# ELECTRIC FIELD

Edited by **Mohsen Sheikholeslami Kandelousi**

**Electric Field**
http://dx.doi.org/10.5772/intechopen.69824
Edited by Mohsen Sheikholeslami Kandelousi

**Contributors**

Dmitry Milovzorov, Eugen Anitas, Liviu Chirigiu, Ioan Bica, Francis Hart, John Palisano, Zhibin Qiu, Jiangjun Ruan, Mirza Sarajlić, Jože Pihler, Nermin Sarajlić, Peter Kitak, Liqiu Wei, Yongjie Ding, Daren Yu, Salvatore Domenic Domenic Morgera, Walter Gustavo Fano, Anatoly Vasilievich Panyukov, Hongbo Liu, Juan Manuel Velazquez Arcos, Ricardo Teodoro Páez Hernández, Samuel Alcántara Montes, Alejandro Pérez Ricardez, Jaime Granados Samaniego, Hadi Nazaripoor, Adham Riad, Mohtada Sadrzadeh, Eyad A. Feilat



First published in London, United Kingdom, 2018 by IntechOpen
IntechOpen is the global imprint of INTECHOPEN LIMITED, registered in England and Wales, registration number: 11086078, The Shard, 25th floor, 32 London Bridge Street
London, SE19SG – United Kingdom
Printed in Croatia

British Library Cataloguing-in-Publication Data
A catalogue record for this book is available from the British Library

Additional hard copies can be obtained from orders@intechopen.com

Electric Field, Edited by Mohsen Sheikholeslami Kandelousi
p. cm.
Print ISBN 978-1-78923-186-1
Online ISBN 978-1-78923-187-8

For EU product safety concerns:
IN TECH d.o.o., Prolaz Marije Krucifikse Kozulić 3, 51000 Rijeka, Croatia,
info@intechopen.com or visit our website at intechopen.com.

# Meet the editor

Dr. Mohsen Sheikholeslami Kandelousi works at the Babol Noshirvani University of Technology's Department of Mechanical Engineering in Iran. His research interests are CFD, nanofluid, mesoscopic modeling of fluid, nonlinear science, magnetohydrodynamics, ferrohydrodynamics, and electrohydrodynamics. He has written several papers and books in various fields of mechanical engineering. According to the reports of Thomson Reuters Corporation and Clarivate Analytics, he has been selected as a Web of Science *Highly Cited Researcher* (Top 0.01%) in 2016 and 2017. He is also the first author of the books *Applications of Nanofluid for Heat Transfer Enhancement*, *Hydrothermal Analysis in Engineering Using Control Volume Finite Element Method*, *Applications of Semi-Analytical Methods for Nanofluid Flow and Heat Transfer*, and *External Magnetic Field Effects on Hydrothermal Treatment of Nanofluid*, which are published in *Elsevier*. Also, he published a book *Nanofluid Heat and Mass Transfer in Engineering Problems* in *IntechOpen*.

# Contents

# Preface

In this book, various applications of electric field have been presented. For the past decades, researchers and scientists have been experimenting and exploring the use of electric fields in a diverse range of applications: in health and biology like treating cancer and cell sorting, in engineering and technological applications like enhancing the heat transfer, in colloidal hydrodynamics, and in stability and lithography. Electric field is defined as a force field arising from the electric charges. Depending on the nature of the material (ability to polarize) and the inherent or attained surface charges, the response in the electric field varies. In Chapter 1, Hall thrusters are considered to satisfy the national demands for developing a long-life satellite platform, lunar exploration, and deep-space exploitation. In the second chapter, the mathematical model describing the dynamics and spatiotemporal evolution of thin liquid film is presented. Electroconductive magnetorheological elastomers (MREs) are studied in the third chapter. Chapter 4 gives a brief introduction on conduction, polarization, dissipation, and breakdown of dielectrics under electric field. In Chapters 5 and 6, local electric fields are studied in dielectric and semiconductors due to the destruction of symmetry, creating the vacancies, point defects, and chemical impurities in the material. Chapter 7 deals with the influence of the electric field on a medium-voltage indoor post insulator (MVIPI) with standard and modified external shapes. A wide range of applications of electric fields in biology and medicine is discussed in Chapter 8. In Chapter 9, a prediction method is proposed based on the electric field features and support vector machine (SVM). The electric and magnetic fields from an electromagnetic wave usually are a matter of interest for regulatory standards, to accomplish electromagnetic compatibility. This topic is investigated in Chapter 10. In Chapter 11, a review of single-stress and multiple-stress life models of electrical insulation is introduced. The origin and spatiotemporal structure of the endogenous (internal) electric near-fields associated with the neurological network activity of the brain are described in Chapter 12. Chapter 13 deals with the problem of the location parameters identifying the equivalent dipole electromagnetic radiation source through measured horizontal magnetic and vertical electric components at some point of the infinite conducting ground. In Chapter 14, Maxwell equations description of the electromagnetic fields with the Fourier transform of the Fredholm integral equations appropriate for a broadcasting process.

**Mohsen Sheikholeslami Kandelousi**
Department of Mechanical Engineering
Babol Noshirvani University of Technology
Babol, Islamic Republic of Iran

Chapter 1

# Application of an Electric Field to Low-Frequency Oscillation Control in Hall Thrusters

Liqiu Wei, Yongjie Ding and Daren Yu

Additional information is available at the end of the chapter

http://dx.doi.org/10.5772/intechopen.71009

**Abstract**

In order to satisfy the national demands for developing a long-life satellite platform, lunar exploration, and deep-space exploitation, Hall thrusters are now considered the preferred candidate for spacecraft propulsion. A Hall thruster is a type of electric propulsion with an annular structure, in which a propellant, usually xenon, is ionized and then accelerated by electrostatic force to create a propulsive thrust. Low-frequency discharge current oscillations, also called breathing mode oscillations in some references, are among the major research topics of Hall thrusters. Low-frequency oscillations in the range of 10–100 kHz might affect the reliability of power processing unit and reduce the efficiency and specific impulse of Hall thrusters. The control of low-frequency oscillations is an essential challenge in the space application of Hall thrusters. It is proved that the electric field is a highly important influence factor for low-frequency oscillations; therefore, application of a dynamic electric field is a practical way to control low-frequency oscillation.

**Keywords:** electric field, low-frequency oscillation, control, filter unit, Hall thruster

## 1. Introduction

In order to adapt the technology requirements of surveying deep space and conducting interplanetary travel, it is an inexorable trend in the development of spacecraft propulsion systems that electric propulsion will take the place of chemical propulsion. A Hall thruster is a type of electric propulsion with an annular structure, in which a propellant, usually xenon, is ionized and then accelerated by electrostatic force to create a propulsive thrust. The typical configuration is shown in **Figure 1**. A mostly transverse magnetic flux is produced by the magnetic circuit and magnetic coils in the discharge channel, with a maximum radial intensity near the channel exit. The discharge voltage is applied between the anode and an external hollow cathode. The propellant, usually Xenon, is injected through the gas distributor. The

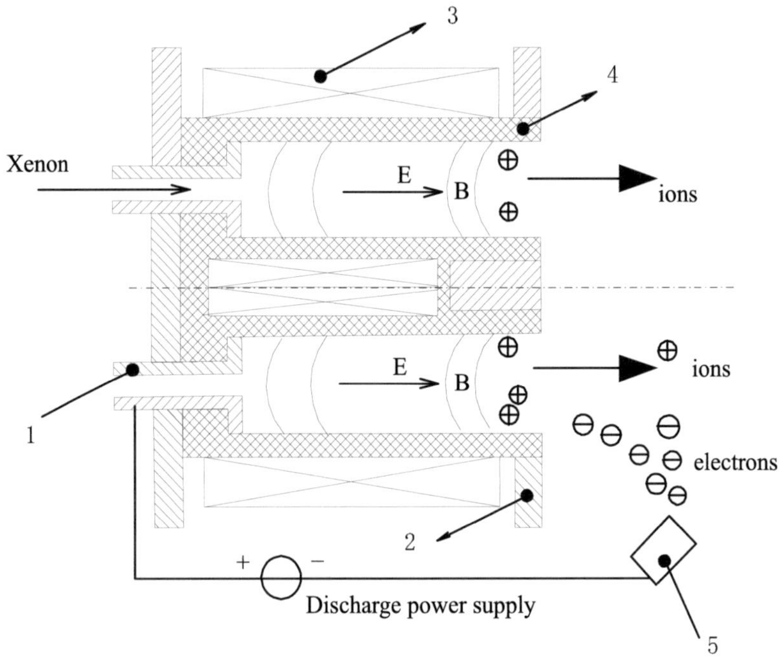


**Figure 1.** Schematic diagram of a Hall thruster: (1) anode, (2) magnetic pole, (3) magnetic coils, (4) accelerating channel, and (5) cathode.

magnetic field limits the axial mobility of electrons and then causes a localized voltage drop of the order of the anode-cathode discharge voltage. Electrons drift in the azimuthal direction, with an average azimuthal velocity $E_z/B_r$. Typical values of the axial electric field $E_z$ are on the order of $10^4$ V/m, and the radial magnetic density $B_r$ is 0.02 T, leading to average electron energies higher than 10 eV. This magnetically confined electron efficiently ionizes the propellant gas. Ions are accelerated toward the exit by the electric field, which produces the thrust for the satellite.

The plasma instability is inevitable in magnetic-confined plasma discharge of Hall thrusters. Past research has shown that regardless of the magnetic topology used, a highly rich and complex wave and noise characteristics over a wide frequency spectrum are inherent in the magnetic-confined plasma discharge. Therefore, to study all kinds of plasma instability, clarify the physical mechanism, understand their characteristics, and seek a stabilization method have always been the central issues of plasma physics [1–6]. Hall thrusters, as typical magnetized plasma discharge devices, exhibit varied oscillations with different lengths and time scales related with various physical processes. These oscillations play a major role in the process of ionization, diffusion, and acceleration of the particles. Low-frequency oscillations in the range of 10–100 kHz, also called breathing mode oscillations in some references, are among the major research topics of Hall thrusters [7, 8]. The typical low-frequency oscillation wave is shown in **Figure 2**. These oscillations might affect the reliability of power processing unit, reduce the efficiency and specific impulse of Hall thrusters, and therefore, its suppression is an essential challenge in the space applications of Hall thrusters not only to relate with the effects on the

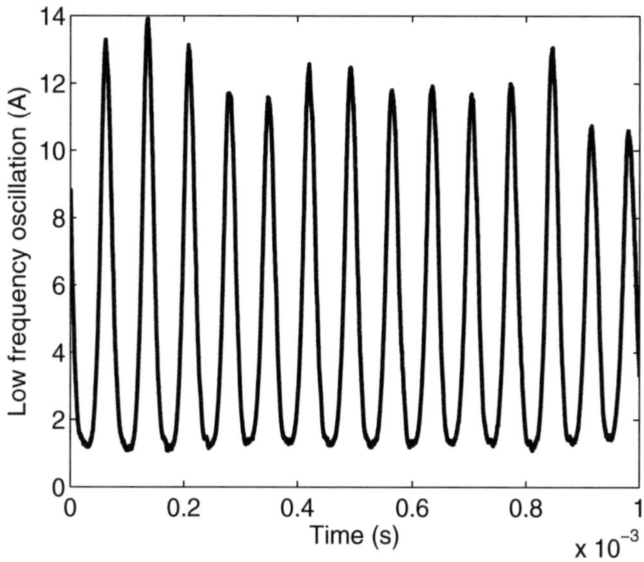


**Figure 2.** Typical low-frequency oscillation wave.

thruster operation but also to control electromagnetic compatibility with other equipment installed on satellites [9, 10]. Researchers have recognized the low-frequency oscillation phenomena with Hall thrusters since the early 1970s [11]. Even now, research on this subject is ongoing [12].

## 2. Research groups for plasma instability

In the world, there are many electric propulsion research groups, such as the Princeton Plasma Physics Laboratory (PPPL) in the U.S., the LAPLACE (Laboratory on Plasma and Conversion of Energy) in France, the Komurasaki-Koizumi Laboratory in Japan, the Moscow Institute of Radio-Engineering in Russia, and other laboratories in Asia and Europe. Some studies of low-frequency oscillation were performed in Russia in the early and mid-1970s; the main research groups continuing to study this subject are located in the U.S., France, Poland, Japan, and China. These groups have researched this subject from different angles and contributed to a full understanding of low-frequency oscillations in Hall thrusters.

In the U.S., studies on low-frequency oscillations are conducted mainly in universities. In PPPL, their research objectives include reducing beam divergence, scaling Hall thrusters, and studying the physics involved in Hall thrusters, as well as plasma instabilities and their control [13–16]. They have studied low-frequency oscillations experimentally and have established some excellent plasma diagnostics devices, included a high-speed positioning system for inside measurements, a movable radial probe, and a segmented probe [17–20]. Another research group at Princeton is EPPDyL (Electric Propulsion and Plasma Dynamics Lab), which has studied the physics and applications of plasma thrusters for more than 30 years. A review of plasma oscillations in Hall thrusters published in *Physics of Plasmas* is their typical oscillation research work [21]. Plasmadynamics and Electric Propulsion Laboratory (PEPL) founded and directed by Gallimore in University of Michigan has operated all thruster types, including electrostatic thrusters, electrothermal propulsion, and electromagnetic thrusters. A majority of the research presently performed at the PEPL lies in Hall thrusters and ion thrusters [22]. In addition, the Stanford Plasma Physics Lab and the Space Propulsion Laboratory at the Massachusetts Institute of Technology are also advancing theoretical and experimental study on plasma physics and space propulsion, including the low-frequency oscillations.

In Europe, the groups focusing on low-frequency oscillations in Hall thrusters are mainly located in Poland and France. In France, Hall thrusters have been studied in the frame of a Research Group on Plasma Propulsion consisting of about 10 academic labs that specialize in all fields of Hall thruster physics (such as plasma physics, optical diagnostics, magnetism, ceramics, and numerical simulation). LAPLACE at the University of Toulouse and the Research Group on Ionized Media (GREMI) at Orléans University, headed by A. Bouchoule, are two of the most important research groups. These two groups have performed many theoretical and experimental studies on discharge instabilities of Hall thrusters in a wide frequency range. Barral is one of the most active researchers in Hall thruster discharge instability for the past few years [8, 23–26]. His work is mainly focused on modeling of plasma instability, discharge instabilities theory, and interactions between thruster and power supply. He has done much research and provided some new perspectives on low-frequency oscillations. Moreover, the Space Propulsion and Plasmas (EP2) research group in Spain, headed by E. Ahedo, has also performed some theoretical studies on the subject of discharge oscillations in Hall thrusters. EP2 has built research relationships with SPL, PPPL, and the Dynamics of Ionized Media group concerning the fundamental physics of Hall thrusters [24, 26–28].

The works of the Komurasaki-Koizumi Laboratory at the University of Tokyo include developing model and improving the characteristics and suppressing low-frequency oscillations [7, 29]. The works of the Gas Discharge Physics Lab in the Korea mainly focus on cylindrical type Hall thrusters and plasma diagnostics [30, 31]. The Harbin Institute of Technology Plasma Propulsion Laboratory (HPPL) in China is a new research group, which began to study Hall thrusters in 2004. They have done some exploratory research on low-frequency oscillations, including the influencing factors, physical mechanism, and stabilization [4, 10, 32–37].

In addition to these research centers, some researchers in Japan (such as Nagoya University and Gifu University) [38], Israel (such as in Technion Israel Institute of Technology and Holon Institute of Technology) [39, 40], and the U.S. (such as the Air Force Institute of Technology) [41], among others, also perform some research work on the subject of low-frequency oscillations in Hall thrusters.

## 3. Model and physical mechanism of low-frequency oscillation

The mechanism of low-frequency oscillations is usually explained using a predator-prey model. In this model, the ions and neutrals in an ionization zone can be expressed as

$$\frac{dn_i}{dt} = \beta n_i n_n - n_i \frac{V_i}{L_{ion}}$$
$$\frac{dn_n}{dt} = -\beta n_i n_n + n_n \frac{V_n}{L_{ion}} \tag{1}$$

where $n_i$ and $n_n$ are the ion number density and neutral number density, respectively. $L_{ion}$ is the length of the ionization zone. Neutrals enter the ionization zone at a rate $n_n V_n$, but few neutrals leave. Ions leave the ionization zone at a rate $n_i V_i$, but few ions arrive. Ionization occurs at the rate of $\beta n_i n_n$. Results show that the $n_i$ and $n_n$ oscillate in an opposed-phase form, which indicates that the ions and neutrals have the relationship of predator and prey.

Although this model can reflect the complex processes of low-frequency oscillations, its boundary conditions are unreasonable. We investigate the meaning of the term $n_n V_n / L_{ion}$. In the original predator-prey model, it means without the existence of the predator, the growth rate of the prey is proportional to its own numbers. In the ionization model, it implies that the rate of neutrals entering the ionization zone is proportional to the number density of neutrals in the ionization zone, when in fact, neutrals are introduced into the channel at a constant rate. If the rate of neutrals arriving is fixed, the system would be stable unconditionally. Detailed calculation results are shown in [42].

It shows that the boundary conditions of the predator-prey model have a significant effect on system stability, which is unreasonable. In order to give reasonable boundary conditions, we can use the simplified equation set to reflect the characteristics of low-frequency oscillations. The predator-prey model is derived from the ion continuity equation and the neutral continuity equation. We begin with the following equation set:

$$\frac{\partial n_n}{\partial t} + \frac{\partial n_n V_n}{\partial x} = -\beta n_n n_i \tag{2}$$

$$\frac{\partial n_i}{\partial t} + \frac{\partial (n_i V_i)}{\partial x} = \beta n_n n_i \tag{3}$$

$$\frac{\partial (n_i V_i)}{\partial t} + \frac{\partial \left(n_i V_i^2\right)}{\partial x} = \frac{e n_i}{m_i} E(x) + \beta n_n n_i V_n \tag{4}$$

where the channel length is $\mathbf{L=3\ cm}$, channel cross section is $\mathbf{A=25\ cm^2}$, neutral velocity is $\mathbf{V_n=200\ m/s}$, and ionization rate is $\boldsymbol{\beta=5\times 10^{-13}}$ $\mathbf{m^3/s}$. With operating conditions of a neutral gas flow rate of 3.0 mg/s and pre-ionization rate of 1% (the ionization rate of the neutral gas before entering the channel), the boundary conditions at the anode (x = 0) are $\mathbf{n_n(0, t)= 2.6\times 10^{19}}$ $m^{-3}$, $V_i(0, t)=\mathbf{2000\ m/s}$, and $\mathbf{n_i(0, t)=2.6\times 10^{16}}$ $m^{-3}$. The values of $n_n(x, 0)$, $n_i(x, 0)$, $V_n(x, 0)$, and $E(x)$ are given according to a one-dimensional steady quasi-neutral hybrid model.

Simulation results show that the system is stable. The traces of $n_i$ and $n_n$ in the ionization zone are shown in **Figure 3** (we define the ionization zone as the region where $n_i$ and $n_n$ changes from 5–90%). The results show that even with equation set (2)–(4), the mechanism of ionization oscillations is incomplete. Further, we introduce the dynamic electric field as follows:

$$E = \frac{en_e V_e}{\sigma(x)} = \frac{I/A - en_i V_i}{\sigma(x)} \tag{5}$$

$$\int_0^L E dx = U_0 \tag{6}$$

where the discharge voltage is $\boldsymbol{U_0 = 300}$ **V**. Electronic conductivity $\boldsymbol{\sigma}$**(x)** depends only on the transverse magnetic field, which is given by $\boldsymbol{\sigma(x) = \sigma_0(H_0/H(x))^2}$, $\boldsymbol{\sigma_0 = 8.3 \times 10^{-3}}$ **S/m**, and $\boldsymbol{H_0 = 0.02}$ **T**. $\boldsymbol{H_0}$ is the magnetic field at the cathode, and $\boldsymbol{H(x)}$ is the profile of the transverse magnetic field. $\sigma_0$ is the mean conductivity. The details of the parameter values and model can be found in [**43**]. With this model, the discharge current fluctuates at a frequency of approximately 30 kHz. In addition, the $\boldsymbol{n_i}$ and $\boldsymbol{n_n}$ change with discharge current (**Figure 4**). The results show that low-frequency oscillations are related with the changes of electric field.

From the view of the physical mechanism, the physical processes of low-frequency oscillations are shown in **Figure 5**. Because the rates of neutrals leaving ($\boldsymbol{n_n V_n}$) and ions arriving ($\boldsymbol{n_{i\_i} V_{i\_i}}$) are small, the main processes in the ionization zone are neutrals arriving at the rate of $\boldsymbol{n_{n\_i} V_n}$, ionization at the rate of $\boldsymbol{\beta n_i n_n}$, and ions leaving at the rate of $\boldsymbol{n_i V_{i\_o}}$ (marked with the oval in **Figure 5**). For a fixed electric field, the ions acceleration characteristic is fixed. If a disturbance occurs in the ionization process and the ionization intensity increases, the $\boldsymbol{n_n}$ would decrease and the rate of ions leaving would be increased; the effects of these two processes cause the system balance again. When the time-dependent electric field is involved, the ionization process disturbance would be followed by a various ion momentum $\boldsymbol{n_i V_{i\_o}}$, and this change would affect the ion momentum in the acceleration zone. From Eq. (5), the change of ion

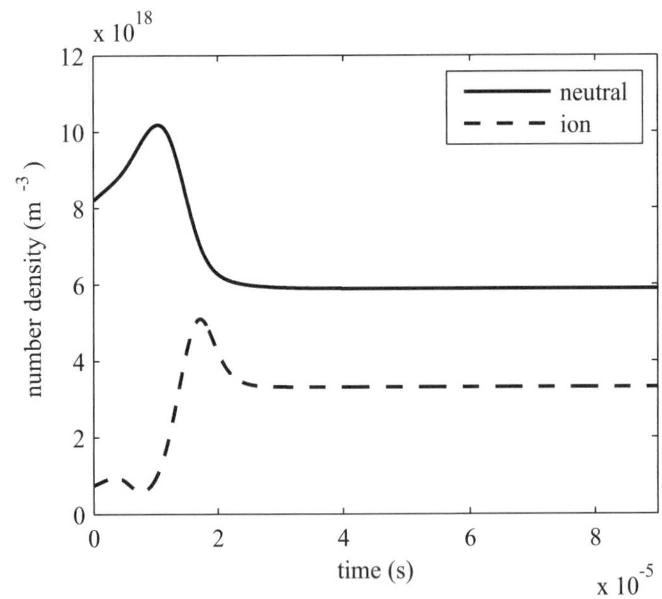


**Figure 3.** Changes in mean neutral number density and mean ion number density in the ionization zone with a constant electric field.

http://dx.doi.org/10.5772/intechopen.71009

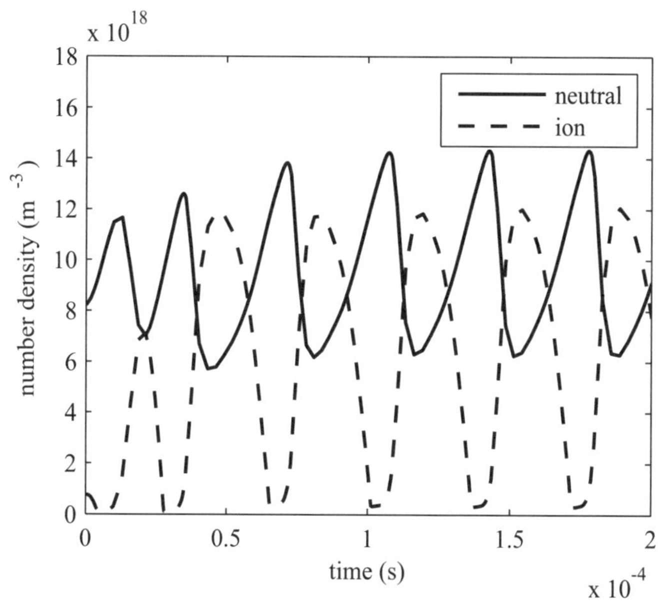


**Figure 4.** Changes in mean neutral number density and mean ion number density in the ionization zone with a dynamic electric field.

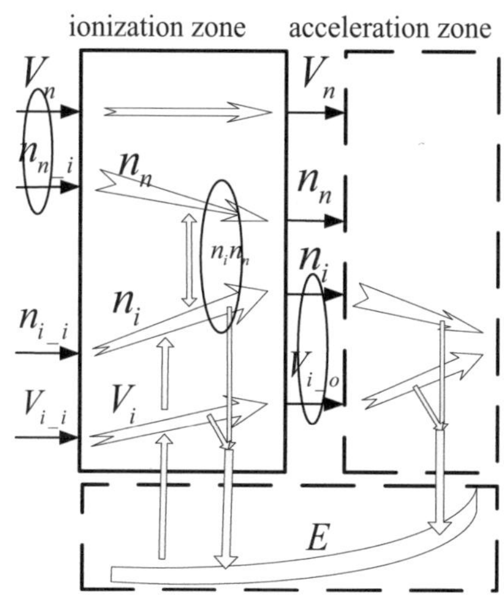


**Figure 5.** Physical processes of low-frequency oscillations.

momentum in the channel would cause the redistribution of the electric field, which influences the ion movement in the ionization zone again. The numerical results show that the feedback of the electric field, together with the replenishment and ionization avalanche of neutral, brings the oscillation. Analyzing particle movement process, the ionization zone is upstream of the acceleration zone. Electrons go into the ionization zone from the acceleration zone. The ionization zone and acceleration zone interact with each other due to this movement. The electric field equation is essentially the electron momentum equation which reflects the interact between the acceleration zone and the ionization zone. Therefore, the import of a dynamic electric field makes the system into whole. The dynamic electric field is a highly important influence factor for low-frequency oscillation. The dynamic electric field is also applied and studied in many research areas, and these research results shown in the electric field have a plentiful scientific connotation [44–58].

## 4. Stabilizing low-frequency oscillation

The stabilizing method of low-frequency oscillation is an essential challenge for the space applications of thrusters. Researchers have tried many different ways to mitigate low-frequency oscillations. A filter, sometimes also called a matching network, is a component mainly applied to reduce the discharge current low-frequency oscillations in the range of 10–100 kHz; it is also an important component in the space mission of a Hall thruster. Previous studies indicated that the low-frequency oscillations are sensitive to filter parameters and can be mitigated to an acceptable level with proper filter values [59]. A filter unit is always involved between the thrusters and the power supply. The traditional filter consists of an inductor and a capacitor, and sometimes a resistor is also applied. This type of filter is a low-pass filter designed to isolate the interfering signal from the thrusters to the power supply. Recently, the role of the filter in the oscillation control was introduced by Yu et al. [4, 10, 33] and Barral et al. [25, 26]. It is noted that the filter regulates the voltage across itself according to the variation of discharge current so as to decrease its fluctuation in the discharge circuit, which is the function of a controller. Therefore, the matching network between the thrusters and power supply has two functions, which are those of a filter and controller. However, it is highly difficult to mix the function of a filter and controller in one network, because the aim and function of these two parts are quite different. This may be the reason that there have been no design methods of the matching network until now. The parameters are in practice always obtained through a trial-and-error method. Therefore, it follows that we should separate the matching network into two stages. The first stage is the filter, which aims to isolate the interfering signal from the thrusters to the power supply. The second stage is the controller, which provides a regulated voltage to decrease the low-frequency discharge current oscillation.

### 4.1. Design of the filter stage

As a matching network between the thruster and power supply, the function of a filter is to protect the power supply, that is, to obtain a stable load for the power supply and, therefore, enable the power supply to operate in its normal electrical state. Another requirement is that the filter has low insert impedance. In a manner of speaking, the design goal is to ensure stable current or voltage on the power supply side, even though the current fluctuates on the thruster side. The filter can be seen as a two-port network: the power supply provides constant input power, and the thruster consumes it with a frequency of 20–40 kHz. Thus, from the point of view of energy conservation, the filter needs a capacitor large enough to store energy and release it to a fluctuating load. Connecting the capacitor between the

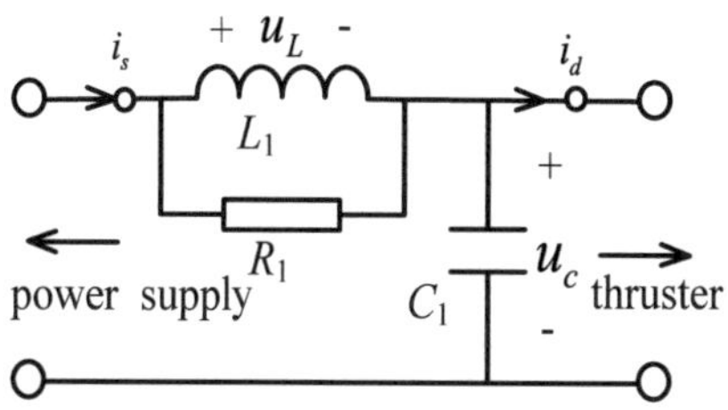


**Figure 6.** Configuration of the filter stage.

positive and negative terminals of the power supply directly may be unreasonable, because there will be a large current peak in the circuit when the power supply is turned on. Therefore, designers must choose a component to connect between the capacitor and the power supply, with the aim of charging the capacitor. This chosen component should have low DC impedance because it is connected in the discharge circuit. Moreover, the component should have high AC impedance. The reason can be understood as follows: The thruster consumes power with a pulsating mode, which would cause fluctuation in the capacitor voltage. The voltage difference between the power supply port and the thruster port would draw a fluctuating current from the power supply, and we do not expect that situation to occur. Thus, the component should have high AC impedance to suppress this kind of current fluctuation. Obviously, the component meeting these criteria is an inductor. Therefore, the filter stage consists of an inductor and a capacitor. On the thruster side, we use specific properties of charge storage to stabilize the discharge voltage, and on the power supply side, we use the inductor to achieve constant current. In order to avoid the LC circuit resonance and electromagnetic interference after extinguishing the thrusters, a resistor is connected parallel with the inductor. That is, the typical configuration of the filter stage is as shown in **Figure 6**.

According to the preceding analysis, the inductor in the filter stage should be sufficiently large to achieve a small power supply current fluctuation. On the other hand, the capacitor should be sufficiently large to ensure a small thruster voltage fluctuation. According to Fourier analysis, a fluctuating signal can be expressed as summation of different frequency components. Therefore, the supply current can be expressed as

$$i_s = \sum_{i}^{n} I_s^{(i)} \sin\left(2\pi f_i t + \varphi_i\right) \tag{7}$$

where $I_s^{(i)}$ is the amplitude of the fluctuating current with frequency $f_i$ and phase angle $\varphi_i$. If we temporarily neglect to consider the existence of the resistor, the inductor voltage can be expressed as

$$u_L(t) = \sum_{i}^{n} 2\pi f_i L_1 \cdot I_s^{(i)} \cos\left(2\pi f_i t + \varphi_i\right) \tag{8}$$

According to electric circuit theory, the capacitor voltage can be expressed as

$$u_C(t) = \int \frac{1}{C_1}(i_d(t) - i_s(t))dt \tag{9}$$

where $i_d(t)$ is the discharge current between the filter stage and the thruster as shown in **Figure 1**. Additionally, we introduce the Fourier analysis method, and therefore, Eq. (9) can be written as

$$u_C(t) = \int \frac{1}{C_1} \sum_i^n \left( I_d^{(i)}(t) \sin\left(2\pi f_i t + \theta_i\right) - I_s^{(i)}(t) \sin\left(2\pi f_i t + \varphi_i\right) \right) dt \tag{10}$$

$$u_C(t) = \sum_i^n \frac{1}{2\pi f_i C_1} \left( I_s^{(i)}(t) \sin\left(2\pi f_i t + \varphi_i\right) - I_d^{(i)}(t) \sin\left(2\pi f_i t + \theta_i\right) \right) \tag{11}$$

According to Kirchhoff's law and taking into account the requirement of a small thruster voltage fluctuation, the sum of capacitor voltage and the inductor voltage fluctuation should be zero. Thus, we obtain

$$|u_C(t)| = \left| \sum_i^n \frac{1}{2\pi f_i C_1} \left( I_s^{(i)}(t) \sin\left(2\pi f_i t + \varphi_i\right) - I_d^{(i)}(t) \sin\left(2\pi f_i t + \theta_i\right) \right) \right| = |u_L(t)| \tag{12}$$

Substituting Eq. (8) into Eq. (12) yields

$$\left| \sum_i^n \frac{\left( I_s^{(i)}(t) \sin\left(2\pi f_i t + \varphi_i\right) - I_d^{(i)}(t) \sin\left(2\pi f_i t + \theta_i\right) \right)}{2\pi f_i C_i} \right| = \left| \sum_i^n 2\pi f_i i_s^{(i)} \cos\left(2\pi f_i t + \varphi_i\right) \right| \tag{13}$$

As is well known, the phase angle of an ideal inductor voltage leads its current by 90°, and the phase angle of an ideal capacitor voltage lags its current by 90°, an ideal voltage source equivalent to a line in an AC circuit. Therefore, the phase relationship between $u_L$ and $u_C$ is opposite. The phase relationship between $i_s$ and $i_C$ is also opposite. Considering that $i_d$ is the sum of $i_s$ and $i_C$, the phase angle of $i_s$ and $i_d$ is equal to 180°. That is, $\theta_i$ and $\varphi_i$ in Eq. (13) have a phase-angle difference of 180°. However, it is highly difficult to simplify Eq. (13) with all frequency components because of the limitation of the mathematical equivalence relation. From the angle of decrease for the low-frequency oscillation, it is also unreasonable to consider waves of all frequencies to set the parameters of the inductor and capacitor. Thus, we simply take into account the main frequency component as low-frequency oscillation $f_0$. The goal of the design is that the power supply current fluctuation is one part per $\alpha$ of the thruster current fluctuation, which means $I_s = \frac{1}{\alpha} I_d$. The inductor and capacitor of the filter stage must satisfy

$$L_1 C_1 = \frac{\left| \frac{1}{2\pi f_0} (1 - \alpha) \cos\left(2\pi f_0 t + \delta_0\right) \right|}{\left| 2\pi f_0 \cos\left(2\pi f_0 t + \varphi_0\right) \right|} \tag{14}$$

where $\delta_0$ is the vector resultant angle of the power supply current and thruster current. $\alpha$ can be represented as the current attenuation ratio. Eq. (14) can be also simplified to

$$L_1C_1 = \frac{|(1-\alpha)\gamma|}{|4\pi^2 f_0^2|}. \tag{15}$$

Here, the parameter $\gamma$ represents the effects of the phase angles $\boldsymbol{\delta_0}$ and $\varphi_0$. Obviously, it is difficult to obtain the value of $\gamma$, but we can estimate the range of its value. $\varphi_0$ is the phase angle of the inductor current, and $\boldsymbol{\delta_0}$ is the vector resultant angle of the power supply current and thruster current. According to the analysis of the phase relationship in [45], the phase angle difference between $\boldsymbol{\delta_0}$ and $\varphi_0$ is equal to 180°. Thus, $\gamma = \frac{\cos\delta_0}{\cos\phi_0}$ and its value is approximately −1. For example, if we want to ensure that the ripple current of the power supply is one-tenth the thruster current fluctuation, then $\boldsymbol{\alpha}$ is larger than or equal to $\boldsymbol{\alpha_0 = 10}$. The frequency of the low-frequency oscillation is assumed to be 20 kHz. Under these assumptions, the product of the inductance and capacitance should be greater than or equal to $5.7 \times 10^{-10}$. If we choose the inductance to be 0.1 mH, the capacitance should be greater than 5.7 μF. The inductance and the capacitance values could be different combinations as long as they satisfy Eq. (9). However, the value of the inductor and capacitor should also take certain factors into consideration, such as the DC power dissipation, volume, and weight. After the determination of the inductance and capacitance, it can be seen that an LC network would have a resonant peak in its characteristic frequency. This would induce undesired electromagnetic interference. Thus, a resistor is usually connected in parallel with the inductor. Under normal conditions, the resistance is in the range 50–200 Ω. Excessive resistance will cause the damping coefficient to be too small, and the electromagnetic interference will increase. Undervalued resistance will weaken the effects of the inductor and cause an increase in power supply current fluctuation. Thus, the resistance of the resistor is often chosen on the basis of experiments.

**Figure 7** shows a contour plot of the current attenuation ratio $\boldsymbol{\alpha}$, with different inductance and capacitance values. It can be seen that the current attenuation ratio increases with the increase in inductance and capacitance in the filter stage. Though the larger inductance and capacitance yields a larger current attenuation ratio, it is also accompanied by an increase in the capacitor volume and inductor weight. Therefore, the reasonable range of current attenuation ratio is approximately 5–10.

### 4.2. Design of the controller stage

From the view of control theory, a thruster, discharge power supply, and filter can be seen as a feedback control system as shown in **Figure 8**. In this system, the Hall thruster is the controlled object, the discharge power supply voltage is the reference signal, and the filter is the controller. Therefore, we designate the filter stage as the controller stage. The controller stage filter regulates the voltage across itself according to the variation of discharge current, so as to affect the electric field distribution in the discharge channel and therefore decrease the discharge current fluctuation in the discharge circuit [8, 10, 12, 26, 33]. The simplest component that can provide a varying voltage with a change in current is an inductor. However, the magnetized plasma of a Hall thruster exhibits varied oscillations ranging from kilohertz to gigahertz. High-frequency current oscillation will cause an undesirable high-amplitude oscillation of the inductor voltage. Thus, we require the controller stage to have sufficient gain in the low band but to decay the signal in the high-frequency band. The required frequency response characteristics

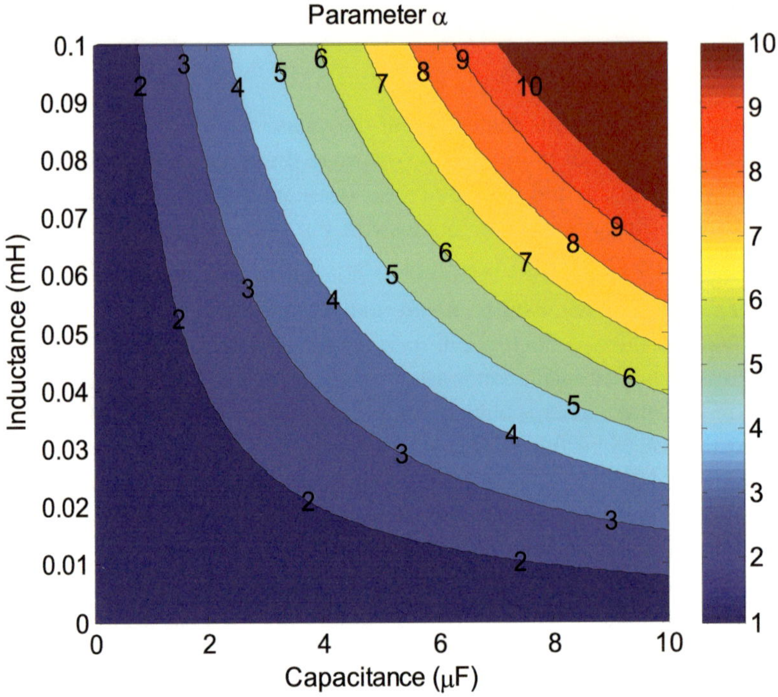


**Figure 7.** Contour plot of the current attenuation ratio $\alpha$, with different inductance and capacitance values.

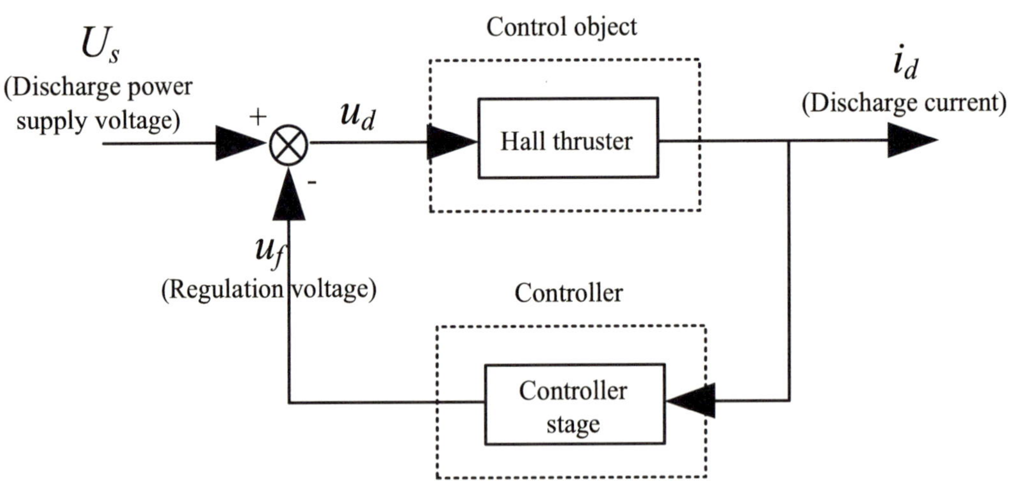


**Figure 8.** Schematic diagram of the feedback control system for Hall thrusters.

would look like that shown in **Figure 9**. According to control theory, it is a typical second-order system that can be realized by the network (controller stage) shown in **Figure 10**. If we consider the oscillation current as the input signal and the voltage of the control filter stage (the voltage at the thruster) as the output signal, the transfer function can be expressed as

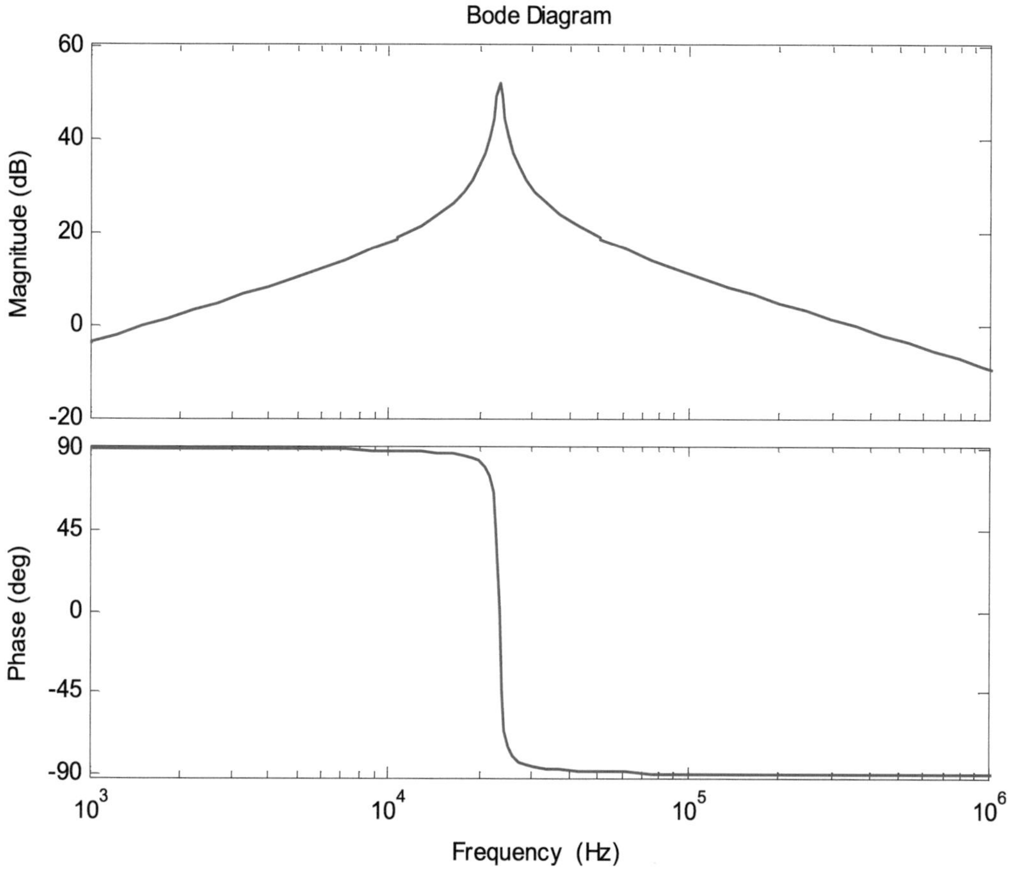


**Figure 9.** Schematic diagram of the ideal frequency response characteristics of the controller stage.

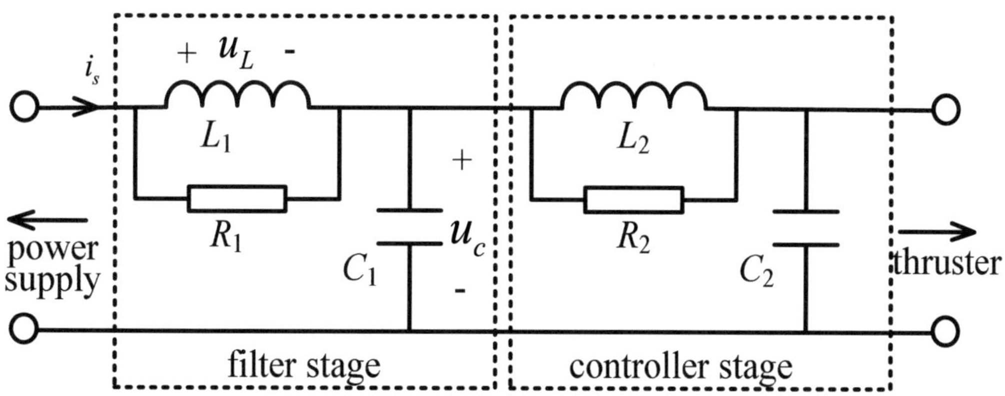


**Figure 10.** Schematic diagram of the two-stage filter for Hall thrusters.

$$\Phi(s) = \frac{K\omega_n^2 s}{s^2 + 2\omega_n \xi s + \omega_n^2} \tag{16}$$

where $K = L_2$, $\omega_n = \frac{1}{\sqrt{L_2 C_2}}$, $\xi = \frac{1}{2R_2}\sqrt{\frac{L_2}{C_2}}$, and $s$ is the Laplacian operator. Obviously, the parameters of the controller stage relate to the required gain of the system. If we suppose the minimum gain is $g_0$ with the low-frequency oscillation frequency $f$, it must satisfy

$$20\ \lg\left(\Phi(s)|_{s=j\omega}\right) = 20\ \lg\left|\frac{K\omega_n^2 j\omega}{(j\omega)^2 + 2\omega_n \xi \cdot j\omega + \omega_n^2}\right| = 20\ \lg\left|\frac{2\pi f L_2 R_2 j}{2\pi f L_2 j - 4\pi^2 f^2 R_2 L_2 C_2 + R^2}\right| \geq g_0 \tag{17}$$

Rewriting Eq. (17), we obtain

$$\frac{2\pi f R_2 L_2}{\sqrt{\left(R_2 - 4\pi^2 f^2 R_2 L_2 C_2\right)^2 + \left(2\pi f L_2\right)^2}} = 10^{\frac{g_0}{20}}. \tag{18}$$

Considering the turnover frequency of the system is

$$f_z = \frac{1}{2\pi\sqrt{L_2 C_2}}, \tag{19}$$

Eq. (18) can be expressed as

$$\frac{\left(1 - \left(\frac{f}{f_z}\right)^2\right)^2}{\left(2\pi f L_2\right)^2} - \frac{1}{R_2^2} = 10^{-\frac{g_0}{10}}. \tag{20}$$

The resistor affects only the gain near the turnover frequency, and thus, we temporarily ignore the effects of the resistor. We then obtain

$$L_2 = 10^{\frac{g_0}{20}} \cdot \frac{1 - \left(\frac{f}{f_z}\right)^2}{2\pi f}. \tag{21}$$

Considering Eq. (19), we therefore know that the capacitance must satisfy

$$C_2 = 10^{-\frac{g_0}{20}} \cdot \frac{f}{2\pi\left(f_z^2 - f^2\right)} \tag{22}$$

A proper filter parameter can provide a regulated voltage with a suitable amplitude and phase angle to control the oscillation of plasma density in the channel so as to decrease the current oscillation in the discharge circuit. Though it is difficult to determine the phase angle of plasma density in the ionization region, the phase relationship between the discharge current and the plasma density in the ionization region can be obtained by analyzing the propellant ionization process in the discharge channel. When the propellant is ionized in the ionization region, the plasma density in the ionization region reaches its highest point. Subsequently, the produced ions are accelerated to their high velocity in the accelerating region, and the discharge current

http://dx.doi.org/10.5772/intechopen.71009

reaches its maximum. From the analysis of the propellant ionization process, it can be seen that there is a lag of time between the discharge current and the plasma density of the ionization region in the time scale of low-frequency oscillation. Thus, when the plasma density in the ionization region increases, the control stage filter should provide a voltage to balance the large ion production. Thus, the regulated voltage of the filter has a phase-angle difference from the discharge current. According to the accelerated time of ions in the discharge channel, the phase-angle, δ, should be in the range of 0–90°. This phase angle can be provided by the resistor paralleled between the inductors. From circuit theory, for the designed filter, the phase angle between the voltage and current can be calculated by

$$\tan\delta = 2\pi f R_2 C_2 - \frac{R_2}{2\pi f L_2}. \tag{23}$$

Substituting Eqs. (21) and (22) into Eq. (23) enables the resistance to be estimated by

$$R_2 = 10^{\frac{g_0}{20}} \cdot \tan\delta, \tag{24}$$

where δ is the phase angle between the regulated voltage and the discharge current. This phase angle is estimated in the range of 0–90°, which would affect the variation of the electric field and therefore help to balance the ion production.

The discharge circuit of a Hall thruster can be seen as a feedback control system. The controller stage regulates the voltage across itself according to the variation of discharge current, so as to decrease its fluctuation in the discharge circuit. As the ion number density increases, the ionization production becomes larger than the number of ions being ejected. A high discharge voltage is helpful for balancing the large ion production by increasing the ion mobility. When the ion number density decreases, a relatively small discharge voltage will decrease the ion mobility and help balance the ion production. According to Eq. (17), the minimum gain is the logarithm transformation of the ratio between the minimum fluctuating discharge voltage and the fluctuating discharge current. Here, the minimum fluctuating discharge voltage is the voltage to balance the fluctuating discharge current at that time. That is, the physical meaning of the minimum gain is the voltage provided by the controller stage to stabilize the unit current oscillation. Obviously, it is almost impossible to determine the exact minimum gain for various discharge operating conditions. However, we also can determine some useful reference values for the minimum gain along with experimental testing and analysis. We can calculate the RLC parameters in the controller stage under different gains and turnover frequencies of the system as shown in **Figure 11** according to Eqs. (21)–(23). It can be seen that the inductance increases with increasing given system gain and turnover frequency, and the capacitance decreases with increasing given system gain and turnover frequency. The choice of turnover frequency has no effect on the resistance of the resistor. The resistance increases only with the increase in given system gain.

Keeping the turnover frequency unchanged, we choose the RLC parameters to obtain different system gains. The relationships between the standard deviation of low-frequency oscillations and system gain with a confidence coefficient of 95% are shown in **Figure 12**. As the system gain increases, the standard deviation of the discharge current low-frequency oscillations measured on the thruster side with a current probe declines and then slightly increases. When

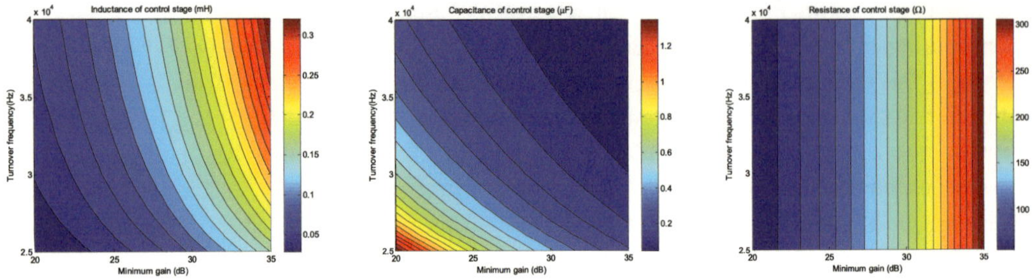


(a) (b) (c)

**Figure 11.** RLC parameters under different gains and turnover frequencies of the system: (a) Calculated results of inductance, (b) calculated results of capacitance, and (c) calculated results of resistance.

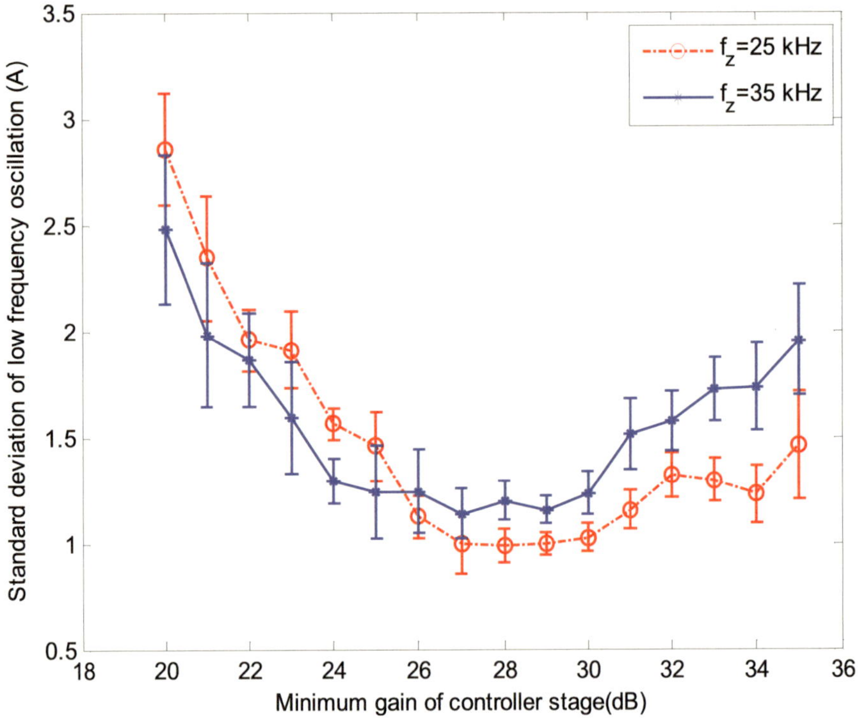


**Figure 12.** Standard deviation of low-frequency oscillation as a function of minimum gain of the controller stage.

the system gain is approximately 25–35 dB, the low-frequency discharge current oscillation reaches its minimum. When the system gain deviates from this range, the low-frequency discharge current oscillations will increase. This can be explained as follows: When the system gain is comparatively low, the provided regulated voltage is too small to stabilize the low-frequency oscillation. The discharge current oscillation thus increases. On the contrary, when the system gain is too high, the provided regulated voltage is excessive. The discharge current oscillation will increase as well. In other words, if the regulated voltage (system gain) is too small or too large, the Hall thruster performs as over-regulated or inadequately regulated. Though the frequency of the low-frequency discharge current oscillations will vary with the discharge parameter, the variation is slight because of the mechanism of low-frequency oscillations. In addition, the controller stage is designed to have enough bandwidth in the range of the frequency of the low-frequency oscillations. Therefore, the variation in frequency may not affect the controllability of the controller stage. [10, 12, 33]

The analysis can also be verified by the experimental data for a turnover frequency 35 kHz as shown in **Figure 12**. The current oscillation in **Figure 12** was measured between the filter and thruster. As seen in **Figure 12**, when the turnover frequency is high, the bandwidth of the controller is widened. More high-frequency components in the discharge current across the control stage networks cause the controller to provide more regulated voltage. Therefore, in the range of low system gain, the low-frequency oscillation with $f_z=35$ kHz is lower than that of $f_z=25$ kHz. However, in the range of high system gain, the relation is the reverse. If we propose $g_0=30$ dB, $\delta=80^{\circ}$, $f=20$ kHz, and $f_z=25$ kHz, we can therefore obtain the parameters of the controller stage as $R_2=179\ \Omega$, $L_2=0.09$ mH , and $C_2=0.447\ \mu$F. According to these experiments and analysis, it can be considered that a reasonable system gain is approximately 25–35 dB per unit current input, the output voltage can be expressed as

$$u=\frac{2\pi f R_2 L_2}{\sqrt{\left(R_2-4\pi^2 f^2 R_2 L_2 C_2\right)^2+(2\pi f L_2)^2}}=10^{\frac{g_0}{20}} \tag{25}$$

If we consider the required gain, $g_0$, is approximately 25–35 dB, the output voltage is in the range 17.78–56.23 V. Supposing that the length of the accelerated region is approximately 10 mm, the equivalent electric field strength increase or decrease for stabilizing a unit discharge current fluctuation is 1700–5600 V/m. We cannot give a precise physical explanation of why it needs this quantity of fluctuating electric field strength, because these are derived average control values. However, it can be evaluated that this fluctuating electric field strength will increase or decrease the average velocity of ions by 3–10%. For the experimental thruster with a 4-A mean discharge current, the unit discharge current fluctuation is roughly equivalent to the discharge current fluctuation in the range 87.5–112.5%. The fluctuation scale of the parameters seems to be reasonable.

Owing to the fact that the effects of the dynamic electric field on plasma instability are a highly complex subject, much work is expected to be done in this field in the future. The contribution of this work lies in the thought process, which is to separate the design of the filter into two stages. The first stage is a filter that aims to isolate the interfering signal from the thrusters to the power supply. This second stage is a controller, which provides a regulated voltage to decrease the low-frequency discharge current oscillation. The range of the parameters in the two stages is given according to the experiments and analysis. Finally, it should be noted that the design method of the two-stage filter works on the premise that the magnetic field of the thruster is reasonably designed. That is, the ionization zone should be controlled in a small region. The action of the controller stage might then be executed effectively [60].

## Acknowledgements

Parts of this chapter are reproduced from authors' recent publication [4, 10, 12]. The authors would like to acknowledge the support of the National Natural Science Foundation of China (No. 51477035 and 51777045).

## Nomenclature

| | |
|---|---|
| $A$ | Channel cross section |
| $B_r$ | Radial magnetic density |
| $C_1$ | Capacitance in filter stage |
| $C_2$ | Capacitance in controller stage |
| $E_z$ | Axial electric field strength |
| $f$ | Fluctuating current frequency |
| $f_0$ | Main frequency of low-frequency oscillation |
| $f_z$ | Turnover frequency |
| $g_0$ | Minimum gain is |
| $H_0$ | Magnetic field at the cathode |
| $i_d$ | Discharge current |
| $I_d^{(i)}$ | Amplitude of discharge current with frequency $\mathrm{f_i}$ and phase angle $\theta_{i.}$ |
| $i_s$ | Power supply current |
| $I_s^{(i)}$ | Amplitude of power supply current with frequency $f_i$ and phase angle $\varphi_i$ |
| $L$ | Channel length |
| $L_1$ | Inductance in filter stage |
| $L_2$ | Inductance in controller stage |
| $L_{ion}$ | Ionization zone length |
| $n_i$ | Ion number density |
| $n_n$ | Neutral number density |
| $R_1$ | Resistance in filter stage |
| $R_2$ | Resistance in controller stage |
| $s$ | Laplacian operator |

| | |
|---|---|
| $u_L$ | Inductor voltage |
| $u_C$ | Capacitor voltage |
| $U_0$ | Discharge voltage |
| $V_i$ | Neutral number velocity |
| $V_n$ | Neutral number velocity |
| $\alpha$ | Current attenuation ratio |
| $\beta$ | Ionization rate |
| $\delta$ | Phase angle between the regulated voltage and the discharge current |
| $\sigma_0$ | Mean conductivity of the channel |
| $\varphi$ | Power supply current phase angle |
| $\theta$ | Discharge current phase angle |

## Author details

Liqiu Wei*, Yongjie Ding and Daren Yu

*Address all correspondence to: weiliqiu@163.com

Harbin Institute of Technology, Harbin, People's Republic of China

## References

[1] Hitendra K. Malik and Sukhmander Singh. Resistive instability in a Hall plasma discharge under ionization effect. Physics of Plasmas. 2013;**20**(5):052115. DOI: http://dx.doi.org/10.1063/1.4804346

[2] Malik HK, Singh S. Conditions and growth rate of Rayleigh instability in a Hall thruster under the effect of ion temperature. Physical Review E. 2011;**83**(3):036406. DOI: 10.1103/PhysRevE.83.036406

[3] Boeuf JP, Garrigues L. Low frequency oscillations in a stationary plasma thruster. Journal of Applied Physics. 1998;**84**(7):3541-3554. DOI: http://dx.doi.org/10.1063/1.368529

[4] Liqiu W, Wang C, Zhongxi N, Weiwei L, ChaoHai Z, Yu D. Experimental study on the role of a resistor in the filter of Hall thrusters. Physics of Plasmas. 2011;**18**(6):063508. DOI: 10.1063/1.3599519

[5] Sukhmander Singh, Hitendra K. Malik. Role of ionization and electron drift velocity profile to Rayleigh instability in a Hall thruster plasma. Journal of Applied Physics. 2012;**112**(1):013307. DOI: http://dx.doi.org/10.1063/1.4733339

[6] Singh S, Malik HK. Growth of low-frequency electrostatic and electromagnetic instabilities in a Hall thruster. IEEE Transactions on Plasma Science. 2011;**39**(11):1910-1911. DOI: 10.1109/TPS.2011.2162652

[7] Yamamoto N, Komurasaki K, Arakawa Y. Discharge current oscillation in Hall thrusters. Journal of Propulsion and Power. 2005;**21**(5):870-876 https://doi.org/10.2514/1.12759

[8] Barral S, Kaczmarczyk J, Kurzyna J, Dudeck M. Closed-loop control of ionization oscillations in Hall accelerators. Physics of Plasmas. 2011;**18**(8):083504. DOI: 10.1063/1.3622655

[9] Tamida T, Nakagawa T, Suga I, Osuga H, Ozaki T, Matsui K. Determining parameter sets for low-frequency-oscillation-free operation of Hall thruster. Journal of Applied Physics. 2007;**102**(4):043304. DOI: 10.1063/1.2771039

[10] Liqiu W, Zhongxi N, Peng E, Daren Y. On the frequency characteristic of inductor in the filter of Hall thrusters. Journal of Vacuum Science and Technology A. 2010;**28**(5):L9-L13. DOI: 10.1116/1.3457152

[11] Morozov AI, Esipchuk YV, Kapulkin AM, Nevrovskii VA, Smirnov VA. Effect of the magnetic field on a closed-electron-drift accelerator. Soviet Physics: Technical Physics. 1972;**17**(3):482-487

[12] Li-Qiu W, Liang H, Yu D-R, Guo N. Low-frequency oscillations in Hall thrusters. Chinese Physics B. 2015;**24**(5):055201. DOI: 10.1088/1674-1056/24/5/055201

[13] Raitses Y, Smirnov A, Fisch NJ. Enhanced performance of cylindrical Hall thrusters. Applied Physics Letters. 2007;**90**(22):221502. DOI: 10.1063/1.2741413

[14] Granstedt EM, Raitses Y, Fisch NJ. Cathode effects in cylindrical Hall thrusters. Journal of Applied Physics. 2008;**104**(10):103302. DOI: 10.1063/1.2999343

[15] Smirnov A, Raitses Y, Fisch NJ. Controlling the plasma flow in the miniaturized cylindrical Hall thruster. IEEE Transactions on Plasma Science. 2008;**30**(5):1998-2003. DOI: 10.1109/TPS.2008.2002148

[16] Raitses Y, Smirnov A, Fisch NJ. Effects of enhanced cathode electron emission on Hall thruster operation. Physics of Plasmas. 2009;**16**(5):057106. DOI: 10.1063/1.3131282

[17] Staack D, Raitses Y, Fisch NJ. Shielded electrostatic probe for nonperturbing plasma measurements in Hall thrusters. Review of Scientific Instruments. 2004;**75**(2):393-399. DOI: 10.1063/1.1634353

[18] Polzin KA, Sooby ES, Raitses Y, Merino E, Fisch NJ. Discharge oscillations in a permanent magnet cylindrical Hall-effect thruster. In: 31st International Electric Propulsion Conference; September 20–24; Ann Arbor, MI. 2009. p. IEPC-2009-122

[19] Raitses Y, Parker JB, Fisch NJ. Plume narrowing and suppression of low frequency oscillations in cylindrical Hall thrusters. In: 31st International Electric Propulsion Conference; September 20–24; Ann Arbor, Michigan. 2009. p. IEPC-2009-123

[20] Parker JB, Raitses Y, Fisch NJ. Transition in electron transport in a cylindrical Hall thruster. Applied Physics Letters. 2010;**97**(9):091501. DOI: 10.1063/1.3486164

[21] Choueiri EY. Plasma oscillations in Hall thrusters. Physics of Plasmas. 2001;**8**(4):1411-1426. DOI: 10.1063/1.1354644

[22] Lobbia RB, Gallimore AD. Two-dimensional time-resolved breathing mode plasma fluctuation variation with Hall thruster discharge settings. In: Presented at the 31st International Electric Propulsion Conference; September 20–24, 2009; University of Michigan Ann Arbor, Michigan, USA. 2009. p. IEPC-2009-106

[23] Barral S, Peradzyński Z. Ionization oscillations in Hall accelerators. Physics of Plasmas. 2010;**17**(1):014505. DOI: 10.1063/1.3292645

[24] Barral S, Ahedo E. Low-frequency model of breathing oscillations in Hall discharges. Physical Review E. 2009;**79**(4):046401. DOI: 10.1103/PhysRevE.79.046401

[25] Barral S, Miedzik J. Numerical investigation of closed-loop control for Hall accelerators. Journal of Applied Physics. 2011;**109**(1):013302. DOI: 10.1063/1.3514151

[26] Barral S, Miedzik J. A model for the active control of low frequency oscillations in Hall. In: 44th AIAA/ASME/SAE/ASEE Joint Propulsion Conference & Exhibit; July 21–23, 2008; Hartford. 2008. p. AIAA 2008-4632

[27] Ahedo E. Plasmas for space propulsion. Plasma Physics and Controlled Fusion. 2011;**53**(18): 124037. DOI: 10.1088/0741-3335/53/12/124037

[28] Parraa FI, Ahedob E, Fifec JM, Martínez-Sánchez M. A two-dimensional hybrid model of the Hall thruster discharge. Journal of Applied Physics. 2006;**100**(2):023304. DOI: 10.1063/1.2219165

[29] Yamamoto N, Nakagawa T, Komurasaki K, Arakawa Y. Discharge plasma fluctuations in Hall thrusters. Vacuum. 2002;**65**(3–4):375-381. DOI: 10.1016/S0042-207X(01)00445-6

[30] Kim H, Lim Y, Choe W, Seon J. Effect of multiply charged ions on the performance and beam characteristics in annular and cylindrical type Hall thruster plasmas. Applied Physics Letters. 2014;**105**(14):144104. DOI: 10.1063/1.4897948], 10.1063/1.4897948]

[31] Lim Y, Kim H, Choe W, Lee SH, Seon J, Lee HJ. Observation of a high-energy tail in ion energy distribution in the cylindrical Hall thruster plasma. Physics of Plasmas. 2014;**21**(10): 103502. DOI: http://dx.doi.org/10.1063/1.4897178

[32] Liqiu W, Wang C, Ke H, Yu D. Effect of ionization distribution on the low frequency oscillations mode in Hall thrusters. Physics of Plasmas. 2012;**19**(1):012107. DOI: 10.1063/1.3676160

[33] Yu D, Wang C, Wei L, Gao C, Guang Y. Stabilizing of low frequency oscillation in Hall thrusters. Physics of Plasmas. 2008;**15**(11):113503. DOI: 10.1063/1.3023150

[34] Yu D, Wei L, Zhao Z, Han K, Yan G. Effect of preionization in Aton-type Hall thruster on low frequency oscillation. Physics of Plasmas. 2008;**15**(4):043502. DOI: 10.1063/1.290 1196

[35] Wang C, Wei L, Ning Z, Daren Y. Effects of magnetic field strength on the low frequency oscillation in Hall thrusters. Physics of Plasmas. 2011;**18**(1):013507. DOI: 10.1063/1.3533915

[36] Liqiu W, Ke H, Chunsheng W, Li H, ChaoHai Z, Daren Y. Study on breathing mode oscillation suppression of self-excited Hall thrusters. Journal of Vacuum Science and Technology A. 2012;**30**(6):061304. DOI: 10.1116/1.4758788

[37] Wang C, Wei L, Daren Y. A basic predator-prey type model for low frequency discharge oscillations in Hall thrusters. Contributions to Plasma Physics. 2011;**51**(10):981-988. DOI: 10.1002/ctpp.201100040

[38] Miyasaka T, Shibata Y, Asato K. Particle simulation of discharge current oscillation in Hall thrusters. Vacuum. 2009;**83**(1):61-66. DOI: 10.1016/j.vacuum.2008.03.023

[39] Kapulkin A, Guelman MM. Low-frequency instability in near-anode region of Hall thruster. IEEE Transactions on Plasma Science. 2008;**36**(5):2082-2087. DOI: 10.1109/TPS.2008.2003359

[40] Fruchtmana A, Cohen-Zur A. Plasma lens and plume divergence in the Hall thruster. Applied Physics Letters. 2006;**89**(11):111501. DOI: 10.1063/1.2349827

[41] Liu D, Huffman RE, Branam RD, Hargus WA Jr. Ultrahigh-speed imaging of Hall-thruster discharge oscillations with Krypton propellant. IEEE Transactions on Plasma Science. 2011;**39**(11):2926-2927. DOI: 10.1109/TPS.2011.2146282

[42] Barral S, Peradzyn'ski Z. A new breath for the breathing mode. In: the 31st International Electric Propulsion Conference; September 20–24, 2009; University of Michigan, Ann Arbor, Michigan, USA. 2009. p. IEPC-2009-070

[43] Morozov AI, Savel'ev VV. One-dimensional hybrid model of a stationary plasma thruster. Plasma Physics Reports. 2000;**26**(10):934-939. DOI: 10.1134/1.1316827

[44] Sheikholeslami M. Influence of magnetic field on nanofluid free convection in an open porous cavity by means of Lattice Boltzmann method. Journal of Molecular Liquids. 2017;**234**:364-374. DOI: 10.1016/j.molliq.2017.03.104

[45] Sheikholeslami M. Magnetohydrodynamic nanofluid forced convection in a porous lid driven cubic cavity using Lattice Boltzmann method. Journal of Molecular Liquids. 2017;**231**:555-565. DOI: 10.1016/j.molliq.2017.02.020

[46] Sheikholeslami M. CuO-water nanofluid free convection in a porous cavity considering Darcy law. The European Physical Journal Plus. 2017;**132**:55. DOI: 10.1140/epjp/i2017-11330-3

[47] Sheikholeslami M. Magnetic field influence on nanofluid thermal radiation in a cavity with tilted elliptic inner cylinder. Journal of Molecular Liquids. 2017;**229**:137-147. DOI: 10.1016/j.molliq.2016.12.024

[48] Sheikholeslami M. Numerical simulation of magnetic nanofluid natural convection in porous media. Physics Letters A. 2017;**381**:494-503. DOI: 10.1016/j.physleta.2016.11.042

[49] Sheikholeslami M. Magnetic source impact on nanofluid heat transfer using CVFEM. Neural Computing and Applications. 2016;**27**:1-10. DOI: 10.1007/s00521-016-2740-7

[50] Sheikholeslami M. Influence of Lorentz forces on nanofluid flow in a porous cylinder considering Darcy model. Journal of Molecular Liquids. 2017;**225**:903-912. DOI: 10.1016/j.molliq.2016.11.022

[51] Sheikholeslami M. CVFEM for magnetic nanofluid convective heat transfer in a porous curved enclosure. The European Physical Journal Plus. 2016;**131**:413. DOI: 10.1140/epjp/i2016-16413-y

[52] Sheikholeslami M. Influence of Coulomb forces on $Fe_3O_4$-$H_2O$ nanofluid thermal improvement. International Journal of Hydrogen Energy. 2017;**42**:821-829. DOI: 10.1016/j.ijhydene.2016.09.185

[53] Kandelousi MS. Effect of spatially variable magnetic field on ferrofluid flow and heat transfer considering constant heat flux boundary condition. The European Physical Journal Plus. 2014;**129**:248. DOI: 10.1140/epjp/i2014-14248-2

[54] Sheikholeslami M. Magnetic field influence on CuO-$H_2O$ nanofluid convective flow in a permeable cavity considering various shapes for nanoparticles. International Journal Of Hydrogen Energy. 2017;**42**:19611-19621. DOI: 10.1016/j.ijhydene.2017.06.121

[55] Sheikholeslami M, Rokni HB. Nanofluid two phase model analysis in existence of induced magnetic field. International Journal of Heat and Mass Transfer. 2017;**107**:288-299. DOI: 10.1016/j.ijheatmasstransfer.2016.10.130

[56] Sheikholeslami M, Sadoughi M. Mesoscopic method for MHD nanofluid flow inside a porous cavity considering various shapes of nanoparticles. International Journal of Heat and Mass Transfer. 2017;**113**:106-114. DOI: 10.1016/j.ijheatmasstransfer.2017.05.054

[57] Sheikholeslami M, Shehzad SA. CVFEM for influence of external magnetic source on $Fe_3O_4$-$H_2O$ nanofluid behavior in a permeable cavity considering shape effect. International Journal of Heat and Mass Transfer. 2017;**115**:180-191. DOI: 10.1016/j.ijheatmasstransfer.2017.07.045

[58] Sheikholeslami M, Bhatti MM. Active method for nanofluid heat transfer enhancement by means of EHD. International Journal of Heat and Mass Transfer. 2017;**109**:115-122. DOI: 10.1016/j.ijheatmasstransfer.2017.01.115

[59] Randolph T, Fischer G, Kahn J, Kaufman H, Zhurin V, Kozubsky K, Kim V. The mitigation of discharge oscillations in the stationary plasma thruster. In: 30th AIAA/ASME/SAE/ASEE Joint Propulsion Conference; June 27–29, 1994; Indianapolis. 1994. p. AIAA-94-2587

[60] Liqiu W, Liang H, Yongjie D, Daren Y, Chaohai Z. Stabilizing low-frequency oscillation with two-stage filter in Hall thrusters. Review of Scientific Instruments. 2017;**88**(7):073502. DOI: 10.1063/1.4990045

Chapter 2

# Electrified Pressure-Driven Instability in Thin Liquid Films

Hadi Nazaripoor, Adham Riad and
Mohtada Sadrzadeh

Additional information is available at the end of the chapter

http://dx.doi.org/10.5772/intechopen.72618

**Abstract**

The electrified pressure-driven instability of thin liquid films, also called electrohydrodynamic (EHD) lithography, is a pattern transfer method, which has gained much attention due to its ability in the fast and inexpensive creation of novel micro- and nano-sized features. In this chapter, the mathematical model describing the dynamics and spatiotemporal evolution of thin liquid film is presented. The governing hydrodynamic equations, intermolecular interactions, and electrostatic force applied to the film interface and assumptions used to derive the thin film equation are discussed. The electrostatic conjoining/disjoining pressure is derived based on the long-wave limit approximation since the film thickness is much smaller than the characteristic wavelength for the growth of instabilities. An electrostatic model, called an ionic liquid (IL) model, is developed which considers a finite diffuse electric layer with a comparable thickness to the film. This model overcomes the lack of assuming very large and small electrical diffuse layer, as essential elements in the perfect dielectric (PD) and the leaky dielectric (LD) models, respectively. The ion distribution within the IL film is considered using the Poisson-Nernst-Planck (PNP) model. The resulting patterns formed on the film for three cases of PD-PD, PD-IL, and IL-PD double layer system are presented and compared.

**Keywords:** thin liquid films, electrohydrodynamic instabilities, electrokinetic, perfect dielectric, ionic liquids, micro- and nano-patterning

## 1. Introduction

For the past decades, researchers and scientists have been experimenting and exploring the use of electric fields in diverse range of applications: In health and biology like treating cancer [1] and cell sorting [2], in engineering and technological applications like enhancing the heat transfer [3–6], colloidal hydrodynamics and stability [7, 8], and lithography [9]. The electric field is defined as a force field arising from the electric charges. Depending on the nature of material (ability to polarize) and the inherent or attained surface charges, the response in the electric field varies.

Surface instabilities can be either triggered using external mechanical, thermal, and electrical forces or via intermolecular interactions like van der Waals forces [9–11]. Development of these instabilities leads to the film disruption and formation of patterns, which is of interest in many applications. In the coating and cooling processes, determining the drainage time (i.e., the time when the film breakdown occurs) and in lithography providing insight regarding different morphological structures of the film interface are few examples in numerous applications. Proper implementation of the patterning process highly depends on the knowledge of the dynamics, instability, and morphological evolution of the interface or film. Interfacial tension and viscosity of liquid film are known as dampening factors for fluctuations on the free surfaces. In small-scale systems, intermolecular forces, which depend on material properties of the substrate, liquid film, and the bounding layer, also play a dominant role in the creation and amplifying the instabilities in thin liquid films. Electrically triggered instability of thin liquid films or often call electrohydrodynamic (EHD) patterning has gained extensive attention because of its ability in the creation of micro- and nano-sized structures ranging from single and bifocal microlens arrays [12], micro and nanochannels [13], and mushroom-shaped microfibers [14].

To investigate the electrically induced instabilities of thin films, it is necessary to have an electrostatic model to find the Maxwell stress acting on the interface as a primary driving force in the system. The molten polymer film is typically assumed to be a perfect dielectric (PD), with no free charge, or a leaky dielectric (LD) which has an infinitesimal amount of charges. This requires assuming the very large and very small electric diffuse layer compared to the film thickness as the main characteristics of the PD and LD models, respectively. In the nanofilms, the film thickness is comparable to the formed diffuse layer during the evolution process which violates the PD and LD assumptions. A general model is needed to bridge the gap between PD and LD models. In ionic liquid (IL) films, with a finite amount of free charges or ions, the diffuse layer thickness is comparable to the film thickness. In this chapter, an electrostatic model is presented for different cases of PD-PD, PD-IL, and IL-PD systems to find the net electrostatic force acting on the interface. Furthermore, the spatiotemporal evolution of the interface is investigated under each condition.

## 2. Deriving nonlinear equation for thin film dynamics

In this part, the nonlinear governing equation is derived for the two-dimensional (2D) thin film. **Figure 1** shows the schematic of the 2D thin liquid film and bounding media sandwiched between two electrodes.

The evolution of the thin film is described using mass and momentum balances for both thin film and bounding media. The boundary conditions are a no-slip condition on the walls, no penetration (two media are immiscible), and stress balance (normal and tangent) at the interface. It is assumed that fluid is Newtonian and incompressible. Detailed mathematical representations are as follows:

http://dx.doi.org/10.5772/intechopen.72618

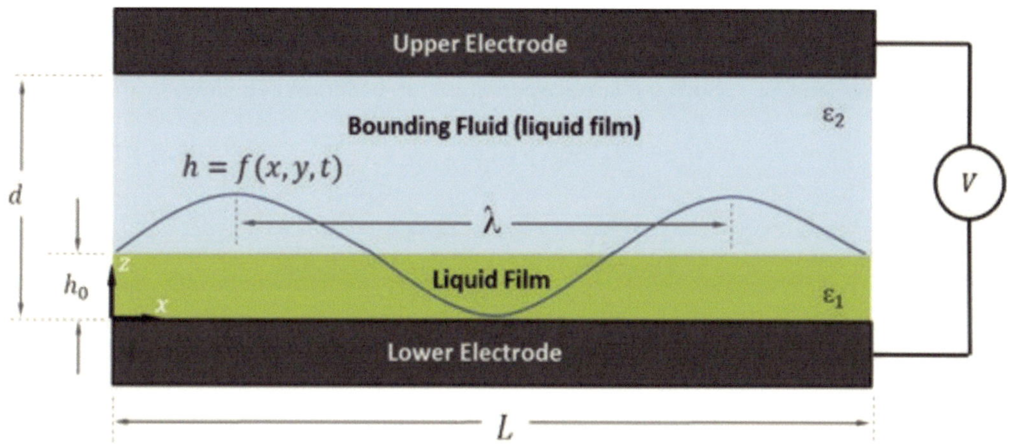


**Figure 1.** Schematic of the thin film sandwiched between two electrodes.

$$\rho_i\left(\frac{\partial \vec{u}_i}{\partial t} + \left(\vec{u}_i.\nabla\vec{u}_i\right)\right) = -\nabla P_i + \nabla.\left[\mu_i\left(\nabla\vec{u}_i + \left(\nabla\vec{u}_i\right)^T\right)\right] + \vec{f}_e \tag{1}$$

$$\nabla.\left(\rho_i\vec{u}_i\right) = 0 \tag{2}$$

where $\vec{f}_e = -\nabla\phi$ accounts for external body force in which $\phi$ is the conjoining/disjoining pressure, $\mu$ is the dynamic viscosity, and $i = 1, 2$ denotes the $i^{th}$ fluid phase. Here, we chose 1 for liquid film (molten polymer) and 2 for bounding media. Above equations are solved by the following boundary conditions:

$$\vec{u}_i = 0 \quad at \quad z = 0 \quad \& \quad z = d \tag{3}$$

No slip condition and penetration at the interface:

$$\vec{u}_{relative} = 0 \quad at \quad z = h(x, y, t) \tag{4}$$

and kinematic boundary condition for the vertical component of velocity:

$$w = \frac{\partial h}{\partial t} + u\frac{\partial h}{\partial x} + v\frac{\partial h}{\partial y} \; at \;\; z = h(x, y, t) \tag{5}$$

Normal and tangent stress balances at the interface:

$$\vec{n}.\left[\sigma_1.\vec{n} - \sigma_2.\vec{n}\right] = \kappa\gamma + \vec{f}_e.\vec{n} \tag{6}$$

$$\vec{t}_i.\left[\sigma_1.\vec{n} - \sigma_2.\vec{n}\right] = \vec{f}_e.\vec{t}_i \tag{7}$$

where $\vec{n}$ and $\vec{t}_i$ are normal and tangent vectors of the interface, respectively. $\gamma$ is surface tension which is assumed constant here (Note that it can vary by location) and $\kappa$ is the mean interfacial curvature of the film interface. Generally, in a 3D analysis, the film interface can be presented mathematically in the form of $G(x, y, z, t) = h(x, y, t) - z$. Using differential geometry, normal and tangent vectors and surface curvature can be found as bellow:

$$\vec{n} = n_x \vec{i_x} + n_y \vec{i_y} + n_z \vec{i_z} \tag{8}$$

$$n_x = \frac{1}{C_n}\left(\frac{\partial h}{\partial x}\right) \quad n_y = \frac{1}{C_n}\left(\frac{\partial h}{\partial y}\right) \quad n_z = \frac{-1}{C_n} \tag{9}$$

The scaling for the unit normal vector is $C_n = \sqrt{1 + \left(\frac{\partial h}{\partial x}\right)^2 + \left(\frac{\partial h}{\partial y}\right)^2}$ and the tangent vector $\vec{t_i}$:

$$\vec{t_i} = t^i{}_x \vec{i_x} + t^i{}_y \vec{i_y} + t^i{}_z \vec{i_z} \tag{10}$$

$$t^1{}_x = \frac{-1}{C_{t1}}\left(\frac{\partial h}{\partial y}\right) \quad t^1{}_y = \frac{1}{C_{t1}}\left(\frac{\partial h}{\partial x}\right) \quad t^1{}_z = 0 \tag{11}$$

$$t^2{}_x = \frac{1}{C_{t2}}\left(\frac{\partial h}{\partial x}\right) \quad t^2{}_y = \frac{1}{C_{t2}}\left(\frac{\partial h}{\partial y}\right) \quad t^2{}_z = \frac{1}{C_{t2}}\left[\left(\frac{\partial h}{\partial x}\right)^2 + \left(\frac{\partial h}{\partial y}\right)^2\right] \tag{12}$$

The scaling for the tangent vectors are $C_{t1} = \sqrt{\left(\frac{\partial h}{\partial x}\right)^2 + \left(\frac{\partial h}{\partial y}\right)^2}$ and $C_{t2} = \sqrt{\left(\frac{\partial h}{\partial x}\right)^2 + \left(\frac{\partial h}{\partial y}\right)^2 + \left(\left(\frac{\partial h}{\partial x}\right)^2 + \left(\frac{\partial h}{\partial y}\right)^2\right)^2}$. Finally, the surface curvature which is presented by two principle radii of curvature $R_1$ and $R_2$ is given by:

$$\kappa = \left(\frac{1}{R_1} + \frac{1}{R_2}\right) \tag{13}$$

Principle radii of curvature for the interface can be defined based on the film thickness as:

$$\frac{1}{R_1} + \frac{1}{R_2} = \frac{\frac{\partial^2 h}{\partial x^2}\left(1 + \left(\frac{\partial h}{\partial y}\right)^2\right) + \frac{\partial^2 h}{\partial y^2}\left(1 + \left(\frac{\partial h}{\partial x}\right)^2\right) - 2\frac{\partial^2 h}{\partial x \partial y}\frac{\partial h}{\partial x}\frac{\partial h}{\partial y}}{\left(1 + \left(\frac{\partial h}{\partial x}\right)^2 + \left(\frac{\partial h}{\partial y}\right)^2\right)^{\frac{3}{2}}} \tag{14}$$

In the case of 2D analysis, $G(x, z, t) = h(x, t) - z$, above relations are simplified as follows:

$$n_x = \frac{1}{C_n}\left(\frac{\partial h}{\partial x}\right); \quad n_y = 0\ ; \quad n_z = \frac{-1}{C_n}; C_n = \sqrt{1 + \left(\frac{\partial h}{\partial x}\right)^2} \tag{15}$$

$$t^1{}_x = 0 \quad t^1{}_y = \frac{1}{C_{t1}}\left(\frac{\partial h}{\partial x}\right) \quad t^1{}_z = 0 \tag{16}$$

$$t^2{}_x = \frac{1}{C_{t2}}\left(\frac{\partial h}{\partial x}\right) \quad t^2{}_y = 0 \quad t^2{}_z = \frac{1}{C_{t2}}\left(\frac{\partial h}{\partial x}\right)^2 \tag{17}$$

$$C_{t1} = \left|\frac{\partial h}{\partial x}\right| \quad \& \quad C_{t2} = \left|\frac{\partial h}{\partial x}\right|\sqrt{1+\left(\frac{\partial h}{\partial x}\right)^2}\,;\, \kappa = \frac{\frac{\partial^2 h}{\partial x^2}}{\left(1+\left(\frac{\partial h}{\partial x}\right)^2\right)^{\frac{3}{2}}} \tag{18}$$

Hence, normal and tangent components of stress balance Eqs. (6) and (7) for the film in the 2D case are given by

$$-(p_1 - p_2) + \frac{2\mu_1 h_x}{1+{h_x}^2}[-(u_{1z}+w_{1x}) + r(u_{2z}+w_{2x})] + \frac{2\mu_1(1-{h_x}^2)}{1+{h_x}^2}[w_{1z} - rw_{2z}]$$
$$= \frac{h_{xx}}{(1+{h_x}^2)^{\frac{3}{2}}}\gamma \tag{19}$$

$$(1-{h_x}^2)[(u_{1z}+w_{1x}) - r(u_{2z}+w_{2x})] + 2h_x[(w_{1z}-u_{1x}) - r(w_{2z}-u_{2x})] = 0 \tag{20}$$

where $r = \frac{\mu_2}{\mu_1} \ll 1$ is the bounding media to film dynamic viscosity ratio which is the small quantity in our case. For instance, water viscosity at $25^{\circ}$C is 0.894 [mPa.s or cP] and for castor oil is 985 [mPa.s or cP]. Therefore, the bounding media can be considered as an inactive media and there is no need to solve the evolution equations (Navier–Stokes and continuity) for both media. Also, as film thickness $h_0 \ll 1$, inertial effects are negligible ($Re \ll 1$, creeping flow). Here and thereafter, we just consider film evolution equations and do not use the subscript 1 for the film velocity components.

The subscripts in above equations $[]_x$and $[]_z$ represent derivatives with respect to x and z, respectively. In what follows, scaling parameters are used to nondimensionalize governing equations and boundary conditions (capital letters are dimensionless):

$$x = \lambda X \; y = \lambda Y \; z = h_0 Z \; u = u_0 U \; v = u_0 V \tag{21}$$

where $\lambda$ is the maximum wavelength of the surface instabilities, $h_0$ is the initial thickness of the film, $u_0$ is the maximum value of lateral velocity, and $k$ is defined as $k = \frac{h_0}{\lambda}$. By using definitions in Eq. (21) and applying to mass continuity equation Eq. (2):

$$w = u_0 k W \tag{22}$$

Time is also re-scaled by employing long length scale $\lambda$ and characteristic velocity $u_0$

$$t = \frac{h_0}{k u_0} T \tag{23}$$

As a result of the long wave approximation theory $k = \frac{h_0}{\lambda} \ll 1$, flow within the film is locally parallel [15]. Pressure and conjoining/disjoining pressure are scaled as

$$(p,\phi)=\frac{kh_0}{\mu_1 u_0}(P,\Phi) \tag{24}$$

$(P,\Phi)$ are dimensionless pressure and conjoining/disjoining pressure.

Stress balance in normal direction, Eq. (19), can be written in the scaled form as follows:

$$\begin{aligned}&-(P_1-P_{ext})+\frac{2H_X}{1+k^2H_X{}^2}k^2[-(U_Z+kW_X)]+\frac{2(1-k^2H_X{}^2)}{1+k^2H_X{}^2}[k^2W_Z]\\&=\left(\frac{H_{XX}(1+k^2H_{YY})+H_{YY}(1+k^2H_{XX})-2k^4H_{XY}H_XH_Y}{(1+k^2H_X{}^2+k^2H_Y{}^2)^{\frac{3}{2}}}\right)\frac{k^3\gamma}{\mu_1 u_0}\end{aligned} \tag{25}$$

and in tangential direction, Eq. (20), is scaled as below:

$$(1-k^2H_X{}^2)[U_Z+k^2W_X]+2k^2H_X[W_Z-U_X]=0 \tag{26}$$

where $P_{ext}$ represents scaled form of $p_2$ in Eq. (19). $P_1-P_{ext}$ can be called gage pressure ($P$) without loss of generality. Rewriting Navier–Stokes and mass continuity equations, Eqs. (1) and (2), by using above definitions we have,

$$kRe(U_T+UU_X+VU_Y+WU_Z)=-(P+\Phi)_X+k^2(U_{XX}+U_{YY})+U_{ZZ} \tag{27}$$

$$kRe(V_T+UV_X+VV_Y+WV_Z)=-(P+\Phi)_Y+k^2(V_{XX}+V_{YY})+V_{ZZ} \tag{28}$$

$$kRe(W_T+UW_X+VW_Y+WW_Z)=-(P+\Phi)_Z+k^4(W_{XX}+W_{YY})+k^2W_{ZZ} \tag{29}$$

$$U_X+W_Z=0 \tag{30}$$

and boundary conditions, Eqs. (3) and (5),

$$\vec{U_i}=0 \quad at \quad Z=0 \quad \& \quad Z=\frac{d}{h_0} \tag{31}$$

$$W=\frac{\partial H}{\partial T}+U\frac{\partial H}{\partial X} at\ Z=H(X,T) \tag{32}$$

In the EHD pattern evolution process, the electric field destabilizes the interface of the fluids, causing the interface deformation with the height of $h=f(x,y,t)$. The characteristic wavelength of the growing instabilities, $\lambda$, is much larger than the initial film thickness ($k\ll 1$), so a "long-wave approximation" [16] is used to simplify the governing equations. To avoid losing the effects of interfacial tension, this assumption is made: $k^3\gamma=O(1)$.

$$(P+\Phi)_X=U_{ZZ} \tag{33}$$

$$(P+\Phi)_Z=0 \tag{34}$$

$$U_X+W_Z=0 \tag{35}$$

$$-P = H_{XX} \quad at \quad Z = H(X,T) \tag{36}$$

$$U_Z = 0 \qquad at \quad Z = H(X,T) \tag{37}$$

$(P + \Phi)$ is called modified pressure $(\overline{P})$. Eq. (34) shows that, the modified pressure does not change across the film thickness. From Eq. (31), (33), and (37), velocity component U is found as follows:

$$U = \overline{P}_X Z\left(\frac{1}{2}Z - H\right) \tag{38}$$

Using Eqs. (38) and (35) leads to velocity component W,

$$W = \frac{1}{2}Z^2\left[-\frac{1}{3}Z(\overline{P}_{XX}) + \left(\overline{P}_X H\right)_X\right] \tag{39}$$

Substituting velocity component values U and W that are evaluated at $Z = H(X,T)$ into Eq. (32) gives the film evolution equation:

$$H_T - \left[\frac{H^3}{3}\overline{P}_X\right]_X = 0 \tag{40}$$

Replacing the modified pressure with pressure and conjoining/disjoining pressure, then using Eq. (36) for pressure results in the following thin film equation,

$$\begin{gathered} H_T + \frac{\partial}{\partial X}\left[H^3\frac{\partial \Psi}{\partial X}\right] = 0 \\ \Psi = \frac{1}{3}(H_{XX} - \Phi) \end{gathered} \tag{41}$$

and in a two-dimensional form and considering y-direction,

$$\begin{gathered} \mu h_t + \frac{\partial}{\partial x}\left[h^3\frac{\partial \psi}{\partial x}\right] + \frac{\partial}{\partial y}\left[h^3\frac{\partial \psi}{\partial y}\right] = 0 \\ \psi = \frac{1}{3}\left(\gamma\left(h_{xx} + h_{yy}\right) - \phi\right. \end{gathered} \tag{42}$$

## 2.1. Interaction potentials

Conjoining pressure, $\phi$, (force acting on the film interface per unit area) is defined as the gradient of excess intermolecular interactions, $\phi = \nabla(\Delta G)$. In EHD, the electric field induces a pressure at the film interface, which is added to the natural interactions to generate excess intermolecular interactions. The conjoining pressure, $\phi$, is a summation of these interaction potentials: van der Waals, electrostatic, and Born repulsive interaction potentials and is given as:

$$\phi = \phi_{vdW} + \phi_{Br} + \phi_{EL} \tag{43}$$

The van der Waals interaction is the summation of Keeson, Debye, and London dispersion forces [17, 18]. This interaction is defined as $\phi_{vdW_L} = A_L/6\pi h^3$ and $\phi_{vdW_U} = -A_U/6\pi(d-h)^3$ for the lower and upper electrodes, respectively. $A_L$ and $A_U$ are effective Hamaker constant lower and upper electrodes, which depend on the materials used for electrodes and fluid layers. For instance, for three-layered systems, it is defined: $A_{213} = \left(\sqrt{A_{33}} - \sqrt{A_{11}}\right)\left(\sqrt{A_{22}} - \sqrt{A_{11}}\right)$ in which 1, 2, and 3 denote substrate, liquid film, and bounding fluid, respectively.

The van der Waals interaction becomes singular as $h \to 0$ and $h \to d$. To avoid nonphysical penetration of liquid to solid phase, in case of film rupture and touching, a cutoff distance, $l_0$, is defined for which a short-range repulsive force, called Born repulsion, acts on the film interface [19]. This is used to maintain a minimum equilibrium liquid thickness on both electrodes and is defined as $\phi_{Br_L} = -8B_L/h^9$ and $\phi_{Br_U} = 8B_U/(d-h)^9$ for the lower and upper electrodes, respectively. Coefficients $B_L$ and $B_U$ are found by setting the net conjoining pressure equal to zero at $h = l_0$ and $h = d - l_0$ for lower and upper surfaces.

The electrostatic conjoining/disjoining pressure depends on the electrostatic property of liquid thin film and bounding layer, which is discussed in the following sections. Throughout this study, it is assumed that electric breakdown does not occur during the EHD patterning process.

### 2.2. Numerical scheme

Different numerical methods are used to track the free interfaces [20, 21], and particularly, in the EHD patterning process [22–25]. The numerical methods are applied as versatile tools to monitor and visualize the transient evolution of liquid film subject to an electric field. Here, the thin film equation, Eq. (42) is solved numerically to obtain the transient behavior using finite difference method and adaptive time step solver. More details about the numerical scheme are available in [26].

## 3. Perfect dielectric film and bounding media

In this part, we consider the film and bounding media as two perfect dielectric media and solve Laplace equation in 1D. Governing equations and applied boundary conditions [27–29] are as follows:

$$\nabla^2 \psi_i = 0 \tag{44}$$

$$\psi_1 = \psi_{low} \; at \quad z = 0 \tag{45}$$

$$\psi_1 = \psi_2 \; at \quad z = h(x, y, t) \tag{46}$$

$$\left(\vec{n} \,.\varepsilon\, \vec{D}\right)_1 - \left(\vec{n} \,.\varepsilon\, \vec{D}\right)_2 = 0 \quad at \quad z = h(x, y, t) \tag{47}$$

$$\psi_2 = 0 \qquad\qquad at \quad z = d \tag{48}$$

Solving Eq. (44) with mentioned boundary conditions give rise to electric potential distribution across the domain as follows:

http://dx.doi.org/10.5772/intechopen.72618

$$\psi_1 = \frac{\varepsilon_2}{\varepsilon_1}\left[\frac{-\psi_{low}}{h\left(\frac{\varepsilon_2}{\varepsilon_1}-1\right)+d}\right]z + \psi_{low} \tag{49}$$

$$\psi_2 = \left[\frac{-\psi_{low}}{h\left(\frac{\varepsilon_2}{\varepsilon_1}-1\right)+d}\right](z-d) \tag{50}$$

The *Maxwell stress* tensor is defined as follows:

$$T = \varepsilon\begin{bmatrix} E_x^2 - \frac{1}{2}E^2 & E_xE_y & E_xE_z \\ E_yE_x & E_y^2 - \frac{1}{2}E^2 & E_yE_z \\ E_zE_x & E_zE_y & E_z^2 - \frac{1}{2}E^2 \end{bmatrix}$$

where $E$ is defined as $E^2 = E_x^2 + E_y^2 + E_z^2$ and electric field components can be found as follows:

$$E_x = -\frac{\partial\psi}{\partial x}\ E_y = -\frac{\partial\psi}{\partial y}\ E_z = -\frac{\partial\psi}{\partial z} \tag{51}$$

Traction force vector acting on the interface because of *Maxwell stress* is found as follows:

$$\vec{F} = \int_S \left(\vec{n}\,.\overline{T}\right) dS \tag{52}$$

where $S$ is the interface area and $\vec{n}$ is the unit normal vector of the interface.

$$\vec{n} = n_x\vec{i}_x + n_y\vec{i}_y + n_z\vec{i}_z$$

where $\vec{n}$ is the unit normal vector into the medium. In the long-wave limit approximation, the traction forces for each layer are given as:

$$\vec{F}_1 = -\frac{1}{2}\varepsilon_1 {E_{1z}}^2 S\vec{i}_z \tag{53}$$

$$\vec{F}_2 = \frac{1}{2}\varepsilon_2 {E_{2z}}^2 S\vec{i}_z \tag{54}$$

Therefore, the net force per unit area, conjoining pressure is

$$\phi_{ES} = -\frac{1}{2}\varepsilon_1\left(\frac{\varepsilon_1}{\varepsilon_2}-1\right)\left[\frac{\psi_{low}}{\frac{\varepsilon_1}{\varepsilon_2}d - h\left(\frac{\varepsilon_1}{\varepsilon_2}-1\right)}\right]^2 \tag{55}$$

The net electrostatic force acting on the interface depends on the electric permittivity of each layer, applied potential, electrodes separation distance, and interface height.

3D snapshots were plotted to investigate the effect that a transverse electric field has on the nondimensional structural height variations over time of a PD film, with an initial thickness of 30 nm. **Figure 2(a)** shows the formation of small random disturbances in the liquid film that are conical in shape. These evolve with time as liquid flows from regions of lower thickness to regions of higher thickness causing them to increase in size and length and become more pillar-shaped as shown in **Figure 2(b)**. However, once the pillars reach the upper electrode, their height ceases to increase, and the pillars begin to increase their cross-sectional area, as shown in **Figure 2(c)**.

A detailed spatiotemporal evolution for the liquid–liquid interface instabilities in a 2D domain highlighting the different patterns formation using different initial film thickness is presented in **Figure 3**. The relative electric permittivity ratio of bilayer system is 0.6 and the initial lower layer film thickness is increased from 20 nm to 85 nm. As a result, four patterns of pillars, bicontinuous, holes, and roll-like features are formed. At lower thickness, 20 nm, pillars are formed (image a(ii)) but merging of neighbor pillars results in coarse final structure (image a (iii)). Bicontinuous structures are formed when the film thickness is increased to 50 nm (image b(i–iv)) which is not desired. The bicontinuous structures behave similarly to an air-in-liquid dispersion that also takes place in air-polymer systems with high filling ratios. The further increase in the film thickness leads to a columnar holes formation in the film (images c(i–iv)). In very thick films, 75 nm, the role-like features form that are spaced with micrometer distance. This type of features can be used as nanochannels in practical applications. The roll-like structures, in d(i), seem to be an organized version of the bicontinuous structures generated by the same phase inversion mechanism shown in b(i).

In the EHL process, both the initial layer thickness and electric permittivity ratio play an important role in the process and can develop different patterns on the film. **Figure 4** shows the different types of patterns formed as a function of the relative electric permittivity ratios of layers '$\epsilon_r$' and nondimensional initial mean film thickness of $h_0/d$, in a liquid–liquid interface. As shown in **Figure 4**, changing the electric permittivity ratios may lead to having two to four kinds of structures depending on the initial layer thickness. In electric permittivity ratios greater than one, two main shapes of structures form, pillars in filling ratios ($h_0/d$) less than 0.5 and bicontinuous structures in relatively thicker films. However, as shown in **Figure 4**, in electric permittivity ratios less than one, four different shapes of structures can form. The generated map shown in **Figure 4** provides details about the threshold values of the filling ratio in which the transition between structures happens at different electric permittivity ratios. From thermodynamics, the shape that leads to the lowest free energy in the system

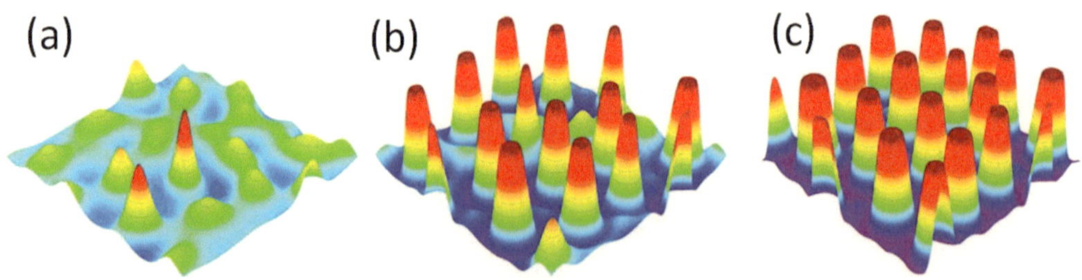


**Figure 2.** PD film with $h_0 = 30\ nm$, 3D spatiotemporal evolution of a PD liquid PD film (images a–c). Nondimensional times, T are: (a) $3 \times 10^5$ (b) $3.5 \times 10^5$ (c) $4.5 \times 10^5$.

http://dx.doi.org/10.5772/intechopen.72618

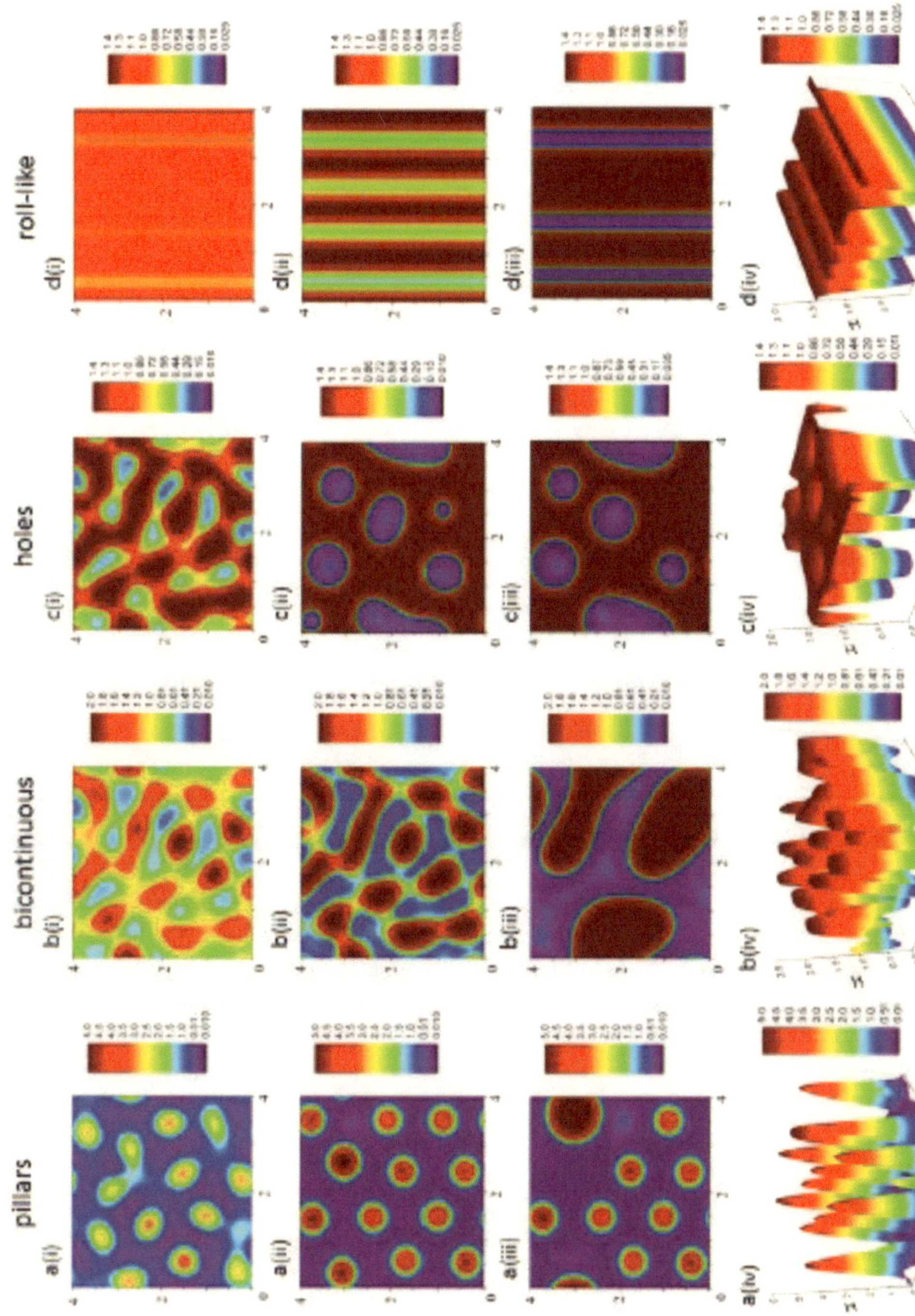


**Figure 3.** Base case PD-PD bilayer, (a–d) images from i to iii show the 2D spatiotemporal evolution for liquid–liquid interface instabilities in a 2 domain when $\epsilon_r$= 0.6 and $\epsilon_l$ = 2.5. Initial mean film thicknesses ($h_0$)are: A(i–iv) 20 nm, b(i–iv) 50 nm, c(i–iv) 70 nm, and d(i–iv) 85 nm. Initial electric field intensities ($E_0$) are 294, 250, 227, and 212 MV/m, respectively. Reprinted with permission from (HADI NAZARIPOOR et al. LANGMUIR 2014, 30, 14734–14744). Copyright (2014) American Chemical Society.

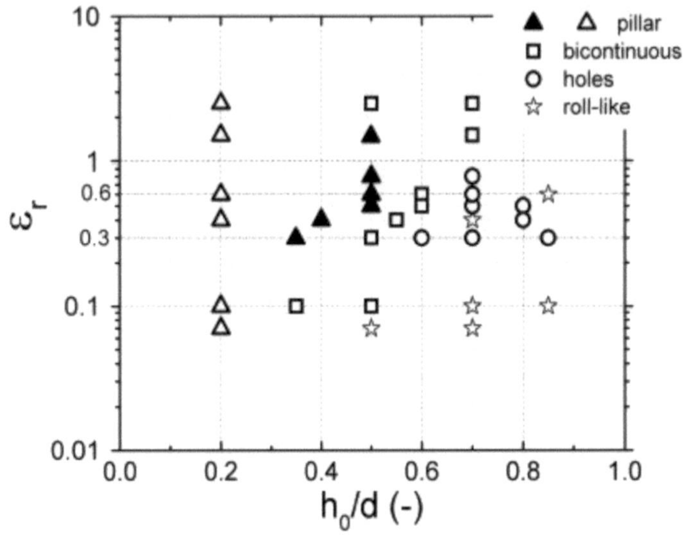


**Figure 4.** A parametric map that shows the different types of patterns formed, in 2D, as a function of the relative electric permittivity ratios of layers '$\epsilon_r$' and nondimensional initial mean film thickness of $h_0/d$, in a liquid–liquid interface. Reprinted with permission from (HADI NAZARIPOOR et al. LANGMUIR 2014, 30, 14734–14744). Copyright (2014) American Chemical Society.

determines the final pattern. **Figure 4** also shows that there is a critical initial film thickness for each value of $\epsilon_r$, where below it only pillars are formed.

## 4. Perfect dielectric film and ionic conductive bounding media

In this section, an ionic conductor is chosen as a bounding media and a dielectric media for thin film, for instance, salt water–oil system. Saltwater behaves like conductors due to having free ions, so Poisson equation is solved to find the electric potential distribution over the domain, and for the oil part similar to the previous section, Laplace equation is considered. Detailed mathematical procedure is

$$\varepsilon_i \nabla^2 \psi_i = -\rho_{f_i} \tag{56}$$

$$\psi_1 = 0 \qquad\qquad at \quad z = 0 \tag{57}$$

$$\psi_1 = \psi_2 \quad at \quad z = h(x, y, t) \tag{58}$$

$$\left(\vec{n}\,.\varepsilon\,\vec{D}\right)_1 - \left(\vec{n}\,.\varepsilon\,\vec{D}\right)_2 = 0 \quad at \quad z = h(x, y, t) \tag{59}$$

$$\psi_2 = \psi_{up} \quad at \ z = d \tag{60}$$

The space charge density of the mobile ions $\rho_f$ is zero for dielectric media as they do not have any ions, but for the IL, $\rho_{f_2}$ is given by [18],

$$\rho_{f_2} = \sum_{i=1}^{N} z_i e n_{i\infty} \exp\left(-\frac{z_i e(\psi - \psi_\infty)}{k_B T}\right) \tag{61}$$

Eq. (61) is called the Boltzmann distribution. Here, $z_i$ is the valence of species $i$, $e$ is the magnitude of electron charge, $1.602 \times 10^{-19}(C)$, $n_{i\infty}$ is the bulk ionic number concentration (say, in $m^{-3}$), $N$ is the number of species in the electrolyte, $k_B$ is the Boltzmann constant, $1.378 \times 10^{-23}(J/K)$, and $T$ is the temperature in $(K)$.

Ionic number concentration, $n_\infty$, is given by

$$n_\infty = 1000 N_A M$$

with Avogadro number, $N_A = 6.022 \times 10^{23} mol^{-1}$ and $M$ being the electrolyte molar concentration (mol/L).

*Proof*

Generally, $\rho_f$ is defined based on ionic number concentration $n_i$ of species $i^{th}$ as follows:

$$\rho_f = \sum_{i=1}^{N} z_i e n_i \tag{62}$$

Ionic number concentration, $n_i$, near a charged surface can be found by considering ion conservation Eq. (63) for the steady-state and equilibrium condition which leads to zero-ion flux, $\vec{j_i} = 0$. Here, $R_i$ represents for reactions that produce/consume species $i$ in the electrolyte and set to zero.

$$\frac{\partial n_i}{\partial t} = -\nabla . \vec{j_i} + R_i \tag{63}$$

Ion flux, $\vec{j_i}$, for species $i$, is defined by the Nernst-Plank relationship Eq. (64)

$$\vec{j_i} = n_i \vec{u} - D_i \nabla n_i - \frac{z_i e n_i D_i}{k_B T} \nabla \psi \tag{64}$$

At the equilibrium, there is no fluid velocity, $\vec{u} = 0$, so for the 1D case (z-direction) it is simplified as follows:

$$\frac{dn_i}{dz} + \left(\frac{z_i e n_i}{k_B T}\right) \frac{d\psi}{dz} = 0 \tag{65}$$

Eq. (65) has an analytical solution with the following boundary conditions:

$$n_i = n_{i\infty} \; and \quad \psi = \psi_\infty at \; |z| \gg 0 \tag{66}$$

$$n_i = n_i \; and \;\; \psi = \psi \;\; at \quad |z| = 0 \tag{67}$$

where|z| shows distance from charged surface, $\psi_\infty$ and $n_{i\infty}$ are bulk (electro neutral) electric potential and ion number concentration, respectively. $\psi_{\infty i}$ is also called reference potential, $\psi_{ref}$. From now we use $\psi_{ref}$ instead of $\psi_\infty$. By using Eqs. (66) and (67), the solution is:

$$n_i = n_{i\infty}\exp\left[-\frac{z_i e\left(\psi - \psi_{ref}\right)}{k_B T}\right] \tag{68}$$

Substituting Eq. (68) into Eq. (62) results in Eq. (61).

*Proof. end*

Using Boltzmann distribution (relation Eq. (61)) for the free space charge density, $\rho_f$, in the Poisson Eq. (56) give rise to well-known Poisson-Boltzmann (PB) equation

$$\varepsilon_2 \nabla^2 \psi_2 = -\sum_{i=1}^{N} z_i e n_{i\infty} \exp\left(-\frac{z_i e\left(\psi_2 - \psi_{ref}\right)}{k_B T}\right) \tag{69}$$

and for the 1D case

$$\varepsilon_2 \frac{d^2\psi_2}{dz^2} = -\sum_{i=1}^{N} z_i e n_{i\infty} \exp\left(-\frac{z_i e\left(\psi_2 - \psi_{ref}\right)}{k_B T}\right) \tag{70}$$

PB equation is simplified for the monovalent, $N = 2$, symmetric, $z : z$, electrolyte solution which is known as the Gouy-Chapman theory

$$\varepsilon_2 \frac{d^2\psi_2}{dz^2} = 2zen_\infty \sinh\left(\frac{ze\left(\psi_2 - \psi_{ref}\right)}{k_B T}\right) \tag{71}$$

Governing equations in the long-wave limit condition [29] is simplified as follows:

$$\frac{d^2\psi_1}{dz^2} = 0 \tag{72}$$

$$\varepsilon_2 \frac{d^2\psi_2}{dz^2} = 2zen_\infty \sinh\left(\frac{ze\left(\psi - \psi_{ref}\right)}{k_B T}\right) \tag{73}$$

and boundary conditions

$$\psi_1 = 0 \qquad at \quad z = 0 \tag{74}$$

$$\psi_1 = \psi_2 \quad at \quad z = h(x, y, t) \tag{75}$$

$$\varepsilon_1 \frac{d\psi_1}{dz} = \varepsilon_2 \frac{d\psi_2}{dz} \quad at \ z = h(x, y, t) \tag{76}$$

$$\psi_2 = \psi_{up} \quad at \;\; z = d \tag{77}$$

The following scaling parameters are defined to nondimensionalize the governing equations and the boundary conditions:

$$\Psi = \frac{ze\psi}{k_B T} \quad , \quad Z = \frac{z}{d} \tag{78}$$

Also, a dimensionless parameter is defined, $\kappa^{-1} = \left(\frac{\varepsilon_2 k_B T}{2e^2 z^2 n_\infty}\right)$, called Debye length. By using these definitions, we can rewrite equations and boundary conditions in scaled form

$$\frac{d^2\Psi_1}{dZ^2} = 0 \tag{79}$$

$$\frac{d^2\Psi_2}{dZ^2} = (\kappa d)^2 \sinh\left(\Psi - \Psi_{ref}\right) \tag{80}$$

and boundary conditions

$$\Psi_1 = 0 \qquad\qquad at \;\; Z = 0 \tag{81}$$

$$\Psi_1 = \Psi_2 \quad at \;\; Z = \frac{h(x,y,t)}{d} = Z^* \tag{82}$$

$$\varepsilon_1 \frac{d\Psi_1}{dZ} = \varepsilon_2 \frac{d\Psi_2}{dZ} \quad at \;\; Z = \frac{h(x,y,t)}{d} = Z^* \tag{83}$$

Solution is as follows,

$$\Psi_1 = \frac{\Psi_s}{Z^*} Z \;\; at \quad (0 \le Z \le Z^*) \tag{84}$$

$$\Psi_2 = \Psi_{up} + 2\ln\left(\frac{1 + \exp(-\kappa d(Z - Z^*))\tanh\left(\frac{\Psi_s - \Psi_{up}}{4}\right)}{1 - \exp(-\kappa d(Z - Z^*))\tanh\left(\frac{\Psi_s - \Psi_{up}}{4}\right)}\right) \quad at \quad (Z^* \le Z \le 1) \tag{85}$$

where $\Psi_s$ results from solution of

$$\Psi_s + \frac{\varepsilon_2}{\varepsilon_1} 2\kappa d \sinh\left(\frac{\Psi_s - \Psi_{up}}{2}\right) Z^* = 0 \tag{86}$$

In dimensional form

$$\psi_1 = \frac{\psi_s}{h} z \;\; at \quad (0 \le z \le h) \tag{87}$$

$$\psi_2 = \psi_{up} + 2\frac{k_B T}{ze} \ln\left(\frac{1 + \exp(-\kappa(z - h))\tanh\frac{k_B T}{ze}\left(\frac{\psi_s - \psi_{up}}{4}\right)}{1 - \exp(-\kappa(z - h))\tanh\frac{k_B T}{ze}\left(\frac{\psi_s - \psi_{up}}{4}\right)}\right) \quad at \quad (h \le z \le d) \tag{88}$$

After finding the electric potential distribution, we can calculate the net electrostatic force acting on the interface. By using definitions in Eqs. (53) and (54), the net force is given by,

$$\frac{1}{S}\vec{i}_z\,.\vec{F}_{net} = \frac{1}{2}\varepsilon_1\left(\frac{\varepsilon_1}{\varepsilon_2}-1\right)\left(\frac{\psi_s}{h}\right)^2 \tag{89}$$

So conjoining pressure becomes [29]:

$$\phi_{ES} = -\frac{1}{2}\varepsilon_1\left(\frac{\varepsilon_1}{\varepsilon_2}-1\right)\left(\frac{\psi_s}{h}\right)^2 \tag{90}$$

**Figure 5(a)** shows the effects of changing the electrolytes molarity on the resultant electrostatic pressure acting on the IL-PD interface. The negative values' curves have an electric permittivity ratio ($\boldsymbol{\epsilon_r} = \mathbf{2.5}$) and are associated with upward disjoining pressure, while the positive values' curves have an electric permittivity ratio ($\boldsymbol{\epsilon_r} = \mathbf{0.6}$) and are associated with downward joining pressure. However, both curves experience a decay in the value of the electrostatic pressure as the interface height increases. **Figure 5(b)** outlines the effect of changing the applied voltage on the electrostatic pressure for both IL-PD bilayer and PD-PD bilayer. The electrolytic molarity and electric permittivity ratio ($\boldsymbol{\epsilon_r} = \mathbf{0.6}$) were kept constant for all applied voltages. The graph shows that the IL-PD bilayer experiences a much larger pressure than that of the PD bilayer for the same applied voltage (Notice at V = 10 V).

**Figure 6** shows a 2D spatiotemporal evolution for uniform electric field liquid–liquid interface instabilities, where (b) is the PD-PD bilayer and (c) is the IL-PD bilayer. When visually compared to one another, the number of pillars in the b(i) 3D representation of the PD-PD bilayer appears to be in the same order of magnitude of that of the c(i) 3D representation of the PD-IL bilayer. However, the difference between them is much higher than that since the physical domain size for PD-IL is actually 10 times smaller than that of the PD-PD. Moreover, **Figure 6** b(ii) and c(ii) confirms this size difference and shows that the average center distance between the pillars for

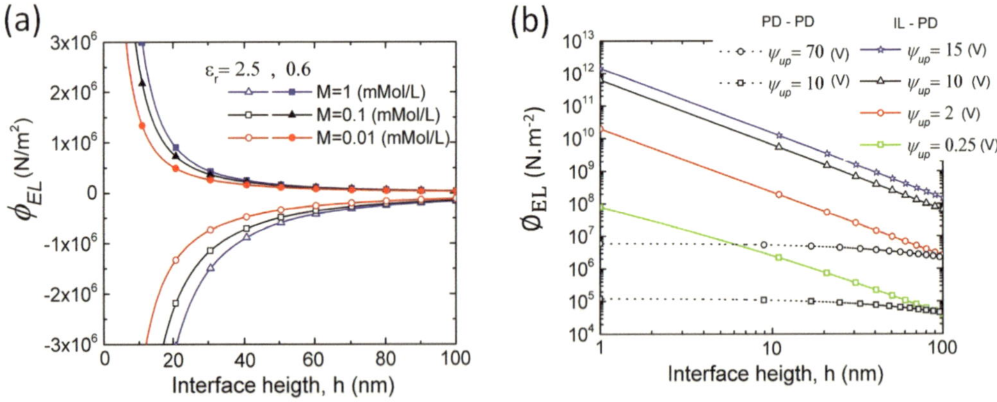


**Figure 5.** Electrostatic pressure distribution versus interface height. (a) Effects of electrolyte molarity in IL-PD bilayers and (b) comparison between PD-PD and PD-IL bilayers. (a) Molarity M = 0.001, 0.0001, and 0.00001 mol/L and applied potential of 0.25 V and $\epsilon_1 = 2.5$. (b) M = 0.0001 mol/L for IL-PD,$\epsilon_1 = 2.5$ and $\epsilon_1 = 4.17$. Reprinted with permission from (HADI NAZARIPOOR et al. LANGMUIR 2014, 30, 14734–14744). Copyright (2014) American Chemical Society.

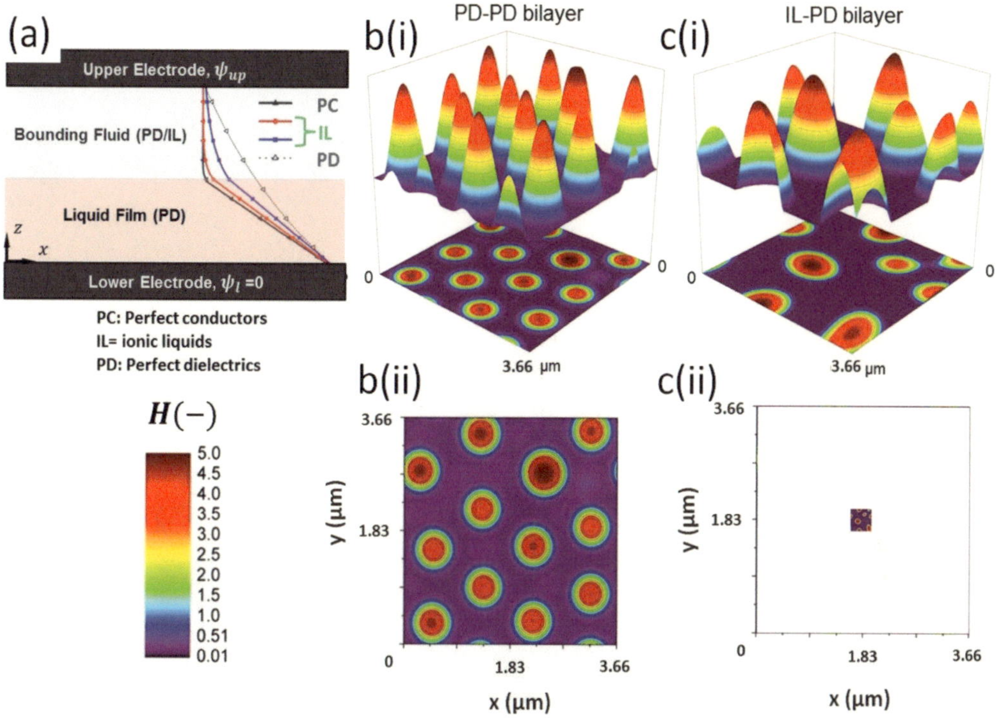


**Figure 6.** (a) Schematic of the IL-PD bilayer. 3D and 2D snapshots of interface morphology of (b) PD-PD, (c) IL-PD bilayer, (i) in a 3D domain, (ii) in physical domain, when $\epsilon_r = 0.6$ and $M = 100\ (mmol/L)$ for IL-PD bilayer. Initial mean film thickness is: $h_0 = 20\ nm$ and nondimensional time for the plots are $T =$ (b) $3.4 \times 10^5$ (c) $13.5.\psi_{up} = 20\ V$ and electric field intensities ($E_0$) are (a) 294 and (b) 995.

PD-IL bilayer is 210 nm while that of the PD-PD bilayer has an average center distance of 1336 nm. The formed pillars also disordered and dispersed randomly in the PD-IL case.

## 5. Ionic conductive film and perfect dielectric bounding media

In the previous section, the complicated case of nonlinear Poisson-Boltzmann equation Eq. (59) was considered for the IL layer. In case of low applied potential (less than 25 mV), the Debye Hückel approximation [10, 30, 31] is used to linearize the Poisson-Boltzmann equation (Eq. (80)).

$$\frac{d^2\psi_1}{dz^2} = \kappa^2\left(\psi_1 - \psi_{ref}\right) \tag{91}$$

and ions conservation within the layer is satisfied as follows:

$$\int_0^h \rho_f dz = \int_0^h \left(\psi_1 - \psi_{ref}\right) dz = 0 \tag{92}$$

and boundary conditions

$$\psi_1 = \psi_l \quad at \quad z = 0 \tag{93}$$

$$\Psi_1 = \Psi_2 \quad at \qquad z = h(x, y, t) \tag{94}$$

$$\varepsilon_1 \frac{d\psi_1}{dZ} = \varepsilon_2 \frac{d\psi_2}{dZ} \quad at \qquad z = h(x, y, t) \tag{95}$$

$$\psi_2 = 0 \quad at \quad z = 1 \tag{96}$$

where $\kappa = \left(\frac{2000 \; e^2 N_A M}{\varepsilon_1 \varepsilon_0 k_B T}\right)^{1/2}$ is the inverse of Debye length. Solving Eqs. (91)–(96) results in the electric potential distribution within PD and IL layers as follows:

$$\psi_2 = \left(\frac{z - d}{h - d}\right)\psi_s \tag{97}$$

$$\psi_1 = (\psi_s - \psi_l)\left[\frac{\cosh(\kappa h) + 1}{2\sinh(\kappa h)}\right]\sinh(\kappa z) \quad - \left(\frac{\psi_s - \psi_l}{2}\right)\cosh(\kappa z) + \frac{\psi_s + \psi_l}{2} \tag{98}$$

where $\psi_s$ is the interface potential and is determined by the following equation:

$$\psi_s = \frac{\psi_l(1 + \cosh(\kappa h))}{1 + \cosh(\kappa h) + \frac{2\varepsilon_2}{\varepsilon_1 \kappa (d-h)} \sinh(\kappa h)} \tag{99}$$

The electrostatic pressure, in this case, is given by,

$$\phi_{EL} = \frac{1}{2}\varepsilon_2 \varepsilon_0 \left(\frac{\varepsilon_2}{\varepsilon_1} - 1\right)\left[\frac{\psi_s}{d - h}\right]^2 \tag{100}$$

A 50-nm thick IL film surrounded by a 50 nm PD media, such as air, was considered to plot the effects of changing the electrolytic molarity have on the potential distribution across the film thickness. The plots obtained for the electric potential distribution are linear as shown in **Figure 7(a)**. For molarity (M) of 0.01 $mol/L$, the reduction in electric potential across the film thickness is almost zero and, thus the electric field is zero. On the other hand, the molarity of 0.00001 $mol/L$ causes the electric potential distribution to behave similarly to a PD.

**Figure 7(b)** shows the changes in the electrostatic pressure and the spinodal parameter across the IL film for different molarities. The graph shows that as the molarity increases the concentration of ions increases and consequently the electrostatic pressure and its corresponding force increase. This relationship is even more significant for higher IL thicknesses. Additionally, the negative values for the pressure indicate that the forces are pushing the interface toward the upper electrode.

**Figure 8(a)** shows the schematic view of the IL film bounded with a PD film while **Figure 8(b)** and **(c)** compare the structural variations between a PD film and IL film. Additionally, **Figure 8(c)** emphasizes the influence of altering the film thickness on the morphology of the IL film. It is found that a faster growth of instabilities is developed by increasing the initial thickness of IL

http://dx.doi.org/10.5772/intechopen.72618

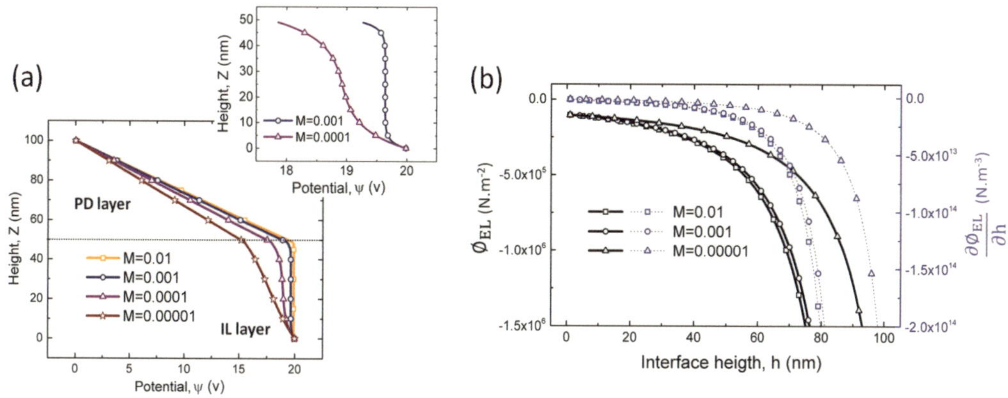


**Figure 7.** (a) Effect of molarity in electric potential distribution. Molarity, M= 0.01, 0.001, 0.0001, and 0.00001$mol/L$ and corresponding Debye lengths, $(\kappa)^{-1}$ are: 0.5, 1.7, 5.4, and 17 nm. (b) Electrostatic pressure, left axis, and spinodal parameter, right axis, distributions versus film thickness for three molarity values of M= 0.01, 0.001, and 0.00001 $mol/L$. $\phi_l = 20\ V$, $\varepsilon_1 = 2.5$ and $\varepsilon_2 = 1$ [31].

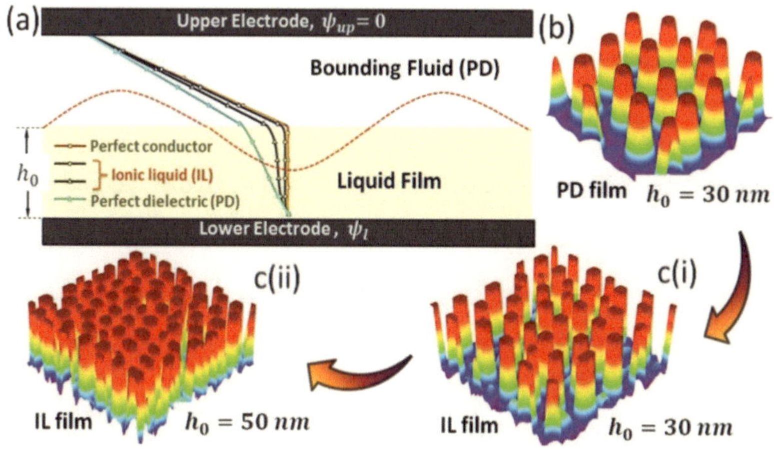


**Figure 8.** (a) A schematic view of the IL (film)–PD (bounding layer) bilayer and electric potential distribution. 3D snapshot of interface morphology of (b) PD film (c) IL film. Initial film thickness, $h_0 =$ (b) 30 $nm$c(i) 30 $nm$ c(ii) 50 $nm$. $\psi_L = 20V$ and $(c)M = 0.0001\ mol/L$ [31].

film from $h_0$= 30 to 50 nm, which mostly leads to more rapid patterning. On the other hand, the formed patterns are less stable as they tend to coalesce at more advanced stages.

As a way of understanding the IL films ability to make smaller sized patterns in the EHD patterning process, molarity is increased. **Figure 9** portrays the influence of molarity and the initial film thickness on the number of pillars in the EHD patterning process. In the IL film interface, the presence of large numbers of pillars indicates the creation of smaller sized pillars, given a fixed area. Changing the molarity affects the conductivity of the IL films, which in turn affect the numbers and speed of formation of the pillars. **Figure 9(a)** shows a plot of the number of pillars with respect to nondimensional time, for a PD film of the same initial

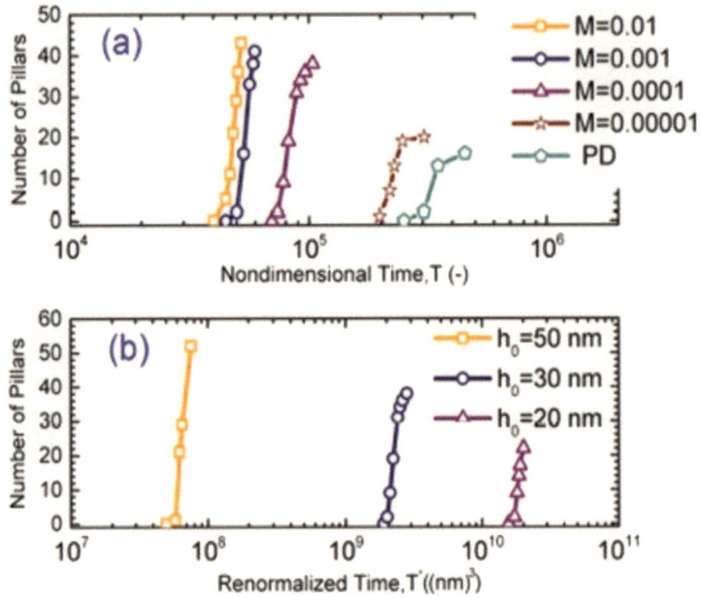


**Figure 9.** Number of formed pillars with (a) changes in molarity with a constant initial film thickness of $h_0$= 30 nm and (b) changes in initial film thickness with a constant molarity of M = 0.0001 mol/L [31].

thickness, $h_0$= 30 nm. The end point of each curve represents the final number of pillars formed when the quasi-stable conditions are established. As shown in **Figure 9(a)**, the final number of pillars formed for the PD film is 16, while that of the IL films is 20, 38, 41, and 43 for the following molarities, respectively: M = 0.00001, 0.0001, 0.001, and 0.01 mol/L.

Therefore, one can deduce that when the molarity is initially increased by 10 times, the number of formed pillars is almost doubled. However, any further increase in molarity causes a slight increase in the number of formed pillars, finally reaching a plateau. Additionally, it can be noted that increasing the molarity decreases the time gap between the initial and final pillar formed. **Figure 9(b)** shows how the initial film thickness affects the number of pillars formed. In order to cancel the effect of changing the film thickness, nondimensional time was normalized with the initial film thickness and was plotted against the number of pillars formed, keeping the molarity constant. The results showed that the higher initial film thicknesses tend to have a faster time evolution when compared to smaller film thicknesses.

## 6. Conclusion

In the confined liquid layer system, the electrically induced instabilities are enlarged leading to interface deformation and pattern generation. This technique has been employed as a fast and inexpensive method for patterning of molten polymer films. In this chapter, the electrified pressure-driven instability of thin liquid films and bilayers is discussed under the long-wave approximation limit. It is shown that the difference between electrical properties of film layers results in a net electrostatic force acting normal to the interface due to Maxell stress. The thin film equation governs the dynamics and instability of film layers, and its derivation is discussed. An extensive numerical study is performed to generate a map, in bilayer systems, based on the filling ratio and the electric permittivity ratio of layers. This map provides a baseline for IL-PD and PD-IL bilayers and is used as a predictive model for the formation of various structures in bilayer systems. To create different morphologies with lower pattern size, the net electrostatic force is increased by introducing an ionic conductive property of liquid layers. Thus, an electrostatic model is developed to find the net electrostatic force ionic liquid films and bilayers. The developed model is integrated to the thin film equation, and the spatiotemporal evolution of the interface is presented to show the compact and smaller sized pattern formation compared to the base cases of perfect dielectric films.

## Nomenclature

| | |
|---|---|
| $A$ | Hamaker constant |
| $d_i$ | Electrodes distance, $i = 1; 2$ |
| $D$ | Diffusion coefficient |
| $e$ | Electron charge magnitude |
| $\vec{f}_e$ | External body force |
| $F_E$ | Electrostatic force |
| $F_{os}$ | Osmotic force |
| $h$ | Interface height |
| $h_0$ | Initial mean film thickness |
| $I$ | Identity tensor |
| $k_B$ | Boltzmann constant |
| $l_0$ | Born repulsion cut off distance |
| $L$ | Domain length |
| $Ls$ | Scaling factor for length |
| $m$ | the mobility of charges/ions |
| M | Molarity |
| $\vec{n}$ | Normal vector to the interface |
| $n_\pm$ | Number concentration of ions/charges, positive or negative |
| $n_\infty$ | Bulk number concentration of ions/charges |

| | |
|---|---|
| $N_A$ | Avogadro number |
| P | Pressure |
| *Pos* | Osmotic pressure |
| $R_\pm$ | Species production rate in chemical reactions |
| $\vec{t_i}$ | Tangential vector to the interface, $i = 1$ *and* 2 |
| T | Temperature |
| *Ts* | Scaling factor for time |
| $\vec{u_i}$ | Velocity vector for film and bounding layer, $i = 1$ *and* 2 |
| $u_{relative}$ | Relative velocity |
| $x$ | $x$ direction in Cartesian coordinate |
| $y$ | $y$ direction in Cartesian coordinate |
| $z$ | $z$ direction in Cartesian coordinate |

## Greek Letters

| | |
|---|---|
| $s$ | Amplitude for growth of instabilities |
| $\varepsilon i$ | Electric permittivity of film and bounding layer, *i*=1 and 2 |
| $\varepsilon r$ | Electric permittivity ratio of layers |
| $\varepsilon 0$ | Free space electric permittivity |
| $\varphi$ | Conjoining/disjoining pressure |
| $\varphi Br$ | Born repulsion pressure |
| $\varphi EL$ | Electrostatic pressure |
| $\Phi s$ | Scaling factor for conjoining pressure |
| $\varphi T$ | Thermocapillary pressure |
| $\varphi vdW$ | van der Waals pressure |
| $\gamma$ | Relative interfacial tension between film and bounding layer |
| $\kappa$ | Inverse of Debye length |
| $\kappa v$ | Wave number for growth of instabilities |
| $\kappa^*$ | Mean interfacial curvature of the film interface |

$\lambda max$ Maximum wavelength for growth of instabilities

$\lambda$ Wavelength for growth of instabilities

$\mu$ Dynamic viscosity

$\nu$ Kinematic viscosity

$\rho$ Density

$\rho f$ Free charge density

$\psi i$ Electric potential within the layers, i=1 and 2

$\psi r$ Reference electric potential

$\psi l$ Electric potential of the lower electrode

$\psi s$ Interface electric potential

$\psi up$ Electric potential of the upper electrode

$\Psi$ Nondimensional electric potential

$\Delta t$ Time step

$\sigma$ Conductivity

$\tau c$ Charge relaxation time

$\tau p$ Process time

## Author details

Hadi Nazaripoor*, Adham Riad and Mohtada Sadrzadeh

*Address all correspondence to: hadi@ualberta.ca

Mechanical Engineering Department, University of Alberta, Edmonton, Canada

## References

[1] Pavesi A, Adriani G, Tay A, Warkiani ME, Yeap WH, Wong SC, Kamm RD. Engineering a 3D microfluidic culture platform for tumor-treating field application. Scientific Reports. 2016

[2] Shields CW, Reyes CD, López GP. Microfluidic cell sorting: A review of the advances in the separation of cells from debulking to rare cell isolation. Lab on a Chip. 2015;**15**(5):1230-1249

[3] Safarnia H, Sheikholeslami M, Ganji DD. Electrohydrodynamic nanofluid flow and forced convective heat transfer in a channel. The European Physical Journal Plus. 2016;**131**:96

[4] Sheikholeslami M, Chamkha AJ. Electrohydrodynamic free convection heat transfer of a nanofluid in a semi-annulus enclosure with a sinusoidal wall. Numerical Heat Transfer, Part A: Applications. 2016;**69**:781-793

[5] Sheikholeslami M, Hayat T, Alsaedi A, Abelman S. Numerical analysis of EHD nanofluid force convective heat transfer considering electric field dependent viscosity. International Journal of Heat and Mass Transfer. 2017;**108**:2558-2565

[6] Sheikholeslami M. Influence of coulomb forces on $Fe_3O_4$-$H_2O$ nanofluid thermal improvement. International Journal of Hydrogen Energy. 2017;**42**(2):821-829

[7] Teubner M. The motion of charged colloidal particles in electric fields. The Journal of Chemical Physics. 1982;**76**(11):5564-5573

[8] de Ven, Van and T. G., M.. Colloidal Hydrodynamics. San Diego, CA: Academic Press; 1989

[9] Ning W, Russel WB. Micro- and Nano-patterns created via Electrohydrodynamic instabilities. Nano Today. 2009;**4**:180-192

[10] Gambaryan-Roisman T. Modulation of Marangoni convection in liquid films. Advances in Colloid and Interface Science. 2015;**222**:319-331

[11] Mukherjee R, Sharma A. Instability, self-organization and pattern formation in thin soft films. Soft Matter. 2015;**11**:8717–8740

[12] Hong H, Tian H, Shao J, Ding Y, Jiang C, Liu H. Fabrication of bifocal microlens arrays based on controlled electrohydrodynamic reflowing of pre-patterned polymer. Journal of Micromechanics and Microengineering. 2014;**24**:095027

[13] Li H, Yu W, Wang Y, Bu H, Liu Z, Abraham E, Desmulliez MPY. Simulation of the electrohydrodynamic instability process used in the fabrication of hierarchic and hollow micro/nanostructures. RSC Advances. 2014;**4**:13774-13781

[14] Hu H, Tian H, Li X, Shao J, Ding Y, Liu H, An N. Biomimetic mushroom-shaped microfibers for dry adhesives by electrically induced polymer deformation. ACS Applied Materials & Interfaces. 2014;**6**:14167–14173

[15] Atta A, Crawford DG, Koch CR, Bhattacharjee S. Influence of electrostatic and chemical heterogeneity on the electric-field-induced destabilization of thin liquid films. Langmuir. 2011;**27**(20):2472-2485

[16] Oron A, Davis SH, Bankoff SG. Long-scale evolution of thin liquid films. Review of Modern Physics. 1997;**69**:931-980

[17] Israelachvili JN. Intermolecular and Surface Forces. Burlington, MA: Academic Press; 2011

[18] Masliyah JH, Bhattacharjee S. Electrokinetic and Colloid Transport Phenomena. Hoboken, NJ: Wiley-Interscience; 2006

[19] Vladimir SM. Dewetting of solid surface: Analogy with Spinodal decomposition. Journal of Colloid and Interface Science. 1993;**156**:491-497

[20] Harlow FH, Eddie Welch J. Numerical calculation of time-dependent viscous incompressible flow of fluid with free surface. Physics of Fluids. 1965;**8**:2182-2189

[21] Hirt CW, Nichols BD. Volume of fluid (VOF) method for the dynamics of free boundaries. Journal of Computational Physics. 1981;**39**:201–225

[22] Craster RV, Matar OK. Electrically induced pattern formation in thin leaky dielectric films. Physics of Fluids (1994-present). 2005;**17**:1131–1198

[23] Verma R, Sharma A, Kargupta K, Bhaumik J. Electric field induced instability and pattern formation in thin liquid films. Langmuir. 2005;**21**(8):3710-3721

[24] Tian H, Shao J, Ding Y, Li X, Liu H. Numerical characterization of electrohydrodynamic micro- or nanopatterning processes based on a phase-field formulation of liquid dielectrophoresis. Langmuir. 2013;**29**(15):4703-4714

[25] Yang Q, Li BQ, Ding Y. A numerical study of nanoscale electrohydrodynamic patterning in a liquid film. Soft Matter. 2013;**9**:3412-3423

[26] Hadi N. Electrohydrodynamic and Thermocapillary Instability of Thin Liquid Films [thesis]. University of Alberta: University of Alberta; 2016

[27] Nazaripoor H, Koch CR, Sadrzadeh M, Bhattacharjee S. Thermo-Electrohydrodynamic patterning in Nanofilms. Langmuir. 2016;**32**(23). DOI: 10.1021/acs.langmuir.6b01810

[28] Nazaripoor H, Koch CR, Sadrzadeh M. Enhanced electrically induced micropatterning of confined thin liquid films: Thermocapillary role and its limitations. Industrial & Engineering Chemistry Research. 2017;**56**. DOI: 10.1021/acs.iecr.7b02814

[29] Nazaripoor H, Koch CR, Bhattacharjee S. Electrical perturbations of ultrathin bilayers: Role of ionic conductive layer. Langmuir. 2014;**30**(49):14734-14744. DOI: 10.1021/la503839x

[30] Nazaripoor H, Koch CR, Sadrzadeh M, Bhattacharjee S. Compact micro/nano electrohydrodynamic patterning: Using a thin conductive film and a patterned template. Soft Matter. 2016;**12**(4)

[31] Nazaripoor H, Koch CR, Sadrzadeh M, Bhattacharjee S. Electrohydrodynamic patterning of ultra-thin ionic liquid films. Soft Matter. 2015;**11**(11):2193-2202

Chapter 3

# Studies of Electroconductive Magnetorheological Elastomers

Eugen Mircea Anitas, Liviu Chirigiu and Ioan Bica

Additional information is available at the end of the chapter

http://dx.doi.org/10.5772/intechopen.72732

**Abstract**

Electroconductive magnetorheological elastomers (MREs) have attracted a wide scientific attention in recent years due to their potential applications as electric current elements, in seismic protection, in production of rehabilitation devices, and sensors or transducers of magnetic fields/mechanical tensions. A particular interest concerns their behavior under the influence of external magnetic and electric fields, since various physical properties (e.g., rheological, elastic, electrical) can be continuously and/or reversibly modified. In this chapter, we describe fabrication methods and structural properties from small-angle neutron scattering (SANS) of various isotropic and anisotropic MRE and hybrid MRE. We present and discuss the physical mechanisms leading to the main features of interest for various medical and technical applications, such as electrical (complex dielectric permittivity, electrical conductivity) and rheological (viscosity) properties.

**Keywords:** electric field, active magnetic materials, magnetorheological elastomers, magnetodielectric effect, dielectric permittivity, dielectric polarization time

## 1. Introduction

Magnetorheological materials is a class of smart materials whose electrical, magnetic, mechanical, or rheological properties can be controlled under the application of an external magnetic field. Since their inception [1], magnetorheological materials have evolved into several main subclasses: magnetorheological fluids (MRFs), elastomers (MREs), and gels (MRGs). Generally, they consist of a non-magnetizable phase in which a second phase is embedded. The non-magnetizable phase is usually an elastic matrix based on natural or silicone rubber, while the second one consists of magnetizable nano/microparticles. For MRFs, the non-magnetizable phase is liquid [2, 3], while for MREs it is solid [4]. Although MRFs are the most common magnetorheological materials, their physical properties change with time due to the sedimentation process of the magnetizable phase. However, for both conventional [5] and hybrid [6] MREs, this phenomenon does not occur since magnetizable particles are linked to the polymer chains and are fixed in the elastic matrix after curing. In addition, MRFs are known to be a source of environmental contamination and they exhibit sealing problems when used in production of brakes, clutches, or variable-friction dampers [7, 8].

Due to this property, MREs have shown many promising technical, industrial, and bio-medical applications, such as in producing of adaptive tuned vibration absorbers [9], in fabrication of devices for varying the stiffness of suspension bushings [10], electric current elements [11, 12], or seismic protection [13]. Furthermore, MREs can be manufactured in the presence or in the absence of an external magnetic field, depending on the required applications. Usually, this is done at a constant temperature (>120°C) so that the flexibility of the magnetic particles is maintained [14]. In the first case, an isotropic MRE is produced, while in the second one, an anisotropic MRE is obtained [15]. The curing process for anisotropic MREs requires a magnetic field above 0.8 T to align the chains [14]. At the end of curing process, the magnetic particles are fixed inside the elastic matrix. By formation of parallel chains of magnetizable nano/microparticles along the magnetic field lines, drastic changes of the elasticity coefficients [16] and of the main shear tensions [17] can be induced in MREs. Therefore, knowledge of the nano/microstructure has fundamental implications in the production of various types of novel MREs with predetermined properties and functions.

The structural analysis of MREs is performed mainly by using electron microscopy, computed tomography (CT) or small-angle scattering [18, 24] (SAS; neutrons or X-rays) techniques. Real-space analysis (i.e., electron microscopy or CT) is in principle more powerful than reciprocal-space analysis (i.e., SAS), but usually it requires extensive sample preparation, and as a consequence the same sample cannot subsequently be used for additional investigations. These disadvantages can be overcome by CT, which gives a three-dimensional (3D) representation of the structure, but it is difficult to resolve structures smaller than few micrometers. A solution to this issue is provided by small-angle scattering techniques [19, 37, 38]. Although the loss of information in scattering is a severe limitation, SAS technique has the advantage that it is suitable for structures with dimensions within 1–1000 nm and the quantities of interest are averaged over a macroscopic volume. In particular, small-angle neutron scattering (SANS) is very useful in studying magnetic properties of materials or in emphasizing certain features [20–25] since neutrons interact with the atomic nuclei and with the magnetic moments in the sample.

In particular, for MREs in which the matrix and/or the fillers form self-similar structures (either exact or statistical) [26, 27], the most important advantage is that scattering methods can distinguish between mass [28] and surface fractals [29]. From an experimental point of view, the difference is accounted through the value of the scattering exponent $\tau$ in the region where the SAS intensity $I(q)$ has a power-law decay, that is,

$$I(q) \propto q^{-\tau}, \tag{1}$$

where $q = (4\pi/\lambda)\sin\theta$ is the scattering vector, $\lambda$ is the wavelength of the incident radiation, and $2\theta$ is the scattering angle. In terms of the fractal dimensions, the scattering exponent in Eq. (1) can be written as

$$\tau = \begin{cases} D_{\mathrm{m}}, & \text{for mass fractals} \\ 6 - D_{\mathrm{s}}, & \text{for surface fractals}' \end{cases} \tag{2}$$

where $D_{\mathrm{m}}$ is the mass fractal dimension with $0 < D_{\mathrm{m}} < 3$, and $D_{\mathrm{s}}$ is the surface fractal dimension with $2 < D_{\mathrm{s}} < 3$. Thus, when the measured absolute value of the scattering exponent is smaller than 3, the sample is considered to be a mass fractal with fractal dimension $\tau$, while if the exponent is between 3 and 4, the sample is considered to be a surface fractal with fractal dimension $6 - \tau$. The mass fractal dimension describes the way in which the mass $M(r)$ of a disk centered on the fractal varies with its radius $r$. The closer the value is to 3, the more compact is the structure. The surface fractal dimension describes the way in which the surface varies with the radius. When its value is close to 3, the fractal is so folded that it fills the space almost completely. However, when its value is close to 2, the surface is almost smooth. Therefore, for a mass fractal $D_{\mathrm{s}} = D_{\mathrm{m}} < 3$, while for a surface fractal, we have $D_{\mathrm{m}} = 3$ [30, 31].

For MREs, both the compactness of the structures formed as well as their roughness is mainly influenced by the volume fraction of the magnetizable phase or additives consisting of electroconductive particles, and by the internal structure of the matrix. In turn, by varying these parameters, various physical characteristics, such as dispersion and absorption characteristics of an electromagnetic field, or the electrical conductivity [32] can be tuned in the presence of an electric field superimposed over a magnetic one. Practically, a common way to perform such tuning is by preparing membranes based on MREs and placing them between two electrodes. In this way, they will act either as electric capacitors or as magnetoresistors whose output signals can be modified in the presence of an electric and/or magnetic field [33–35].

In this chapter, we present the fabrication process of few types of magnetorheological and magnetic elastomers based on silicone rubber (SR), silicone oil (SO), turmeric, carbonyl iron (CI), cotton fabric, Stomaflex, and Nivea creme. By using MREs, we fabricate membranes which are introduced between two plates and thus forming electrical capacitors. The structural properties are determined both in real and reciprocal space, by using scanning electron microscopy (SEM) and SANS. For the MREs, we investigate the behavior of the equivalent electrical capacitance $C_p$, and respectively of the equivalent resistance $R_p$ in a magnetic field with or without an electric field. The effect of volume concentration of various microparticles on the components of complex relative permittivity, electrical conductivity, and on viscosity is investigated by using the dipolar approximation. The components of elasticity, of mechanical tensions, and the modulus of elasticity are presented in the framework of a model based on elasticity theory.

## 2. Preparation of electroconductive magnetorheological elastomers

In this section, we present the materials and the fabrication methods used for preparation of several new classes of elastomers nanocomposites. For this purpose, we use materials with possible applications in biomedicine, consumer goods, and rubber engineering. The internal structure of elastomers is revealed with the help of SEM images.

### 2.1. Magnetic elastomers based on Stomaflex Creme and silicone oil

In this example, the base matrix is a complex medium in which the magnetic phase is dispersed and consists of a condensation curing a polysiloxane impression material, which is conventionally used in dental medicine. It is a low-viscosity, light-bodied, syringeable, and reline material for functional impressions, commonly known as Stomaflex Creme. The chemical components are polydimethylsiloxane, calcium carbonate, taste ingredients, and catalyst (dibutyitindilaureate, benzyl silicate and pigments). The base matrix is then mixed with silicone oil, and **Figure 1** (left part), presents an SEM of the final polymer matrix. The image shows that the matrix has a rough surface. Quantitatively, this is characterized by a surface fractal dimension of about $D_s = 2.47$ [23]. Generally, the surface fractal dimension lies between 2 and 3. A value close to 2 indicates a surface almost completely planar, while a value close to 3 indicate a surface so folded that it almost fills all the available space.

The magnetic phase consists of Fe particles with a mean radius of about 2 μm. **Figure 1** (right part) presents an SEM of the polymer matrix together with the magnetizable particles. From a structural point of view, the presence of the magnetizable particles alters the roughness of the matrix and thus the value of the surface fractal dimension is changed [23].

Depending on the required applications, the magnetic elastomer can be polymerized with or without magnetic field at various concentrations of Fe microparticles [23]. Furthermore, additives can also be used to keep magnetizable particles apart from each other. However, this issue arises mainly in MRFs, where additives have to hinder the sedimentation process.

### 2.2. Magnetic elastomers based on silicone rubber and magnetic liquid

For this class of magnetic elastomers, the matrix is a silicone rubber (SR) reinforced with silicone oil (SO), stearic acid (SA), catalyst (C), and magnetic liquid (ML) at various concentrations. ML contains magnetizable nanoparticles, oleic acid, and crude oil, and it has a density of 1.465 g/cm$^3$. Initially, an SA solution is prepared, which is then heated for homogeneity up to 350 K. This solution has the property that at about 300 K, it is in liquid phase, while at lower temperatures it crystallizes. Then, a homogeneous mixture is prepared by using SR, SA solution, C, and ML at

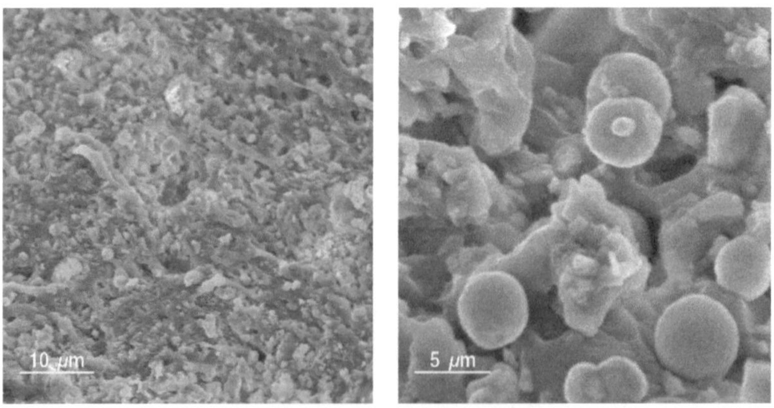


**Figure 1.** SEM images from MREs based on Stomaflex Creme and Fe microparticles [23]. Left part: The polymer matrix. Right part: The polymer matrix together with magnetizable phase.

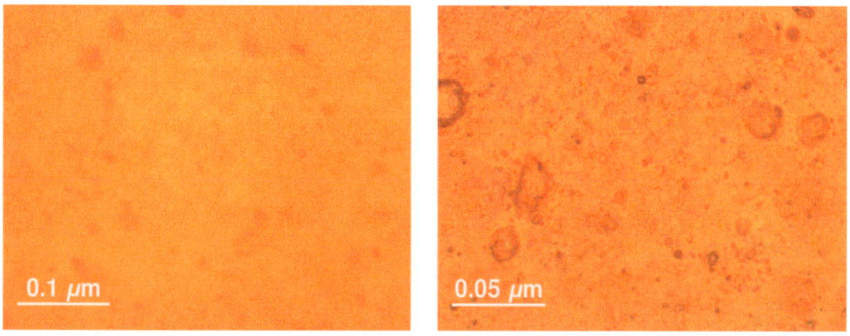


**Figure 2.** Electron microscopy images from magnetic elastomers based on silicone rubber and reinforced with silicone rubber, silicone oil, catalyst and magnetic liquid [27]. Left part: 1%vol. conc. Right part: 2%vol. conc.

various concentrations. Further, each new mixture is poured between two polyethylene thin foils and pressed in-between two parallel plates. After polymerization of about 24 h, we obtain the elastomeric magnetic membranes [27]. **Figure 2** (Left and Right) parts show the electron microscopy images for 1%vol. conc. and 2%vol. conc. of ML, respectively.

### 2.3. Magnetorheological elastomers based on silicone oil and Fe nano-/microparticles

An important class of MREs is based on silicone oil and reinforced with Fe nano/micro particles, since this gives electrical properties of importance for various particular applications. The materials used for producing this class of MRE are SO (with viscosity 200 mPa s at 260 K), Fe content of min. 97% and mass density 7860 kg/m$^3$, CI spherical micro particles with diameters between 4.5 and 5.4 μm, graphene nanoparticles (nGr) with granulation between 6 and 8 nm, cotton fabric gauze bandage (GB) (**Figure 3**, left part), and copper-plated textolite (*TCu*).

Based on these materials we prepare membranes (**Figure 3**, right part), following similar procedures as presented first in [6, 36]. First, homogeneous solutions are prepared by keeping constant the volume concentration of SO and CI, and by varying the volume concentration of SR and nGr. Each solution has a volume of 5 cm$^3$. Pieces of GB and copper-plated textolite are cut in rectangular shapes with sides 5×4 cm, and 5×5 cm respectively. Each GB cotton fabric is

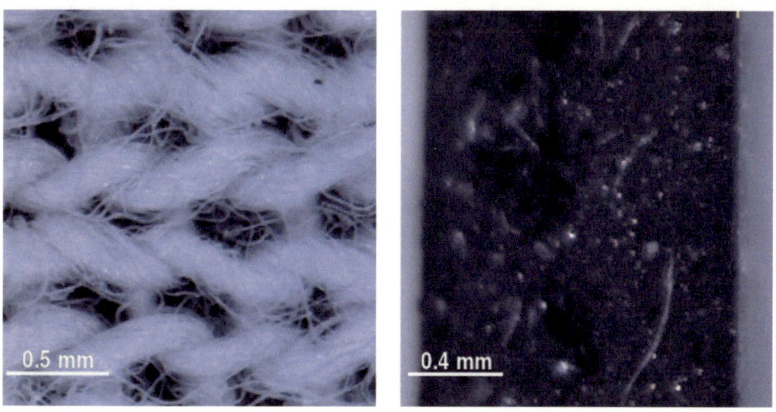


**Figure 3.** Left part: Electron microscopy of cotton fabric. Right part: Electron microscopy images from magnetic elastomers based on silicone rubber and reinforced with silicone rubber, and graphene nanoparticles [6, 36].

impregnated with solutions, which are then deposited on the copper-side plates of copper-plated textolite. Then, the impregnated cotton fabric is introduced between the copper-side plates in such a way that an edge of 0.5 cm of TCu remains uncovered. Finally, the whole system is compressed.

## 3. Structural properties using SANS

### 3.1. Theoretical background

In a SANS experiment, a beam of neutrons is emitted from a source and is directed toward the sample. In a scattering process, a small fraction is deviated from its initial path and is recorded by the detector. By considering that the scattering objects have the length $b_j$, then the scattering length density SLD can be written as $\rho(\boldsymbol{r}) = \sum_j b_j(\boldsymbol{r} - \boldsymbol{r}_j)$ [38], where $\boldsymbol{r}_j$ are the positions vectors. In a particulate system where the particles have density $\rho_m$ and the matrix has density $\rho_p$, the excess scattering SLD is d $\Delta\rho = \rho_m - \rho_p$. We consider also that the objects are randomly distributed and their positions are uncorrelated. Thus, the scattering intensities can be written as [37, 38] $I(q) = n|\Delta\rho|^2 V^2 \left\langle |F(\boldsymbol{q})|^2 \right\rangle$, where $n$ is the concentration of objects, $V$ is the volume of each object, and $F(\boldsymbol{q}) = 1/V \int_V e^{-i\boldsymbol{q}\cdot\boldsymbol{r}} \mathrm{d}^3 r$.

Many experimental SANS curves are characterized by a simple power-law of the type

$$I(q) = Aq^{-\tau} + B, \tag{3}$$

where $B$ is the background. Depending on the value of scattering exponent $\tau$, reflects the dimensionality of the object. **Figure 4** (left part), shows the corresponding simple power-law decay at fixed values of $A$ and $B$, for $\tau = 1$ (this rods), $\tau = 2$ (thin disks), $\tau = 2.5$ and 3.5 for fractals, and $\tau = 4$ ($3D$ objects).

However, a large number of SANS data obtained from MRE shows a succession of two power-law decays with arbitrarily exponents on a double logarithmic scale. This indicates the formation of a complex structure with two hierarchical levels. Therefore, for these data the Beaucage

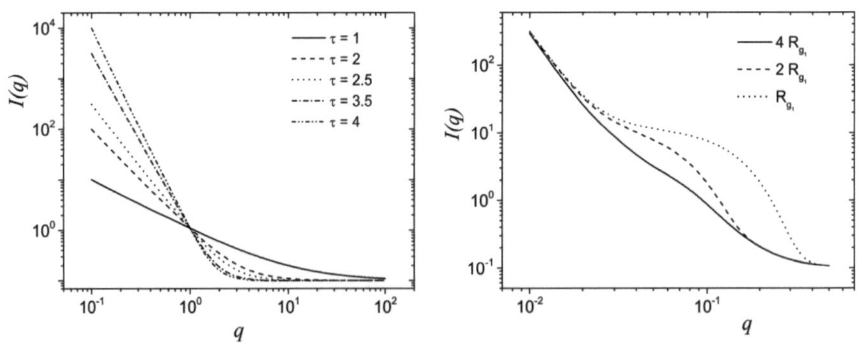


**Figure 4.** Left part: Simple power-law decays given by Eq. (3) for various values of the scattering exponent $\tau$. Continuous curves – Scattering from regular objects. Dashed curves – Scattering from fractal objects. Right part: Scattering intensity given by Beaucage model in Eq. (4) for various values of the radius of gyration of the smallest hierarchical level.

http://dx.doi.org/10.5772/intechopen.72732

model [39] is most commonly used. However, under some conditions a SANS experiment can show only a single power-law decay, and therefore only a single structural level can be investigated. For a two-level structure, the intensity can be approximated by [39]:

$$I(q) \propto G_1 e^{\frac{-q^2 R_{g1}^2}{3}} + B_1 e^{\frac{-q^2 R_{g2}^2}{3}} \left(\frac{h_1}{q}\right)^{D_1} + G_2 e^{\frac{-q^2 R_{g2}^2}{3}} + B_2 \left(\frac{h_2}{q}\right)^{D_2} + B. \tag{4}$$

Here, $h_1 = \text{erf}\left(qR_{g1}/\sqrt{6}\right)^3/q$, $h_2 = \text{erf}\left(qR_{g2}/\sqrt{6}\right)^3/q$ and $B$ is the background. The first term in Eq. 4 describes a large-scale structure of overall size $R_{g1}$ composed of small-scale structures of overall size $R_{g2}$, written in the third term. The second term allows for mass or surface fractal power-law regimes for the large structure. $G_1$ and $G_2$ are the classic Guinier prefactors, and $B_1$ and $B_2$ are the prefactors specific to the type of power-law scattering, specified in the regime in which the exponents $D_1$ and $D_2$ fall. **Figure 4** (right part) shows the scattering given by Beaucage models when the radius of gyration of the smaller structural level is varied, while all the other parameters are kept constant. For small values of the radius of gyration, the curves clearly show the appearance of a plateau (at about $4 \times 10^{-2} \lesssim q \lesssim 10^{-1}$), which may indicate that the sizes of the scattering units are much smaller than the distances between them. Similar behavior of the scattering curve has been observed also in [40–48].

### 3.2. SANS from magnetic elastomers based on Stomaflex Creme and silicone oil

**Figure 5** shows the SANS curves on a double logarithmic scale, corresponding to a polymer matrix consisting of Stomaflex creme and silicone oil (Left part), as well as the polymer matrix in which were embedded Fe particles (Right part) (see Ref. [24]). For the case of scattering from the polymer matrix, the experimental data are characterized by the presence of two successive simple power-law regions, with different values of the scattering exponents. Thus, these data can be modeled by using the simple power-law given by Eq. (3). At low values of the scattering wavevector $q\left(8 \times 10^{-3} \lesssim q\left(\text{Å}^{-1}\right) \lesssim 2.2 \times 10^{-2}\right)$, the absolute value of the scattering exponent is 3.88. This indicates that in this range the scattering signal arises from a surface fractal with fractal dimension $D_1 = 6 - 3.88 = 2.12$, which represents a very smooth surface. However, at higher values of the scattering wavevector ($2.2 \times 10^{-2} \lesssim q\left(\text{Å}^{-1}\right) \lesssim 2 \times 10^{-1}$), the absolute value of the

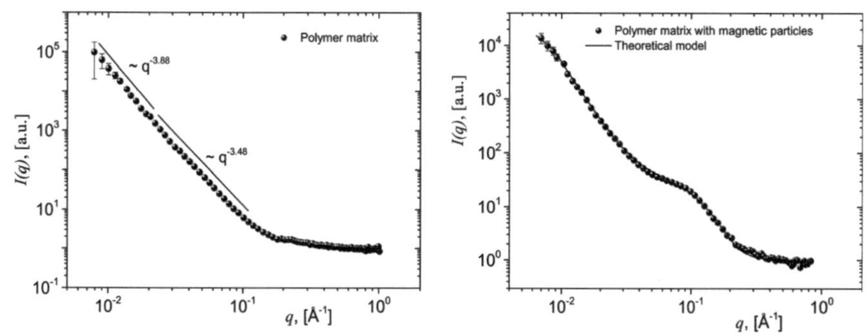


**Figure 5.** Left part: Scattering from the polymer matrix based on Stomaflex creme and silicone oil. Right part: Scattering from the polymer matrix with magnetic Fe particles.

scattering exponent decreases to 3.48, which indicates the presence of a more rough surface, with the fractal dimension $D_2 = 6 - 3.48 = 2.52$.

When magnetic particles are added, the SANS curve shows the presence of two simple power-law decays but with an additional "knee" in-between them, which arise at $q \simeq 1.5 \times 10^{-1}$ $\left(\text{Å}^{-1}\right)$ (**Figure 5**, right part). This shows the formation of multilevel structures in the magnetic elastomers and the Beaucage model (Eq. (4)) is generally used to extract information about the fractal dimensions and the overall size of each structural level. For the data shown in **Figure 7**, a fit with the Beaucage model reveals that the fractal dimensions of the two structural levels are $D_1 = 2.7$ and $D_2 = 2.4$, while the overall sizes are $R_{g_1} \geq 79$ nm and $R_{g_2} = 4.5$ nm, respectively.

### 3.3. SANS from magnetic elastomers based on silicone rubber and magnetic liquid

The SANS data from a polymer matrix based on silicone rubber (sample $M_1$) reinforced with magnetic particles at various concentrations are shown in **Figure 6** (left part) (see Ref. [27]). As in the previous case, the scattering data from the polymer matrix show a single power-law decay with the fractal dimension $D_1 = 2.62$. This shows the formation of a mass fractal with very ramified branches. Addition of magnetic particles leads to a two-level structure. For the sample with the lower concentration of particles (sample $M_2$), the Beaucage model gives $D_1 = 2.68$, $D_2 = 3.21$, $R_{g_1} = 34$ nm, and $R_{g_2} = 3$ nm. For the sample with the higher concentration of magnetic particles (sample $M_3$), one obtains $D_1 = 2.68$, $D_2 = 3.21$, $R_{g_1} = 34$ nm, and $R_{g_2} = 3$ nm. This complex arrangement shows that a mass fractal of nearly constant fractal dimension is composed of surface fractals whose fractal dimension increases significantly with increasing concentration of magnetic phase.

Another useful visual representation is by plotting the data after subtracting the contribution of the polymer matrix. This leads to the appearance of pronounced peaks, from which an estimation of the overall sizes can be inferred (**Figure 6**).

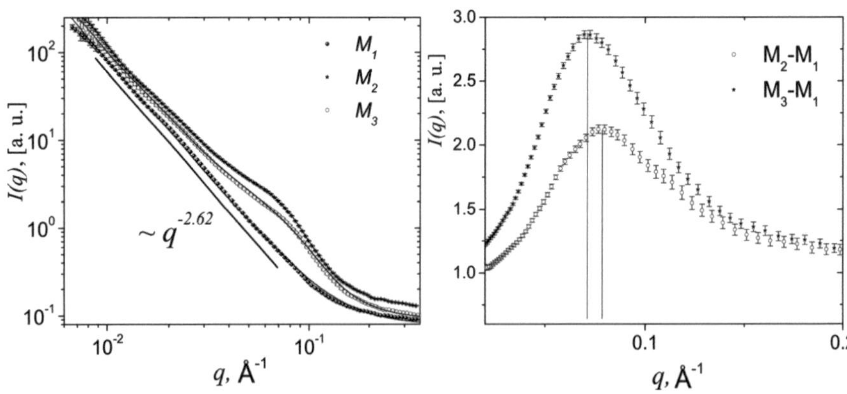


**Figure 6.** Left part: Scattering from polymer matrix based on silicone rubber (sample $M_1$) and scattering from polymer matrix with various concentrations of magnetic particles (samples $M_2$ and $M_3$). Right part: Contribution only of the particles. The vertical lines indicate the maximum size of the second structural level.

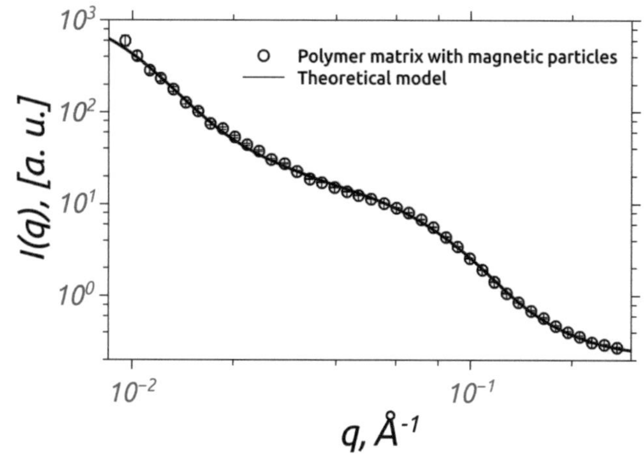


**Figure 7.** Scattering from the polymer matrix based on silicone rubber with graphene nanoparticles.

### 3.4. SANS from magnetorheological elastomers based on silicone oil and *Fe* nano-/microparticles

Another example of materials displaying a complex hierarchical organization in which fractal structures are formed, is presented for magnetorheological elastomers based on silicone rubber and graphene nanoparticles. Since, as expected, the SANS data of the matrix reinforced with graphene particles show a "knee" (**Figure 7**), the main structural characteristics are obtained using Eq. (4). Fitting the data reveals a complex fractal structure of size $R_{g1} \simeq 290$ Å, which consists of smaller structures of size $R_{g2} \simeq 35$ Å. The corresponding fractal dimensions are $D_1 = 2.51$ and $D_2 = 2.38$. This reveals the formation of fractal structures consisting of structural units, which are also fractals.

## 4. Physical properties of MRE-based membranes

The values of the corresponding fractal dimensions in such systems may help to understand the aggregation process and the interaction processes between particles, on the one hand, and between the polymer matrix, on the other hand. In turn, this may provide a better understanding of the physical properties, of various types of materials, such as membranes, based on these components. As an application, we shall present below in more detail the experimental setup and the fabrication process of a class of MRE-based membranes and study the electrical and rheological properties in the presence of a magnetic field.

### 4.1. Experimental setup and fabrication method

The materials used for manufacturing MRE-based membranes doped with graphene nanoparticles (nGr) are: silicone oil (SO), AP200 type, with viscosity 200 m Pa s at temperature 260 K; carbonyl iron (CI) spherical particles, C3518 type with diameters between 4.5 and 5.4 μm, Fe content of min. 97% and mass density 7860 kg/m$^3$; nGr with granulation between

6 and 8 nm; cotton fabric gauze bandage (GB), with the texture shown in **Figure 3** (left part); silicone rubber (SR) (black color); copper-plated textolite (TCu) with a thickness of 35 μm.

The MRE-based membranes are fabricated by performing the following steps: first, three homogeneous solutions are prepared with the chemical composition shown in **Table 1**. Each solution has a volume of 5 cm$^3$; three pieces of GB are cut in rectangular shapes with sides 5×4 cm, and six plates of TCu are cut also in rectangular shapes, with dimensions 5×5 cm; each GB cotton fabric is impregnated with solutions $S_i$, $i = 1, 2, 3$. Solutions are also deposited on the copper-side plates of TCu; the impregnated cotton fabric is introduced between the copper-side plates such that an edge of 0.5 cm of TCu is not covered, and then the whole system is compressed.

Using these solutions, we build a plane electrical capacitor $FC_1$ that corresponds to $S_1$, a second capacitor $FC_2$ that corresponds to $S_2$, and a third capacitor that corresponds to $FC_3$. We use GB for strengthening the membranes. The polymerization time is approximately 24 h. At the end of the polymerization process, one obtains three capacitors having as dielectric materials the membranes *Ms* with sizes 5×4 cm whose composition and thickness $d_0$ are listed in **Table 2**.

**Figure 3** shows the experimental configuration used for studying the MRE-based membranes in magnetic field. It consists of an electromagnet with a coil and a magnetic core current source, an RLC-meter, and a current source. The electric capacitor is placed between the poles of the electromagnet. The magnetic field intensity is measured with a gaussmeter. Magnetic field intensity can be continuously modified, through the intensity of the electric current provided by the source (*A*).

By changing the frequency $f$ of the electric current, we measure the equivalent capacitance ($C_p$) and resistance ($R_p$) of *FC*. We determine that for $f = 10$ kHz, the measured values are stable in time. Then we fix the frequency $f$ at this value and we vary the magnetic field intensity $H$ from 0 to 200 kA/m.

### 4.2. Electrical and rheological properties

**Figure 8** shows that the capacitance increases while the resistance decreases with increasing magnetic field intensity. The volume fraction Φ of nGr greatly influences the values of the equivalent capacitance and of resistance, respectively, for fixed values of magnetic field intensity.

In the presence of a magnetic field, the carbonyl iron microparticles become magnetized, and form magnetic dipoles which attract each other along the direction of $H$. The equation of motion is [49]

| $S_i$ | SR (cm$^3$) | SO (cm$^3$) | CI (cm$^3$) | nGr (cm$^3$) |
|---|---|---|---|---|
| $S_1$ | 3.0 | 1.0 | 1.0 | 0.0 |
| $S_2$ | 2.6 | 1.0 | 1.0 | 0.4 |
| $S_3$ | 2.0 | 1.0 | 1.0 | 1.0 |

**Table 1.** Chemical compositions and the volumes (in cm$^3$) for each component $S_i, i = 1, 2, 3$.

| *Ms* | SR (vol.%) | SO (vol.%) | CI (vol.%) | nGr (vol.%) | $d_0$ (mm) |
|---|---|---|---|---|---|
| $M_1$ | 60 | 20 | 20 | 0 | 4.30 |
| $M_2$ | 52 | 20 | 20 | 8 | 3.90 |
| $M_3$ | 40 | 20 | 20 | 20 | 4.46 |

**Table 2.** Chemical compositions and volume concentrations (%vol.) for membranes $M_i, i = 1, 2, 3$.

$$M\ddot{x} + 2\beta\dot{x} + \frac{3\mu_0}{\pi}\frac{\mu_s}{x^4}m^2 = 0, \tag{5}$$

where $M$ is the mass of the magnetic dipole, $2\beta$ is the particle friction coefficient, $x$ is the distance between two neighboring dipoles, $m$ is the magnetic dipole moment, $\mu_0$ and $\mu_s$ are the magnetic permeability of vacuum, and of SR, respectively. At $t = 0$, when the magnetic field is applied, the distance between the magnetic dipoles is [50] $X_0 = d_m/\sqrt[3]{\phi}$, where $\phi$ is the volume fraction of magnetic dipoles, and which coincides with the volume fraction of CI. For membranes with CI and nGr, the distance between the magnetic dipoles at $t = 0$ can be calculated using [50] $X_0^{nGr} = d_m/\sqrt[3]{\phi(1+\Phi)}$. Here, $\Phi$ is the volume fraction of nGr.

For membranes without nGr, the following conditions shall be satisfied:

$$x = \begin{cases} X_0, & \text{at } t = 0 \text{ and } H \neq 0, \\ x, & \text{at } t \neq 0 \text{ and } H \neq 0, \end{cases} \tag{6}$$

while for membranes with *nGr*, we have the conditions

$$x = \begin{cases} X_0^{nGr}, & \text{at } t = 0 \text{ and } H \neq 0, \\ x_{nGr}, & \text{at } t \neq 0 \text{ and } H \neq 0. \end{cases} \tag{7}$$

By using numerical values $d_m = 5\ \mu m$ for the mean diameter of CI particles, and $\rho_m = 7860\ \text{kg/m}^3$, the mass of magnetic dipole becomes $M \equiv \pi\rho_m d_m^3/6 = 0.514 \times 10^{-12}$ kg. Further, the quantity $2\beta$ in Eq. (5) can be approximated by $2\beta = 3\pi\eta d_m$, where $\eta$ is the viscosity. Thus, for $0.1 \le \eta(Pa\,s) \le 100$, we obtain the variation of the particle friction coefficient: $47.1 \times 10^{-7} \le 2\beta(N) \le 47.1 \times 0^{-4}$. Knowing that [49] $m = \pi d_m^3 \chi H/6$, where $\chi = 3(\mu_p - \mu_s)/(\mu_p + \mu_s)$ and

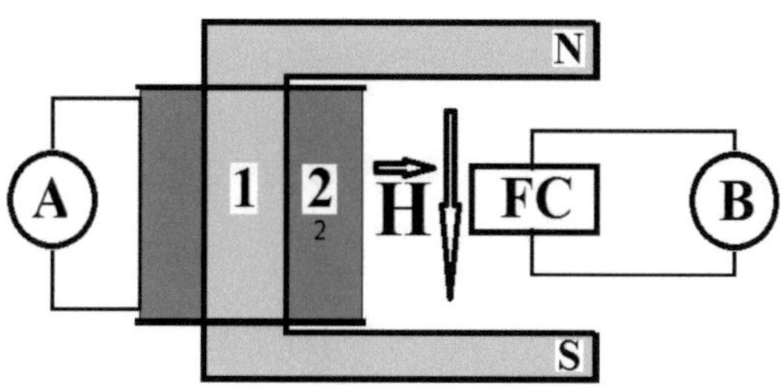


**Figure 8.** Experimental setup (overall configuration): *FC*–plane capacitor, *H*–magnetic field intensity, 1–magnetic core, 2–coil, A–source of continuous current, B–*RLC*–meter, *N* and *S*–magnetic poles [6].

since $\mu_p \gg \mu_s$, we can write in a good approximation that $\chi = 3$, where $\mu_p$ is the magnetic permeability of CI microparticles.

Since $8 \leq H(kA/m) \leq 200$, then $1.57 \times 10^{-8} \leq m\ (A\ cm^2) \leq 3.92 \times 10^{-7}$. Further, by using $d_m = 5\ \mu m$ and $\phi = 20\%$, we obtain the distance between magnetic dipoles $X_0 = 8.55\ \mu m$. Taking into account that $\mu_s = 1$ and $\mu_0 = 4\pi 10^{-7}$H/m, we obtain $0.55 \times 10^{-9} \leq 3\mu_0\mu_s m^2/(\pi x^4)\ (N) \leq 0.345 \times 10^{-6}$. Therefore, Eq. (5) becomes

$$2\beta\dot{x} + \frac{3\mu_0}{\pi}\frac{\mu_s}{x^4}m^2 = 0. \tag{8}$$

Using the numerical values $\chi = 3$ and $x = d_m$ into Eq. (8), we can write the equation of motion of magnetic dipoles along the magnetic field, such as

$$2\beta\dot{x} + \frac{3\pi\mu_0}{4}\mu_s d_m^2 H^2 = 0. \tag{9}$$

By integrating the last equation, and using the initial and boundary conditions, we obtain

$$x = X_0 - \frac{3\pi\mu_0}{8\beta}\mu_s d_m^2 H^2 t. \tag{10}$$

Then, from Eq. (10), the distance between two magnetic dipoles can be written as

$$x = d_m 1\sqrt[3]{\frac{}{\phi}}\left(1 - 3\pi\sqrt[3]{\phi}\frac{\mu_0}{8\beta}\mu_s d_m^2 H^2 t\right). \tag{11}$$

Following a similar procedure for the membrane with nGr, the equation of motion of magnetic dipoles becomes

$$x_{nGr} = d_m 1\sqrt[3]{\frac{}{\phi(1+\Phi)}}\left(1 - 3\pi\sqrt[3]{\phi(1+\Phi)}\frac{\mu_0}{8\beta}\mu_s d_m^2 H^2 t\right). \tag{12}$$

The number of magnetic dipoles from the membrane can be calculated according to [36] $N \equiv V\phi V_d = 6Lld_0/(\pi\phi(1+\Phi)d_m^3)$, where $V$ is the membrane's volume, $V_d$ is the average volume of *CI* microparticles, and $L, l$, and $d_0$ are the length, width, and the thickness of the membrane, respectively. The maximum number of dipoles from each chain is $N_1 = d_0 d_m$, and the total number of dipoles is $N_2 \equiv N/N_1 = 6Ll/(\pi\phi(1+\Phi)d_m^2)$. In a magnetic field, the membrane's thickness can be approximated by the relation $d = (N-1)x$, which together with Eq. (12) gives (for $N_1 \gg 1$):

$$d = d_0 1\sqrt[3]{\frac{}{\phi(1+\Phi)}}\left(1 - 3\pi\sqrt[3]{\phi(1+\Phi)}\frac{\mu_0}{8\beta}\mu_s d_m^2 H^2 t\right). \tag{13}$$

On the other hand, the electrical capacitance of the capacitor can be calculated from $C_p = \varepsilon_0\varepsilon_r' Ll/d$, where $\varepsilon_0$ is the absolute permittivity of vacuum, $\varepsilon_r'$ is the real component of the complex relative permittivity and $d$ is the thickness of the membrane for plane capacitors in magnetic field.

The equivalent resistance of $FC$ is modeled by the resistance of a linear resistor, namely, $R_p = d/(\sigma Ll)$, where $\sigma$ is the electrical conductivity of the membrane, and can be approximated by [51] $\sigma = 2\pi f \varepsilon_0 \varepsilon_r''$. Here, $f$ is the frequency of the electric field and $\varepsilon_r^*$ is the imaginary part of the complex relative permittivity. Thus, the variation of the equivalent electrical capacitance, with the magnetic field intensity becomes

$$C_p = \frac{C_{p_0}}{1 - 3\pi\sqrt[3]{\phi(1+\Phi)}\frac{\mu_0}{8\beta}\mu_s d_m^2 H^2 t,} \tag{14}$$

where, $C_{p_0} = \varepsilon_0 \varepsilon_{r_0}' Ll\sqrt[3]{\phi(1+\Phi)}/d_0$ is the equivalent capacitance of $FC$ at $H = 0$, and $\varepsilon_{r_0}'$ is the real component of the complex relative permittivity at $H = 0$. Similarly, the equivalent resistance can be obtained from

$$R_p = R_{p_0}\left(1 - 3\pi\sqrt[3]{\phi(1+\Phi)}\frac{\mu_0}{8\beta}\mu_s d_m^2 H^2 t\right), \tag{15}$$

where $R_{p_0} = 1/(\sigma Ll) \times d_0/\sqrt[3]{\phi(1+\Phi)}$ is the equivalent electrical resistance of $FC$ at $H = 0$.

The imaginary part of complex relative permittivity can be obtained using Eq. (13): $\varepsilon_r'' = d/\left(2\pi f R_p Ll\right)$. Eqs. (14) and (15) show that $C_p$ increases and $R_p$ decreases with increasing $H^2$. Then, the real component of the complex relative permittivity is obtained as:

$$\varepsilon_r' = \frac{C_p d}{\varepsilon_0 Ll}. \tag{16}$$

For $C_p = C_p(H)_\Phi$ shown in **Figure 9** (left part), $d \approx d_0$, $L = 0.05$ m, and respectively for $l = 0.04$ m, we obtain $\varepsilon_r' = \varepsilon_r'(H)_\Phi$, as shown in **Figure 10** (left part). From the dependence $R_p = R_p(H)_\Phi$ shown in **Figure 9** (right part), we obtain the variation $\varepsilon_r'' = \varepsilon_r''(H)_\Phi$ as shown in **Figure 10** (right part).

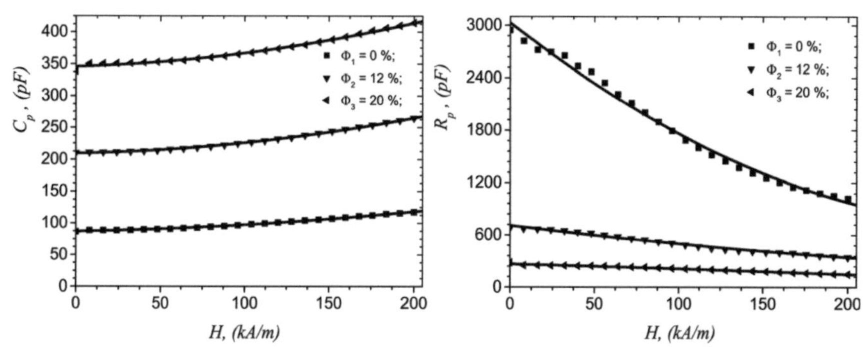


**Figure 9.** Left part: Capacitance. Right part: Resistance of *FC*, as a function of magnetic field intensity *H*. Discrete points – Experimental data, continuous line – Polynomial fit.

From the dependencies $\varepsilon_r'' = \varepsilon_r''\ (H)_\phi$ and $\varepsilon_r' = \varepsilon_r'(H)_\phi$ shown in **Figure 5** we obtain:

$$\varepsilon_r'' = \begin{cases} -11.5 + 0.8\varepsilon_r' - 0.0094\left(\varepsilon_r'\right)^2, & \text{for } M_1, \\ -52.0 + 1.720\varepsilon_r' - 0.014\left(\varepsilon_r'\right)^2, & \text{for } M_2, \\ -58.0 + 0.998\varepsilon_r' - 0.015\left(\varepsilon_r'\right)^2, & \text{for } M_3, \end{cases} \tag{17}$$

The points $\left(\varepsilon_r', \varepsilon_r''\right)$ are found on continuous semicircles (see **Figure 11**; left part) [6], and their values at a fixed $H$ depend on the volume fraction of $nGr$. For fixed electrical frequency $f = 10$ kHz of the electric field, and for $\varepsilon_r'' = \varepsilon_r''(H)_\Phi$ we obtain the dependency of electrical conductivity $\sigma = \sigma(H)_\Phi$ as shown in **Figure 11** (right part). The results show that the electrical conductivity increases with magnetic field intensity, and the obtained values increase sensibly with increasing the quantity of graphene nanoparticles.

With the help of Eq. (15) we can write that: $2\beta = 3\pi\sqrt[3]{\phi(1+\Phi)}\mu_0\mu_s d_m^2 H^2 t/8\left(1 - R_p/R_{p_0}\right)$, and using the relation $2\beta = 3\pi\eta d_m$, we obtain the viscosity of the membranes:

$$\eta = \sqrt[3]{\phi(1+\Phi)}\frac{\mu_0}{8\left(1 - \frac{R_p}{R_{p0}}\right)}\mu_s d_m^2 H^2 t. \tag{18}$$

Using $\Phi$ listed in **Table 2**, together with $\phi = 20\%$, $d_m = 5\ \mu m$, $t = 5$ s, and the functions $R_p/R_{p_0} = R_p/R_{p_0}(H)_\Phi$, into Eq. (18), we obtain:

$$10^7\ \eta_1(\text{Pa s}) = 22.922\ H^2(\text{kA/m})\frac{1}{\left(1 - \frac{R_p}{R_{p0}}\right)_{\Phi_1}}, \tag{19}$$

$$10^7\ \eta_2(\text{Pa s}) = 23.820\ H^2(\text{kA/m})\frac{1}{\left(1 - \frac{R_p}{R_{p0}}\right)_{\Phi_4}}, \tag{20}$$

$$10^7\ \eta_3(\text{Pa s}) = 24.374\ H^2(\text{kA/m})\frac{1}{\left(1 - \frac{R_p}{R_{p0}}\right)_{\Phi_6}}, \tag{21}$$

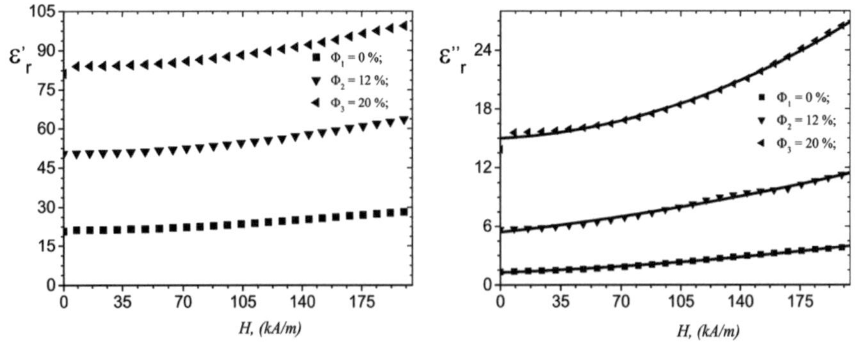


**Figure 10.** Variation of dielectric permittivity (left part), and of dielectric loss factor (right part), with magnetic field intensity.

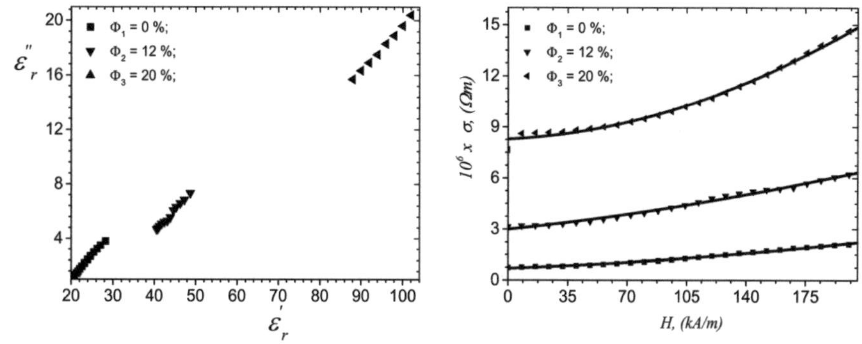


**Figure 11.** Left part: Cole-Cole diagrams for the membranes *Ms*; right part: Variation of electrical conductivity $\sigma$ with magnetic field intensity *H*.

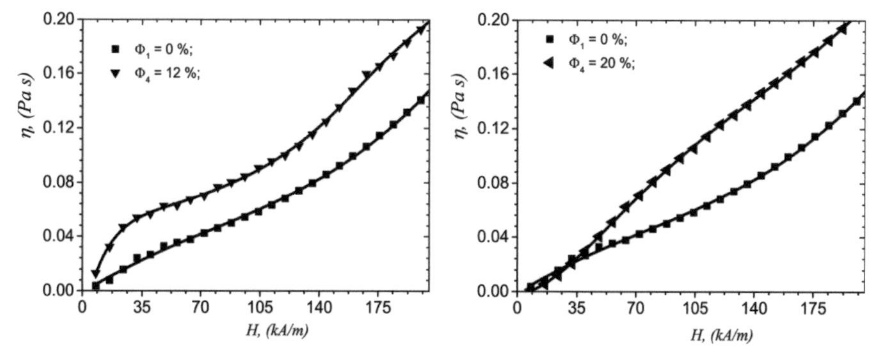


**Figure 12.** Variation of viscosity with magnetic field. Points – Experimental data, line – Polynomial fit.

This shows that viscosity increases with increasing the magnetic field intensity and is sensibly influenced by the nGr concentration as shown in **Figure 12**. By increasing the volume concentration of nGr, the space between the polymer molecules is saturated with nGr. Depending on the available space provided by the matrix, this increase may lead to formation of complex multilevel structures, in which one fractal (either mass or surface) consists of another fractal, as discussed in Section 3.

## 5. Conclusions

In this chapter, by using the EM and SANS techniques, we have presented the main structural properties of various classes of MREs produced with and without a magnetic field. We have shown that the fillers together with the matrix in which they are embedded form usually hierarchically (two-level) fractal structures. We have presented the corresponding theoretical models and have obtained the main parameters from scattering data, namely, the fractal dimensions and the overall sizes of each structural level.

For a recently obtained MRE based on cotton fabric reinforced with a solution containing silicone rubber, carbonyl iron microparticles, silicone oil, and graphene nanoparticles as fillers with various volume concentrations, we have shown in detail the fabrication process of a new class of smart membranes. The electrical and rheological properties of the membranes are greatly influenced by the quantity of fillers. In addition, they can be continuously modified in a magnetic field superimposed on an electrical field. We have shown that the components of the complex relative dielectric permittivity, the electrical conductivity, and the viscosity of the membranes are the result of complex interparticle interaction of the fillers inside the fractal polymer matrix. When the fillers themselves form fractal aggregates, this usually leads to formation of complex two-level hierarchical structures.

We have explained the electrical and rheological properties in the framework of dipolar approximation, where we have proposed a model based on the assumption that carbonyl iron microparticles become magnetic dipoles which attract each other. We have shown that the suggested theoretical model can describe with good accuracy the physical processes leading to the observed effects.

## Acknowledgements

We are grateful to our partners from JINR-Dubna, UMF-Craiova, and UVT-Timisoara for the fruitful collaboration.

## A. Appendix

In the following, we explain in more detail some important terms used throughout the chapter.

**Small-angle scattering** (SAS; neutrons, X-rays, light) is an experimental technique for investigating the structural properties of matter up to about 1 $\mu m$. This technique yields the elastic cross section per unit solid angle as a function of the momentum transfer ($q$) and describes, through a Fourier transform, the spatial density-density correlations of the system.

**Radius of gyration** ($R_g$) is the square root of the average density weighted squared distance of the scatterers from the center of the object and it is a measure of its overall size. In a SAS experiment, $R_g$ is determined from the end of plateau at low $q$ (Guinier region) on a double logarithmic scale.

**erf** is the error function (known also as Gauss error function) defined as

$$f(x) = \frac{1}{\sqrt{\pi}} \int_{-x}^{x} e^{-t^2} \mathrm{d}t$$

**Fractal dimension** (known also as Hausdorff dimension) is a measure of the degree of ramification (for mass fractals) or of the roughness of a surface (for surface fractals). Mathematically, it is defined by considering first a subset A of an n-dimensional Euclidean space, and $\{V)i\}$ the covering of A, with $a_i = \mathrm{diam}(V_i) \leq a$. Then, the $\alpha$-dimensional Hausdorff measure of A is

$$m^{\alpha}(A) \equiv \lim_{a \to 0} \inf_{\{V_i\}} \sum_i a_i^{\alpha}, \quad \alpha > 0,$$

where the infimum is on all possible coverings. The fractal dimension $D$ of the set A is defined by

$$D \equiv \inf\{\alpha : m^{\alpha}(A) = 0\} = \sup\{\alpha : m^{\alpha}(A) = +\infty\},$$

and it gives the value of $\alpha$ for which the Hausdorff measure jumps from zero to infinity. When $\alpha = D$, the measure can take any value between zero and infinity. However, for practical applications, this equation is difficult to use, and one usually resorts to other methods, such as mass–radius relation. For example, in the case of a deterministic (exact self-similar) fractal of length $L$, whose first iteration consists of elements of size $\beta_s L$, we have $M(L) = kM(\beta_s L)$, where $M$ is the "mass" (i.e., mass, volume, surface, etc.). Then, using the last equation we have $k\beta_s^D = 1$, from which the fractal dimension can be obtained. In a SAS experiment, $D$ is given by the absolute value of the scattering exponent in the fractal region.

## Author details

Eugen Mircea Anitas[1,2]*, Liviu Chirigiu[3] and Ioan Bica[4]

*Address all correspondence to: anitas@jinr.ru

1 Joint Institute for Nuclear Research, Dubna, Moscow Region, Russian Federation

2 Horia Hulubei National Institute of Physics and Nuclear Engineering, Bucharest-Magurele, Romania

3 University of Medicine and Pharmacy, Craiova, Romania

4 West University of Timisoara, Timisoara, Romania

## References

[1] Rabinow J. The magnetic fluid clutch. AIEE Trans. 1948;**67**:1308-1315. DOI: 10.1109/EE.1948.6444497

[2] Bica I, Liu YD, Choi HJ. Physical characteristics of magnetorheological suspensions and their applications. Journal of Industrial and Engineering Chemistry. 2013;**19**:394-406. DOI: 10.1016/j.jiec.2012.10.008

[3] Bica I, Balasoiu M, Bunoiu M, Iordaconiu L. Microparticles and electrodonductive magnetorheological suspensions. Romanian Journal of Physics. 2016;**61**:926-945

[4] Bunoiu M, Bica I. Magnetorheological elastomer based on silicone rubber, carbonyl iron and Rochelle salt: Effects of alternating electric and static magnetic fields intensities.

Journal of Industrial and Engineering Chemistry. 2016;**37**:312-318. DOI: 10.1016/j.jiec.2016.03.047

[5] Ginder JM, Nichols ME, Elie LD, Tardiff JL. Magnetorheological elastomers: Properties and applications. Proc. SPIE 3675, Smart Structures and Materials 1999. Smart Materials technologies. 1999. DOI: 10.1117/12.352787

[6] Bica I, Anitas EM. Magnetic field intensity and graphene concentration effects on electrical and rheological properties of MREs-based membranes. Smart Materials and Structures. 2017;**26**:105038(1-11). DOI: 10.1088/1361-665X/aa8884

[7] Deng HX, Gong XL. Application of magnetorheological elastomer to vibration absorber. Communications in Nonlinear Science and Numerical Simulation. 2008;**13**:1938-1947. DOI: 10.1016/j.cnsns.2007.03.024

[8] Shen Y, Golnarghi MF, Heppler GR. Experimental research and modeling of magnetorheological elastomers. Journal of Intelligent Materials Systems and Structures. 2004;**15**:2735. DOI: 10.1177/1045389X04039264

[9] Ginder JM, Schlotter WF, Nichols ME. Magnetorheological elastomers in tunable vibration absorbers. Proc. SPIE 4331, Smart Structures and Materials: Damping and Isolation. 2001;**4331**:103110. DOI: 10.1117/12.432694

[10] Watson JR, Canton M. Method and apparatus for varying the stiffness of a suspension bushing 1997. US Patent 5,609,353

[11] Bica I. Influence of the transverse magnetic field intensity upon the electric resistance of the magnetorheological elastomer containing graphite microparticles. Materials Letters. 2009;**63**:2230-2232. DOI: 10.1016/j.matlet.2009.07.032

[12] Bica I. Influence of magnetic field upon the electric capacity of a flat capacitor having magnetorheological elastomer as a dielectric. Journal of Industrial and Engineering Chemistry. 2009;**15**:605-609. DOI: 10.1016/j.jiec.2009.02.005

[13] Eem SH, Jung HJ, Koo JH. Application of MR elastomers for improving seismic protection of base-isolated structures. IEEE Transactions on Magnetics. 2011;**47**:2901-2904. DOI: 10.1109/TMAG.2011.2156771

[14] Liao GJ, Gong XL, Xuan SH, Kang CJ, Zong LH. Development of a real-time tunable stiffness and damping vibration isolator based on magnetorheological elastomer. Journal of Intelligent Materials Systems and Structures. 2012;**23**:25-33. DOI: 10.1177/1045389X11429853

[15] Chen L, Gong XL, Jiang WQ, Yao JJ, Deng XX, Li WH. Investigation on magnetorheological elastomers based on natural rubber. Journal of Materials Science. 2010;**42**:5483-5489. DOI: 10.1007/s10853-006-0975-x

[16] Bica I. Damper with magnetorheological suspension. Journal of Magnetism and Magnetic Materials. 2002;**241**:196-200. DOI: 10.1016/S0304-8853(02)00009-4

[17] Bica I, Anitas EM, Averis LME. Tensions and deformations in composites based on polyurethane elastomer and magnetorheological suspension: Effects of the magnetic

field. Journal of Industrial and Engineering Chemistry. 2015;**28**:86-90. DOI: 10.1016/j.jiec.2015.02.003

[18] Borin D, Gunther D, Hintze C, Heinrich G, Odenbach S. The level of cross-linking and the structure of anisotropic magnetorheological elastomers. Journal of Magnetism and Magnetic Materials. 2012;**324**:3452-3454. DOI: 10.1016/j.jmmm.2012.02.063

[19] Guinier A, Fournet G. Small-Angle Scattering of X-rays. New York: John Wiley & Sons; 1955. p. 263

[20] Brumberger H, editor. Modern Aspects of Small-Angle Scattering. New York: NATO ASI Series; 1995. p. 463

[21] Feigin LA, Svergun DI. Structure Analysis by Small-Angle X-ray and Neutron Scattering. New York and London: Plenum Press; 1987. p. 335

[22] Balasoiu M, Craus ML, Kuklin AI, Plestil J, Haramus V, Islamov AH, Erhan R, Anitas EM, Lozovan M, Tripadus V, Petrescu C, Savu D, Savu S, Bica I: Microstructure of magnetite doped elastomers investigated by SAXS and SANS. Journal of Optoelectronics and Advanced Materials 2008;**10**(11):2932-2935

[23] Balasoiu M, Anitas EM, Bica I, Erhan R, Osipov VA, Orelovich OL, Savu D, Savu S, Kuklin AI. SANS of interacting magnetic micro-sized Fe particles in a Stomaflex creme polymer matrix. Optoelectronics and Advanced Materials – Rapid Communications. 2008;**2**(11):730-734

[24] Anitas EM, Balasoiu M, Bica I, Osipov VA, Kuklin AI: Small-angle neutron scattering analysis of the microstructure of Stomaflex Creme - Ferrofluid based elastomers. Optoelectronics and Advanced Materials – Rapid Communications 2009;**3**(6):621-624

[25] Craus ML, Islamov AK, Anitas EM, Cornei N, Luca D: Microstructural, magnetic and transport properties of La0.5Pr0.2Pb0.3-xSrxMnO3 manganites. Journal of Alloys and Compounds 2014;**592**:121-126. DOI: 10.1016/j.jallcom.2014.01.002

[26] Balasoiu M, Craus ML, Anitas EM, Bica I, Plestil J, Kuklin AI: Microstructure of stomaflex based magnetic elastomers. Physics of the Solid State 2010;**52**(5):917-921. DOI: 10.1134/S1063783410050070

[27] Anitas EM, Bica I, Erhan RV, Bunoiu M, Kuklin AI. Structural properties of composite elastomeric membranes using small-angle neutron scattering. Romanian Journal of Physics. 2015;**60**(5-6):653-657

[28] Martin JE, Hurd AJ. Scattering from fractals. Journal of Applied Crystallography. 1987;**20**:61-78. DOI: 10.1107/S0021889887087107

[29] Schmidt PW. Small-angle scattering studies of disordered, porous and fractal systems. Journal of Applied Crystallography. 1991;**24**:414-435. DOI: 10.1107/S0021889891003400

[30] Bale HD, Schmidt PW. Small-angle X-ray-scattering investigation of submicroscopic porosity with fractal properties. Physical Review Letters. 1984;**53**:596-599. DOI: 10.1103/PhysRevLett.53.596

[31] Teixeira J. Small-angle scattering by fractal systems. Journal of Applied Crystallography. 1988;**21**(6):781-785. DOI: 10.1107/S0021889888000263

[32] Bica I, Anitas EM, Bunoiu M, Vatzulik B, Juganaru I. Hybrid magnetorheological elastomer: Influence of magnetic field and compression pressure on its electrical conductivity. Journal of Industrial and Engineering Chemistry. 2014;**20**:3994-3999. DOI: 10.1016/j.jiec.2013.12.102

[33] Bica I, Anitas EM, Averis LME, Bunoiu M. Magnetodielectric effects in composite materials based on paraffin, carbonyl iron and graphene. Journal of Industrial and Engineering Chemistry. 2015;**21**:1323-1327. DOI: 10.1016/j.jiec.2014.05.048

[34] Bica I, Anitas EM, Chirigiu L, Bunoiu M, Juganaru I, Tatu RF: Magnetodielectric effects in hybrid magnetorheological suspensions. Journal of Industrial and Engineering Chemistry 2015;**22**:53-62. DOI: 10.1016/j.jiec.2014.06.024

[35] Bica I, Anitas EM, Averis LME. Influence of magnetic field on dispersion and dissipation of electric field of low and medium frequencies in hybrid magnetorheological suspensions. Journal of Industrial and Engineering Chemistry. 2015;**27**:334-340. DOI: 10.1016/j.jiec.2014.09.047

[36] Bica I, Anitas EM, Chirigiu L. Magnetic field intensity effect on plane capacitors based on hybrid magnetorheological elastomers with graphene nanoparticles. Journal of Industrial and Engineering Chemistry. 2017;**56**:407-412. DOI: 10.1016/j.jiec.2017.07.039

[37] Brumberger H, editor. Modern Aspects of Small-Angle Scattering. New York: NATO ASI Series; 1995. p. 463

[38] Feigin LA, Svergun DI. Structure Analysis by Small-Angle X-ray and Neutron Scattering. New York and London: Plenum Press; 1987. p. 335

[39] Beaucage G. Approximations leading to a unified exponential/power-law approach to small-angle scattering. Journal of Applied Crystallography. 1995;**28**:717-728. DOI: 10.1107/S0021889895005292

[40] Anitas EM. Small-angle scattering from fat fractals. European Physical Journal B. 2014;**87**:139. DOI: 10.1140/epjb/e2014-41,066-9

[41] Cherny AY, Anitas EM, Osipov VA, Kuklin AI. Small-angle scattering from multiphase fractals. Journal of Applied Crystallography. 2014;**47**:198-206. DOI: 10.1107/S1600576713029956

[42] Anitas EM. Microscale fragmentation and small-angle scattering from mass fractals. Advances in Condensed Matter Physics. 2015;**2015**, ID 501281. DOI: 10.1155/2015/501281

[43] Anitas EM. A structural model for scattering intensities with multiple fractal regions of variable length. Journal of Optoelectronics and Advanced Materials. 2015;**17**:1122-1127

[44] Anitas EM. Scattering structure factor from fat fractals. Romanian Journal of Physics. 2015;**60**:647-652

[45] Cherny AY, Anitas EM, Osipov VA, Kuklin AI. The structure of deterministic mass, surface and multi-phase fractals from small-angle scattering data. Romanian Journal of Physics. 2015;**60**:658-663

[46] Anitas EM, Osipov VA, Kuklin AI, Cherny AY. Influence of randomness on small-angle scattering from deterministic mass fractals. Romanian Journal of Physics. 2015;**61**:457-463

[47] Anitas EM, Slyamov A, Todoran R, Szakacs Z. Small-angle scattering from nanoscale fat fractals. Nanoscale Research Letters. 2017;**12**:389. DOI: 10.1186/s11671-017-2147-0

[48] Anitas EM, Slyamov A. Structural characterization of chaos game fractals using small-angle scattering analysis. PLOS One. 2017;**12**(7):e0181385. DOI: 10.1371/journal.pone.0181385

[49] Melle S. Study of the Dynamics in Magnetorheological Suspensions Subject to External Fields by Means of Optical Techniques. Madrid: University of Madrid; 1995. PhD Thesis

[50] Jones TB. Electromechanics of Particles. New York, USA: Cambridge University Press; 1995

[51] Liu J. Models of Electromagnetic Properties of Composite Media. Iowa, USA: Iowa State University; 2012. http://lib.dr.iastate.edu/etd/12695. PhD Thesis

# Chapter 4

# Dielectrics under Electric Field

Liu Hongbo

Additional information is available at the end of the chapter

http://dx.doi.org/10.5772/intechopen.72231

**Abstract**

The chapter first gives a brief introduction on conduction, polarization, dissipation, and breakdown of dielectrics under electric field. Then, two of electric field-related applications, dielectrics for electrical energy storage and electrocaloric (EC) effect for refrigeration are discussed. Conclusion and perspectives are given at last.

**Keywords:** dielectrics, electrical energy storage, electrocaloric refrigeration

## 1. Introduction

Dielectrics are materials that can be polarized by an applied electric field. Polarizability is the essential property for dielectrics. The term is closely related to insulator. In electrical phenomena, insulator is commonly used especially in electronic engineering and electrical engineering, that is, in electronic packaging printed circuit board, electrical wire, high voltage system, and so on. It has a longer history than "dielectrics." The main property of an insulator is to prevent the flow of current when it is not desired. This means that insulator must have low electrical conductivity and can resist breakdown under high electric field. Physically, insulator is a subgroup of dielectrics because of the existence of polarization. And dielectrics can include insulator, semiconductor, and other materials with polarizability. Nevertheless, poor insulation could screen the polarizability of dielectrics under electric field, which makes the polarization hard to be "seen" by electrical measurement. And in most cases, poor insulation makes dielectrics useless. Thus, insulating property is commonly expected for dielectrics. Although insulating dielectrics has been applied for long time, the study on the area is continually progressing, that is, using cold sintering processing [1], developing broadband dielectric spectroscopy [2, 3], optoelectrical effect [4, 5], and multiferroics [6].

The subgroups of insulating dielectric include piezoelectric, pyroelectric (electrocaloric), ferroelectric, and multiferroic. Electrocaloric is the reverse effect of pyroelectric. Because of interesting physics and technological importance, there are enormous books dedicated to these areas.

Thus, we only introduce the fundamental phenomena of dielectric under electric field in the chapter. Then, two state-of-the-art applications for dielectrics are included, which are electrical energy storage [7, 8] and electrocaloric refrigeration [9–11]. We show that electric field plays a critical role in both applications.

## 2. Dielectrics under electric field

### 2.1. Conductivity

The main properties for dielectric under electric field are conduction, polarization, dissipation, and breakdown. Although normally a dielectric material is a good insulator, there are still some charge carries flowing through the whole material under electric field, which is called leakage current. Under a direct current (dc) field, the current could be a constant for a material. If the detected current is ***I*** when a voltage ***U*** is applied to a material, the conduction ***G*** of the materials is ***I/U***. If the material has a regular plate structure with surface *A* and thickness *t*, the conductivity $\sigma$ of the material is:

$$\sigma_{DC} = \frac{I}{U}\frac{d}{A} \tag{1}$$

The property can also be expressed by bulk resistivity $\rho$, which is reciprocal of conductivity:

$$\rho = \frac{U}{I}\frac{A}{d} \tag{2}$$

Both conductivity and resistivity are electrical properties for all materials, that is, conductor, semiconductor, and insulator. Because of different mechanism of conduction, there is a huge gap between conductivities of dissimilar materials, which has been well understood by energy band theory. The conductivity of a conductor could be big as $10^9$ S/m; the conductivity of good insulating dielectrics could be small as $10^{-18}$ S/m.

The DC conductivity in dielectrics is related to the hopping transition of defect charge carries, and thus, it is strongly temperature dependent:

$$\sigma_{DC} = \sigma_0 e^{-Ea/k_B T} \tag{3}$$

Here $k_B$ is Boltzmann constant, $T$ is temperature, and $E_a$ is thermal activity energy. By fitting experimental date with Eq. 3, the fitted activation energy $E_a$ gives information about current carriers.

The conductivity of dielectrics under alternative current (ac) field is related to the imaginary part of dielectric constant, which is introduced in Section 3. Since insulating is prerequired for dielectrics, the conductivity is less considered in the study of dielectrics. However, in some cases, it is crucial for dielectric properties [7].

http://dx.doi.org/10.5772/intechopen.72231

### 2.2. Polarization

Polarization is the unique electrical properties for dielectrics, depending on which concept is defined. There are diverse types of polarization, that is, electronic polarization, ionic polarization, orientation polarization, interface polarization, spontaneous polarization, and so on.

In dielectrics, if one positive charge $+q$ and the corresponding negative charge $-q$ are separated by a distance $\boldsymbol{d}$, an electric dipole is produced with magnitude $\boldsymbol{p} = q\boldsymbol{d}$. It is a vector with direction from negative charge toward positive charge. This dipole could be the one induced by electric field like in electronic polarization or ionic polarization, but it could also be one permanent like in polar molecules and ferroelectrics. After introducing electric dipole, the polarization in a material can be defined as the sum of individual dipole moments $\boldsymbol{p}$ divided by volume $V$ of the material:

$$P = \frac{\sum p}{V} \tag{4}$$

From this definition, we can see that the magnitude of polarization equals to the surface charge density, in fact depending on which the spontaneous polarization of ferroelectrics is measured by Sawyer-Tower electrical bridge.

For all dielectrics under electric field, the mutual shifts of nuclei and electrons induce electronic polarization, which occur within very short time, that is, $10^{-14}$–$10^{-16}$ s. Electronic polarization exists for all materials. In ionic crystals, the relative separation of cation and anion is induced by electric field, which is called ionic polarization. The ionic polarization also builds at a very short time, that is, $10^{-12}$–$10^{-13}$ s.

In some molecules, the center of positive charge and negative charge does not coincide, which is called polar molecule. In polar molecules, permanent dipoles exist and can somehow rotate under electric field. In thermal equilibrium state, the dipoles randomly orient, and thus, there is no net polarization. When an external electric field applied, these dipoles align to it in some extent. Thus, orientation polarization is induced. The response time of these permanent dipoles changes from material to material, that is, $10^{-2}$–$10^{-12}$ s. For example, in water at 300 K, the response time is around $5 \times 10^{-11}$ s.

Interface polarization, which is also called space charge polarization, builds at interface between two or more heterogeneous medium, i.e., between electrodes and dielectrics, between different dielectrics, between grain boundaries in ceramics, etc. The electrons, ions, and other defects trap at these sites, which have a slow response toward AC electric field. The response time varies from case to case, that is, $10^{7}$–$10^{-6}$ s.

In ferroelectrics, from high-temperature paraelectric phase to low-temperature ferroelectric phase, a permanent dipole is present because of symmetry break without electric field, which is called spontaneous polarization. The spontaneous polarization can be reversed by an external electric field, which makes ferroelectrics a smart material with lots of interesting properties.

In general, the relation between polarization ***P*** and electric field ***E*** follows [12]:

$$P = \varepsilon\chi E + \text{higher terms in E} \tag{5}$$

where $\varepsilon_0$ is the permittivity of free space, and $\chi$ is the susceptibility. The equation does not include spontaneous polarization of ferroelectrics. For most dielectrics, the first term is dominant. Higher terms are commonly omitted except for nonlinear dielectrics.

From Eq. (5), we can see that $\chi$ represents the polarizability. But the most widely used parameter is dielectric permittivity $\varepsilon$:

$$\varepsilon = \frac{D}{E} \tag{6}$$

Here ***D*** is electric displacement. And in engineering, the relative dielectric permittivity $\varepsilon_r = \varepsilon/\varepsilon_0$, or more generally called dielectric constant, is used because $\varepsilon$ is too small in SI unit.

For a capacitor, electric displacement equals to the surface charge density. Assume a planar capacitor consisting of two parallel electrodes with surface *S* and distance *d*. When a voltage ***V*** is applied and there is no dielectric material between electrodes, the surface charge density $\boldsymbol{Q_0} = \varepsilon_0 \boldsymbol{V}/t$ appears according to Gauss' law. If a dielectric material with susceptibility $\chi$ is filled in between two electrodes, it contributes surface charge density $\boldsymbol{Q_d} = \boldsymbol{P} = \varepsilon_0\chi \boldsymbol{V}/t$. As a result, the total surface charge equals the sum of two: $\boldsymbol{Q} = \varepsilon_0(1+\chi)\boldsymbol{V}/t$. As a result, combining Eqs. (5) and (6), we can get the relation between relative dielectric permittivity $\varepsilon_r$ and susceptibility $\chi$:

$$\varepsilon_r = (1 + \chi) \tag{7}$$

Both $\varepsilon_r$ and $\chi$ are parameters describing the polarizable property of dielectrics under electric field.

### 2.3. Dielectric dissipation

Under AC electric field, there are two types of current flowing through a capacitor, the so-called polarization current $\boldsymbol{I_P}$ and conduction current $\boldsymbol{I_R}$. In this case, dielectric constant is expressed by a complex number:

$$\varepsilon_r = \varepsilon_r^{'} - i\varepsilon_r^{''} \tag{8}$$

The first term is still called dielectric constant, while the second term is called the imaginary part of dielectric constant. When an AC electric field $\boldsymbol{E} = \boldsymbol{E_0}e^{i\omega t}$ ($\boldsymbol{E_0}$ is a constant, $\omega = 2\pi f$, and $f$ is frequency) applied to a dielectric material, its equivalent circuit can be expressed as shown in **Figure 1**. The current $\boldsymbol{I_P}$ flows through capacitance saving energy, while the current $\boldsymbol{I_R}$ flows through resistance dissipating energy. If the surface of the capacitor is *S* and thickness of it is *d*, the polarization current is $I_P = i\omega\varepsilon_0\varepsilon_r^{'}SE_0e^{i\omega t}$, while conduction current is $I_C = \omega\varepsilon_0\varepsilon_r^{''}SE_0e^{i\omega t}$, which means that the phase of polarization current advances applied electric field by 90° while conduction current keeps the same phase as that of applied electric field. The ratio between the two defines dielectric dissipation, also called dielectric loss, or loss tangent $tan\delta$:

http://dx.doi.org/10.5772/intechopen.72231

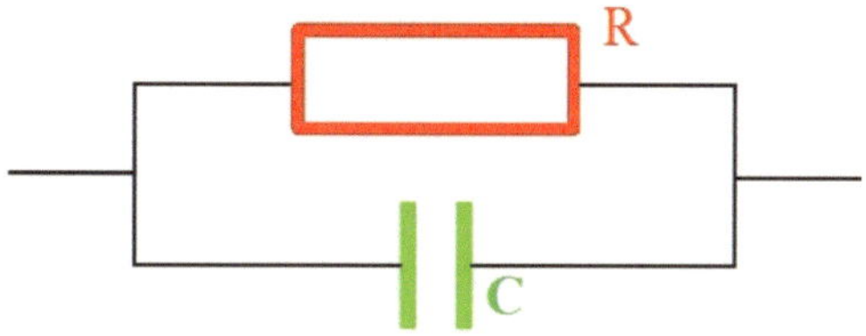


**Figure 1.** Equivalent circuit for a dielectric capacitor.

$$\tan\delta = \frac{I_C}{I_P} = \frac{\varepsilon_r''}{\varepsilon_r'} \tag{9}$$

The conduction current is normally harmful for applications. Thus, there are lots of work on how to decrease dielectric loss.

### 2.4. Breakdown

Under very high electric field, the conduction current increases significantly. Finally, a dielectric enters a high conduction nonequilibrium state, which is called breakdown. Mathematically, it can be expressed as:

$$\frac{dI}{dU} \rightarrow \infty \text{ or } \frac{dU}{dI} \rightarrow 0 \tag{10}$$

If a material with thickness $d$ and breakdown at voltage $U_B$, the breakdown field of the material is $U_B/d$, which changes for material to material, that is, ~MV/m for ceramics, ~10 MV/m for ceramic thin film, and ~100 MV/m for polymers. The breakdown field is also called dielectric strength. For most materials, the breakdown is normally induced by pores, impurities, and other defects because of electric inhomogeneities. Thus, processing control is critical for obtaining a high breakdown field. Covalent compounds commonly have larger breakdown field than ionic compounds. So far, 1.2 GV/m breakdown field gotten in $SiO_2$-BaO-$B_2O_3$-$Al_2O_3$ glass perhaps is the highest value for dielectrics [13].

Except electric field, heat also induces breakdown. Thermal breakdown occurs when a material cannot efficiently dissipate the produced heat. In many cases, a breakdown is companied by both electric and thermal breakdowns. Breakdown also has randomicity, which can be described by Weibull distribution.

## 3. Dielectric relaxation

One essential problem in dielectrics is the response time of a dielectric material toward to a periodic external electric field. When polarization delays with respect to an external oscillating electric field, so-called dielectric relaxation occurs. Dielectric relaxation may happen at an intrinsic time (so-called characteristic relaxation time) for a particular polarization, so it can help to identify the specific polarization mechanism. It also induces significant energy loss (or, in some case, energy conversion like oscillator, and so on), which is very important for engineering applications. Because of mutual inverse between time $t$ and frequency $f$ ($f = 1/t$), the dielectric relaxation is more conventionally studied in frequency domain by dielectric spectroscopy technology. The typical impedance analyzer (or also called LCR meter) usually performs from few Hz to 10 MHz. From MHz to GHz, vector network analyzer can finish the job. For higher frequency, more sophisticated spectroscopy technologies may be applied, that is, waveguided system, THz time-domain spectroscopy, and so on. However, these methods in many cases are in-house and not commercialized.

For clarifying a relaxation mechanism, different electric parameters may be applied, that is, complex dielectric constant $\varepsilon^*=\varepsilon'-i\varepsilon''$, complex impedance $Z^* = Z'-iZ''$, complex admittance $Y^* = Y'-iY''$, and complex electric modulus $M^*=M'-iM''$. Electric modulus is the reciprocal of dielectric permittivity $M^* = 1/\varepsilon^*$, corresponding to the relaxation of electric field in a material when electric displacement remains constant.

### 3.1. Debye relaxation

Debye relaxation equation is the most simple and elegant mathematic equation so far used to describe relaxation phenomenon:

$$\varepsilon_r = \varepsilon_\infty + \frac{\varepsilon_s - \varepsilon_\infty}{1 + i\omega\tau} \tag{11}$$

Here $\varepsilon_s$ and $\varepsilon_\infty$ are the static and high-frequency dielectric permittivity, respectively. $\tau$ is the characteristic relaxation time. This relaxation was first introduced by physicist Peter Debye for noninteracting dipoles in an alternating external electric field with unique relaxation time. In complex plane, $\varepsilon'$-$\varepsilon''$ plots an ideal semicircle for Debye relaxation. Because of the mutual interactions of dipoles in real materials, the semicircle changes to an ellipse, which is called Cole-Cole circle. The relaxation is called Cole-Cole relaxation, which is described by the following equation:

$$\varepsilon_r = \varepsilon_\infty + \frac{\varepsilon_s - \varepsilon_\infty}{1 + (i\omega\tau)^{1-\alpha}} \tag{12}$$

Here exponent parameter $\alpha$ takes a value between 0 and 1. When $\alpha = 0$, the Cole-Cole equation reduces to Debye equation. Cole-Cole relaxation is seen in many polymers.

### 3.2. Maxwell-Wagner relaxation

Maxwell-Wagner relaxation is because of electric inhomogeneous of materials, that is, interface between dielectrics and electrodes, grain boundaries, and so on. It is usually represented by a two-layer equivalent electric circuit as shown in **Figure 2**.

The dielectric permittivity of Maxwell-Wagner relaxation can also be expressed as Debye relaxation style:

http://dx.doi.org/10.5772/intechopen.72231

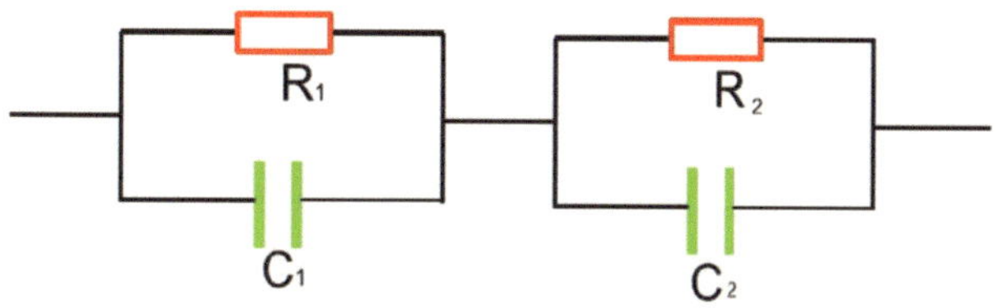


**Figure 2.** A two-layer equivalent electric circuit for Maxwell-Wagner relaxation.

$$\varepsilon_r = \varepsilon_\infty + \frac{\varepsilon_s - \varepsilon_\infty}{1 + i\omega\tau} \text{-} i\frac{\sigma}{\omega} \tag{13}$$

if we set the following parameters: $C_0 = \varepsilon_0 S/d$; $\varepsilon_\infty = C_1 C_2/[C_0(C_1 + C_2)]$; $\varepsilon_0 = [R^2{}_1 C_1 + R^2{}_2 C_2]/[C_0(R_1 + R_2)^2]$; $\delta = 1/[C_0(R_1 + R_2)]$; $\tau = R_1 R_2(C_1 + C_2)/(R_1 + R_2)$. It is obvious that the relaxation would be observed when there is a significant difference between two resistances. In **Figure 2**, each RC unit represents an ingredient in studied dielectrics, that is, grain, grain boundary, or electrode interface. To separate these different ingredient relaxations, plotting date in both impedance and electric modulus is commonly applied. The impedance of the circuit is as follows:

$$Z' = \frac{R_1}{1 + (\omega R_1 C_1)^2} + \frac{R_2}{1 + (\omega R_2 C_2)^2} \tag{14}$$

$$Z'' = R_1 \frac{\omega R_1 C_1}{1 + (\omega R_1 C_1)^2} + R_2 \frac{\omega R_2 C_2}{1 + (\omega R_2 C_2)^2} \tag{15}$$

The electric modulus of the circuit is as follows:

$$M' = \frac{C_0}{C_1} \frac{(\omega R_1 C_1)^2}{1 + (\omega R_1 C_1)^2} + \frac{C_0}{C_2} \frac{(\omega R_2 C_2)^2}{1 + (\omega R_2 C_2)^2} \tag{16}$$

$$M'' = \frac{C_0}{C_1} \frac{\omega R_1 C_1}{1 + (\omega R_1 C_1)^2} + \frac{C_0}{C_2} \frac{\omega R_2 C_2}{1 + (\omega R_2 C_2)^2} \tag{17}$$

From these equations, we can find that the impedance normalization highlights big resistance relaxation, while the electric modulus normalization highlights small capacitance relaxation. It should be noted that the equivalent electric circuit changes from case to case, and more complex circuit model may be applied for treating dielectric behavior of Maxwell-Wagner relaxation [7]. Here, we only give the simplest case in which there only exist two dielectric ingredients.

### 3.3. Universal dielectric relaxation

Universal dielectric relaxation is also called universal dielectric response. To have a better understanding of it, we first introduce complex conductivity $\sigma$.

$$\sigma = \sigma_0 + \sigma^{'} - i\sigma'' \tag{18}$$

Here, $\sigma_0$ is the DC conductivity and $\sigma'$ is the AC conductivity. The existence of $\sigma_0$ in the formula is an open question. The relationship between $\sigma_0$ and $\sigma$ can be looked as: $\sigma_0 = \lim \sigma_{'}$ as $f \to 0$. Generally, from dielectric measurement, the AC conductivity is calculated by:

$$\sigma = \varepsilon_0 \varepsilon'' \omega \tag{19}$$

For many dielectrics, A. K. Jonscher found that the frequency-dependent AC conductivity satisfies the following exponent relation [14]:

$$\sigma = A\omega^n (0 < n < 1) \tag{20}$$

which is called universal dielectric relaxation. If $n = 1$, it is called nearly constant loss. Universal dielectric response is understood as ion hopping effect. But, electric heterogeneity can also induce this behavior at high frequency. Nearly constant loss happens at sufficiently low temperature or high frequency, and the origin was understood as vibration relaxation or off-center relaxations swear out into a broad distribution [15].

So far, to identify a specific relaxation is still not easy. In most cases, the characteristic relaxation time constant $\tau$ is significantly influenced by temperature, which can be expressed as an exponent function of temperature:

$$\tau = \tau_0 e^{E_a/k_B T} \tag{21}$$

Here $E_a$ is activation energy, $k_B$ is Boltzmann constant, and $T$ is temperature. The activation energy reflects the energy needed to hop from one potential well to another for dipoles. Thus, it gives the information for possible relaxation mechanism.

## 4. Dielectrics for energy storage

Energy and environment protection are the challenges we are facing. Effective storing energy, reducing loss, and environment pollution are the hot topics for researchers. Electrical energy perhaps is the most convenient energy form to be applied. For applications in mobile electronic devices, stationary power systems, and hybrid electric vehicles, electric energy must be stored first.

There are many ways to storage electrical energy such as battery, supercapacitor, and dielectric capacitor. Battery has very high energy storage density (10–300 W·h/kg), but the power density is very low (<500 W/kg) and hazard for environment. Supercapacitor has average energy storage density (<30 W·h/kg) and power density (10–$10^6$ W/kg). But it has shortages such as complex configuration, low operating voltage, large leakage, and short recycling period. In comparison, dielectric capacitor has the highest power density ($10^8$ W/kg), broad operating temperature, fast charging and discharging, and long recycling period. The only disadvantage is low energy storage density ($10^{-2}$–$10^{-1}$ W·h /kg). Thus, there is a crucial need for dielectric capacitor with improved energy storage density [7].

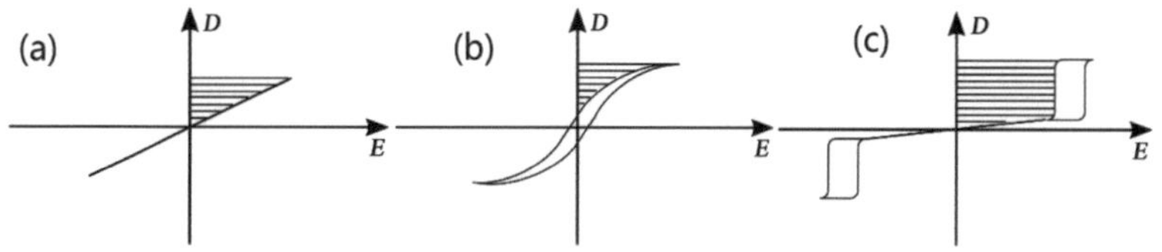


**Figure 3.** Electric displacement versus electric field for linear dielectrics (a), ferroelectrics (b), and antiferroelectrics (c).

The energy stored in a dielectric material under an electric field can be expressed by the shadow area in **Figure 3** in which different relationships between electric field *E* and electric displacement *D* are presented. The energy storage density *J* is defined as the stored energy per unit volume:

$$J = \int E\mathrm{d}D \tag{22}$$

According to Eq. (22), for linear dielectrics as shown in **Figure 3(a)**, energy storage density is proportional to the product of permittivity and the square of the applied electric field:

$$J = \frac{1}{2}\varepsilon_0\varepsilon_r E^2 \tag{23}$$

Thus, high dielectric permittivity and high breakdown field are highly desirable for the dielectrics used in energy storage devices. For linear dielectrics, a very high electric field must be applied to obtain large *J*. As shown in **Figure 3(b)**, ferroelectrics have much higher polarization so more energy can be stored at low field. But because of quick saturation of polarization under high electric field, increasing electric field does not contribute much energy density. At the same time, ferroelectrics exhibit significant hysteresis which corresponds to energy loss. The antiferroelectrics, as shown in **Figure 3(c)**, have a very large area between *D* and *E* because of zero remnant polarization; thus, it can store much more energy than linear dielectrics and ferroelectrics. It is therefore indispensable for high-energy and high-power capacitive storage system [16]. However, few materials have antiferroelectric properties like $PbZrO_3$, $PbHfO_3$, and $NaNbO_3$. $PbZrO_3$ is probably the only choice, but it contains environmental hazard element lead.

### 4.1. Single-phase materials for energy storage

As we have stressed, ferroelectric and antiferroelectric materials have better energy storage density. However, lots of them are inorganic materials prepared by solid reaction method. Various defects, like pores, impurities, and vacancies are produced during sintering, deteriorating the dielectric strength. For instance, the theoretical antiferroelectric-ferroelectric phase transition happens at $2.2 \times 10^7$ V/m for $PbZrO_3$, but the breakdown field for $PbZrO_3$ ceramic is about $1 \times 10^7$ V/m. The typical energy storage density for ferroelectric ceramic is around 1 J/cm$^3$. If materials are prepared in thin-film forms, the energy storage density could increase 10–50 times. However, reducing thickness also means reducing weight or volume, an impractical method for large energy storage application. The configuration of multilayer ceramic capacitor may be employed in this case. On contrary to ceramics, polymers have high breakdown field ($10^8$ V/m), although the dielectric constant is very small (<10). The energy storage density of polymer is around 10 J/cm$^3$. In addition, polymers are flexible, which can be used in flexible electronics.

It is should be noted that many already developed dielectrics may have super electric energy storage density. For instance, $BiFeO_3$-$SrTiO_3$ is a multiferroic system [17]. Recently, it was shown that $BiFeO_3$-$SrTiO_3$ thin film can get an ultrahigh energy density of ~51 J cm$^{-3}$. The breakdown field of the film is 360 MV/m, which is comparable with polymer dielectrics [18].

### 4.2. Ceramic-polymer composites for energy storage

Combining high dielectric constant ceramic with high breakdown field polymer to prepare composites has been extensively studied for energy storage, which is usually called 0–3 composites because zero-dimensional ceramic particles embedded in three-dimensional polymer matrix.

By filling ceramic particles with dielectric constant $\varepsilon_f$ with volume fraction $\varphi$ in polymer matrix with dielectric constant $\varepsilon_m$, the effective dielectric constant $\varepsilon_{eff}$ of 0–3 composite increases with the increase in $\varphi$. Its value is in between parallel and series connections of the two phases, which is well known as Wiener limits:

$$\left[\varphi\varepsilon_f^{-1} + (1-\varphi)\varepsilon_m^{-1}\right]^{-1} \le \varepsilon_{eff} \le \varphi\varepsilon_f + (1-\varphi)\varepsilon_m \tag{24}$$

There are enormous works of 0–3 composites. For example, in $BaTiO_3$-polymer composites, filling 40%, 70%, 70%, 40%, and 70% of $BaTiO_3$ in epoxy, PVDF, polystyrene, PVC, and polyamide, the corresponding dielectric constant is 44 [19], 152 [20], 100 [21], 18 [22], and 80 [23], respectively. It is obvious that dielectric constant limited increase although high volume of ceramic particle is filled in polymer. We propose that this is probably because polymer matrix normally has higher resistivity than ceramic filler [7]. By filling high-resistivity ceramic fillers in polymer matrix, a great increasing of dielectric constant may be achieved. As a result, the overall energy storage density can also be improved.

## 5. Dielectric for refrigeration

Electrocaloric (EC) effect is defined as the isothermal entropy or adiabatic temperature change of a dielectric material when an electric field is applied or removed. It could be used for efficient refrigeration and for the conversion of heat flows into electrical power. As the reverse effect of pyroelectric, EC effect has been explored before but the conclusion was that it was too small for practical applications. Recent interest on it is partly intrigued by the findings of 12 K temperature change in $PbZr_{0.95}Ti_{0.05}O_3$ [24] and more than 12 K temperature change in P(VDF-TrFE) (55/45 mol %) [25]. More importantly, EC refrigeration has been paid great attention for its environment-friendly and solid-state characters. Nowadays, because of highly integrated, a chip may be comprised of thousands of components, which results in large heat generation.

The future of IC may rely on the development of on-chip refrigeration technology, particularly with solid-state character. EC cooler is greatly expected for on-chip refrigeration.

### 5.1. Characterization of electrocaloric effect

At here, we first derive the equation for evaluating EC effect. The thermodynamic Gibbs free energy $G$ of dielectrics under constant stress condition can be expressed as:

$$G = U - TS - EDV \tag{25}$$

where $U$ is the internal energy of the system, $T$ and $S$ are temperature and entropy, $\mathbf{E}$ and $\mathbf{D}$ are electric field and electric displacement, and $V$ is the volume of the system. The differential form of the Gibbs free energy is:

$$dG = -VDdE - SdT \tag{26}$$

According to Maxwell relation, we have

$$\left(\frac{\partial S}{\partial E}\right)_T = V\left(\frac{\partial D}{\partial T}\right)_E \tag{27}$$

Thus, the isothermal entropy change $\Delta S$ and adiabatic temperature change $\Delta T$ are as follows:

$$\Delta S = -V\int_{E_1}^{E_2}\left(\frac{\partial D}{\partial T}\right)_E \mathrm{d}E \tag{28}$$

$$\Delta T = -\int_{E_1}^{E_2}\frac{T}{\rho C_\mathrm{p}}\left(\frac{\partial D}{\partial T}\right)_E \mathrm{d}E \tag{29}$$

in which $\rho$ and $c_p$ are density and specific heat capacity of material. These equations are applied to estimate $\Delta S$ and $\Delta T$ according to the experimental data of temperature dependency of dielectric displacement under constant electric field, which is commonly called indirect method.

It is obvious that the EC effect can also be directly measured by modified differential scanning calorimeter, which has not yet commercialized. Thus, the main method is still the indirect method.

Recently we find that in fact it is possible to evaluate EC temperature variation at paraelectric phase by a simple equation because of dielectric nonlinear behavior of ferroelectric at paraelectric phase [10]:

$$\Delta T = \frac{1}{2}\frac{\varepsilon_0}{\rho C_\mathrm{p}}\frac{\beta^3}{\gamma}T(T-T_0)^2\left\{\frac{1}{\varepsilon_0\beta(T-T_0)} - \frac{1}{\left[\varepsilon_0{}^3\beta^3(T-T_0)^3 + 3\varepsilon_0{}^3\gamma E^2\right]^{1/3}}\right\} \tag{30}$$

where $T_0$ is Curie temperature, $\beta$ is related to the Curie-Weiss law of $\varepsilon_r(T) = 1/[\varepsilon_0\beta(T\text{-}T_0)]$, $\gamma$ is related to the dielectric nonlinearity under electric field:

$$\varepsilon_r(E) \approx \frac{\varepsilon_r(T,0)}{\left\{1+3\gamma[\varepsilon_0\varepsilon_r(T,0)]^3E^2\right\}^{1/3}} \tag{31}$$

In fact, it is also the biquadratic coefficient in Landau-Devonshire Gibbs free energy density. Because dielectric constant of relaxor follows power law, we also can get corresponding EC temperature variation for relaxor [10]. Since EC effect normally gets maximum at phase transition point, our method is practical for fast evaluating EC effect for materials with available Landau-Devonshire Gibbs free energy density with biquadratic term. For materials, the free energy density is not available; measuring dielectric nonlinearity under electric field also can save time than measuring temperature-dependent dielectric displacement under different constant electric fields.

More generally, EC effect over broad temperature range can also be predicted if the complete Landau-Devonshire potential of a material is available. For example, in monodomain state, the potential of Landau-Devonshire including up to eighth-power term can be written as:

$$g = g_0 + \alpha_1(T-T_0)P^2 + \alpha_{11}P^4 + \alpha_{111}P^6 + \alpha_{1111}P^8 - EP \tag{32}$$

where $\alpha_1$, $\alpha_{11}$, $\alpha_{111}$, and $\alpha_{1111}$ are the Landau coefficients and $g_0$ is the Gibbs free energy density without considering the contribution of polarization. The polarization in ($T$, $E$) field can be obtained by finding the minimum of the potential. Then, the EC $\Delta T$ can be calculated according to Eq. (29) since dielectric displacement $\boldsymbol{D}$ equals to polarization $\boldsymbol{P}$ for ferroelectrics.

### 5.2. The influence of electric field for electrocaloric effect

Enormous materials have been studied for their electrocaloric effect from inorganic to organic. Most of them are ferroelectrics. Using model ferroelectric $BaTiO_3$ as an example, we can see the influence of electric field for electrocaloric effect. In single crystal, the reported $\boldsymbol{\Delta T}$ for $BaTiO_3$ changes from 0.9 K@1.2MV/m [26], ~1.2 K@1.2MV/m [27], 1.6 K@1MV/m [28], and 4.8 K@10 MV/m [29]. It is obvious that high electric field is the key factor to induce large electrocaloric effect.

Materials in thin film can withstand high electric field; thus, large EC $\boldsymbol{\Delta T}$ can be measured. However, the overall thermal mass is low. This problem can be overcome by combining many films to form a multilayer capacitor, which is a well-developed commercial product. In multilayer, the $\boldsymbol{\Delta T}$ of BaTiO3 varies from 1.8 K@17.6MV/m [30] to 7.1 K@80MV/m [31]. Again, improving breakdown field is essential for inducing large EC effect.

### 5.3. Device development

**Figure 4** is a schematic diagram for EC cooling cycle. One cycle includes four steps. (1) By applied electric field under adiabatic condition, EC unit changes from state 1 ($\boldsymbol{E_1}$, $\boldsymbol{T_c}$, $\boldsymbol{S_h}$) to state 2 ($\boldsymbol{E_2}$, $\boldsymbol{T_h}$, $\boldsymbol{S_h}$). The temperature of the unit increases. (2) Continuing increasing electric field under isothermal condition, EC unit changes from state 2 ($\boldsymbol{E_2}$, $\boldsymbol{T_h}$, $\boldsymbol{S_h}$) to state 3 ($\mathbf{E_3}$, $\mathbf{T_h}$, $\mathbf{S_h}$). The

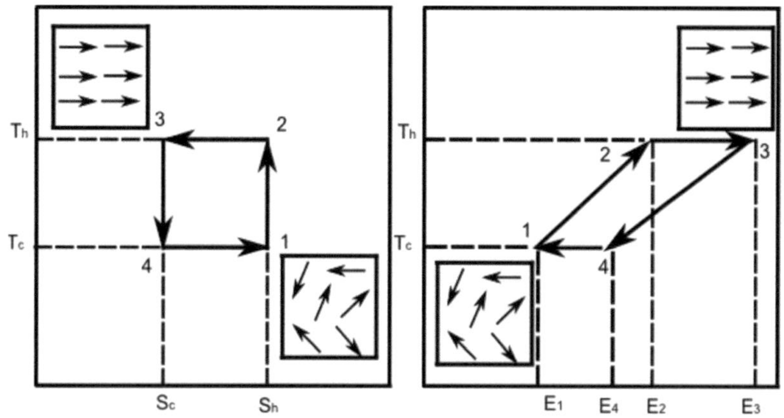


**Figure 4.** A schematic diagram of EC cooling cycle.

heat is expelled to a sink. (3) Decreasing electric field under adiabatic condition, EC unit changes from state 3 ($E_3$,$T_h$, $S_c$) to state 4 ($E_4$, $T_c$, $S_c$). The temperature of it decreases. (4) Continuing decreasing electric field under isothermal condition, EC unit changes from state 4 ($E_4$, $T_c$, $S_c$) to state 1 ($E_1$, $T_c$, $S_h$). The unit absorbs heat from a source. By one cycle, the heat of $T_C \Delta S$ is transferred from the source to the sink.

The critical point for an EC device is to achieve the unidirectional flow of entropy from a source to a sink. Some attempts to achieve it include: (1) shifting EC unit between the source and the sink [32, 33], (2) adding thermal switcher between EC unit with the sink and the source [34, 35], and (3) using liquid for heat transferring [36–38]. It is obvious that these designs increase the complexity of cooling device, which is not suitable for minimization and for chip-scale cooling. A new scheme is realized recently in which the cooler is composed of a flexible electrocaloric polymer film and an electrostatic actuation mechanism [9]. The device works for cooling battery of smartphone, although if it will work for chip cooling is still open. We stress here the sophisticated device development for chip-scale cooling based on EC effect is strongly expected, particularly adoptable for integrated fabrication.

## 6. Conclusion and perspectives

Conductivity, polarization, dissipation, and breakdown are the main electric responses of dielectrics under electric field. The weak electric field response of dielectrics is mainly studied by dielectric spectroscopy technology, particularly for dielectric relaxation. Equivalent electric circuit is the main technique to separate different relaxation mechanism by combining different normalization. The shortage of dielectric spectroscopy technology is that it is largely restricted by frequency limitation. Some relaxations may not enter the window of dielectric spectrum. Developing broadband dielectric spectroscopy is therefore strongly expected.

Dielectric for electrical energy storage is highly desired, but the energy storage density is still low. Theoretically, antiferroelectric can store more electric energy. The problem is we are shortage of practicable lead-free antiferroelectric materials. In 0–3 composite, we show that filling higher resistivity ceramics in polymer matrix has very potential to get higher energy storage density.

EC effect is very important for future cooling especially for microelectronic industry. The large EC effect normally happens around the phase transition. We have developed a practical and simple equation for predicting electrocaloric effect above the phase transition point, which can be applied for fast determining EC effect of a material. The future development in the area is to find new materials with high EC effect and practical device developments for chip-scale refrigeration.

In both applications, withstanding high electric field is essential. Both energy storage density and EC effect can be significantly improved if the breakdown field can be increased. Since breakdown field largely depends on processing, developing new preparation process for dielectric is also very important.

## Lists of symbols

**$I$** current

**$U$** voltage

**$G$** conduction

**$E$** electric field

**$D$** electric displacement

**$P$** polarization

$C$ capacitance

$Z$ impedance

$Y$ admittance

$M$ electric modulus

$\sigma$ conductivity

$\rho$ resistivity

$\varepsilon$ permittivity

$\varepsilon_r$ relative dielectric permittivity or (relative) dielectric constant

$\chi$ susceptibility

$A$ surface of capacitor

$t$ thickness of capacitor

$T$ temperature

$U$ internal energy

$S$ entropy

$V$ volume

$E_a$ thermal activation energy

$\tau$ characteristic relaxation time

$\omega$ angular frequency

$\varepsilon_0$ permittivity of free space

$k_B$ Boltzmann constant

## Author details

Liu Hongbo

Address all correspondence to: bohongliu@gmail.com

School of Materials Engineering, Shanghai University of Engineering Science, China

## References

[1] Guo J, Baker AL, Guo H, Lanagan M, Randall CA. Cold sintering process: A new era for ceramic packaging and microwave device development. Journal of the American Ceramic Society. 2017;**100**:669-677

[2] Kalmykov YP. Recent Advances in Broadband Dielectric Spectroscopy. Netherlands: Springer; 2013

[3] Svirskas Š, Banys J, Kojima S. Broadband dielectric spectroscopy of Pb-based relaxor ferroelectric (1-x)Pb(Mg1/3Nb2/3)O3-xPbTiO3 with intermediate random fields. Journal of Applied Physics. 2017;**121**:134101

[4] Schultze M et al. Controlling dielectrics with the electric field of light. Nature. 2013;**493**:75-78

[5] Paillard C et al. Photovoltaics with Ferroelectrics: Current Status and Beyond. Advanced Materials. 2016;**28**:5153

[6] Liu H, Yang X. A brief review on perovskite multiferroics. Ferroelectrics. 2017;**507**:69-85

[7] Liu H, Dkhil B. Effect of resistivity ratio on energy storage and dielectric relaxation properties of 0–3 dielectric composites. Journal of Materials Science. 2017;**52**:6074-6080

[8] Huan TD et al. Advanced polymeric dielectrics for high energy density applications. Progress in Materials Science. 2016;**83**:236-269

[9] Ma R et al. Highly efficient electrocaloric cooling with electrostatic actuation. Science. 2017;**357**:1130

[10] Liu H, Yang X. Theoretical prediction of electrocaloric effect based on non-linear behaviors of dielectric permittivity under temperature and electric fields. AIP Advances. 2015;**5**, 117134-1–5

[11] Correia T, Zhang Q. Electrocaloric Materials. Springer, 2014

[12] Jonscher AK, Dielectric Relaxation in Solids. Chelsea Dielectrics Press, 1983

[13] Smith NJ, Rangarajan B, Lanagan MT, Pantano CG. Alkali-free glass as a high energy density dielectric material. Materials Letters. 2009;**63**:1245-1248

[14] Jonscher AK. Universal Relaxation Law. Chelsea Dielectrics Press, 1996

[15] León C et al. Origin of Constant Loss in Ionic Conductors. Physical Review Letters. 2001;**86**:1279-1282

[16] Liu H, Dkhil B. A brief review on the model antiferroelectric PbZrO3 perovskite-like material. Zeitschrift für Kristallographie. 2011;**226**:163-170

[17] Liu H, Yang X. Structural, dielectric, and magnetic properties of BiFeO3-SrTiO3 solid solution ceramics. Ferroelectrics. 2016;**500**:310-317

[18] Pan H et al. BiFeO 3–SrTiO 3 thin film as a new lead-free relaxor-ferroelectric capacitor with ultrahigh energy storage performance. Journal of Materials Chemistry A. 2017;**5**: 5920-5926

[19] Kuo DH, Chang CC, Su TY, Wang WK, Lin BY. Dielectric properties of three ceramic/epoxy composites. Materials Chemistry and Physics. 2004;**85**:201-206

[20] Dang ZM, Wang HY, Peng B, Nan CW. Effect of BaTiO 3 size on dielectric property of BaTiO 3 /PVDF composites. Journal of Electroceramics. 2008;**21**:381-384

[21] Dang ZM, Zheng Y, Xu HP. Effect of the ceramic particle size on the microstructure and dielectric properties of barium titanate/polystyrene composites. Journal of Applied Polymer Science. 2008;**110**:3473-3479

[22] Olszowy M, Cz P, Markiewicz E, Kułek J. Dielectric and pyroelectric properties of BaTiO3-PVC composites. Physica Status Solidi. 2010;**202**:1848-1853

[23] Kakimoto MA et al. Polymer-ceramic nanocomposites based on new concepts for embedded capacitor. Materials Science and Engineering B. 2006;**132**:74-78

[24] Mischenko A, Zhang Q, Scott J. Whatmore, R. & Mathur, N. Giant electrocaloric effect in thin-film PbZr0. 95Ti0. 05O3. Science. 2006;**311**:1270-1271

[25] Neese B et al. Large electrocaloric effect in ferroelectric polymers near room temperature. Science. 2008;**321**:821-823

[26] Moya X et al. Giant Electrocaloric Strength in Single-Crystal BaTiO3. Advanced Materials. 2013;**25**:1360-1365

[27] Novak N, Kutnjak Z, Pirc R. High-resolution electrocaloric and heat capacity measurements in barium titanate. EPL Europhysics Letters. 2013;**103**:47001

[28] Bai Y, Ding K, Zheng G-P, Shi S-Q, Qiao L. Entropy-change measurement of electrocaloric effect of BaTiO3 single crystal. Physica Status Solidi -Applied Material Science. 2012;**209**: 941-944

[29] Bai Y, Han X, Zheng X-C, Qiao L. Both High Reliability and Giant Electrocaloric Strength in BaTiO3 Ceramics. Scientific Reports. 2013;**3**:2895

[30] Bai Y, Zheng G, Shi S. Direct measurement of giant electrocaloric effect in BaTiO3 multilayer thick film structure beyond theoretical prediction. Applied Physics Letters. 2010;**96**

[31] Bai Y et al. The giant electrocaloric effect and high effective cooling power near room temperature for BaTiO3 thick film. Journal of Applied Physics. 2011;**110**

[32] Jia Y, Sungtaek Y. A solid-state refrigerator based on the electrocaloric effect. Applied Physics Letters. 2012;**100**

[33] Gu H et al. A chip scale electrocaloric effect based cooling device. Applied Physics Letters. 2013;**102**:122904

[34] Epstein RI, Malloy KJ. Electrocaloric devices based on thin-film heat switches. Journal of Applied Physics. 2009;**106**:064509

[35] Wang YD et al. A heat-switch-based electrocaloric cooler. Applied Physics Letters. 2015;**107**:134103

[36] Plaznik U et al. Bulk relaxor ferroelectric ceramics as a working body for an electrocaloric cooling device. Applied Physics Letters. 2015;**106**

[37] Blumenthal P, Molin C, Gebhardt S, Raatz A. Active electrocaloric demonstrator for direct comparison of PMN-PT bulk and multilayer samples. Ferroelectrics. 2016;**497**:1-8

[38] Sette D et al. Electrocaloric cooler combining ceramic multi-layer capacitors and fluid. APL Mater. 2016;**4**:091101

Chapter 5

# Local Electric Fields in Dielectric and Semiconductors

Dmitry E. Milovzorov

Additional information is available at the end of the chapter

http://dx.doi.org/10.5772/intechopen.74310

**Abstract**

Local electric fields are appeared in dielectric and semiconductors due to the destruction of symmetry, creating the vacancies, point defects and chemical impurities in material. By increasing in external electric field value there are numerous structural changes will be generated. Point defects in silicon films were characterized by using electron-paramagnetic resonance spectroscopy and laser picoseconds spectroscopy. Chemical bonding properties was investigated by means of Fourier-transformed infrared spectroscopy. The possible mechanism of phase destruction was proposed.

**Keywords:** local field, point defect, dangling bonds, Raman spectroscopy, nanocrystals, second-harmonic generation, silicon films

## 1. Introduction

Local electric fields are appeared in dielectric and semiconductors due to the destruction of symmetry, creating the vacancies, point defects and chemical impurities in material. By increasing in external electric field value there are numerous structural changes will be generated. Some of them will produce such great local fields that will destroy all material or change its physical properties. The studying the nature of local electric fields will open new tendency in electronic device producing, from one side, and, help to change materials' properties according to our needs, from another side.

Description of local electromagnetic fields is a continuously durable through the all history of physics and was began with publication of first articles written by Maxwell Garnett which were devoted to colors in metal glasses and metallic films [1], Lorentz [2], and later in works of Brugeman [3] was developed by Edmund Stoner from University of Leeds [4] and Osborn from Naval Research Laboratory [5]. For a complicated medium such as the binary system with components A and B the dielectric function can be estimated as following [5]:

$$\varepsilon = \varepsilon_a \left[1 + 2C_A\left(\frac{\varepsilon_a - \varepsilon_b}{\varepsilon_a + 2\varepsilon_b}\right)\right]\left[1 - C_A\left(\frac{\varepsilon_a - \varepsilon_b}{\varepsilon_a + 2\varepsilon_b}\right)\right]^{-1} \tag{1}$$

In a case that one component is included in another dielectric component and polarized media with averaged value of polarizability of $\langle\alpha(w)\rangle = \frac{\varepsilon_A(w)-\varepsilon_B(w)}{\varepsilon_A(w)+2\varepsilon_B(w)}$ the Maxwell-Garnet formula is given by $\varepsilon(w) = \varepsilon_B(w)\left[1 + 2C_A\left(\frac{\varepsilon_A(w)-\varepsilon_B(w)}{\varepsilon_A(w)+2\varepsilon_B(w)}\right)\right]\left[1 - C_A\left(\frac{\varepsilon_A(w)-\varepsilon_B(w)}{\varepsilon_A(w)+2\varepsilon_B(w)}\right)\right]^{-1}$.

Structural properties of material may be strictly different as for surface and thin films, as for nanostructures such as clusters or nanocrystals, as for bulk material. However, it is obviously that most amounts of media in the universe is nanostructured or even in nanocrystal phase. For example, interplanetary dust was observed charged coupled devices (CCD) detectors and infrared space telescope [6]. They observed a cometary coma of Hale-Bopp comet. The dust destruction in space (Cygnus Loop supernova) was observed by using a Spitzer Telescope tuned in infrared (IR) range from 22 to 36 μm [7] and shows us the chemical properties, such as dust chemical compositions. These observations show the great fraction of silicon in all space dust. Space dust destruction and ion formation was studied by Mann and Czechovsky [8], which results from model calculations in silicate grains, carbon and ice grains. Grain destruction in a supernova remnant shock wave was investigated by astronomers of Harvard University [9]. It was observed by Spitzer telescope in IR 24 μm range of wave length. The case of impact of nanoscopic dust grain with solar wind of spacecraft already was estimated by using the dimensionless parameter equals to ratio between Debye length and radius of dust cloud spherical shell with radius R [10]. It was shown that the dust particle with mass $10^{-20}$ kg produces by impact $10^7$ charged particles.

From the other physical scientific trends we have an observations of local field by a nonlinear spectroscopic experiments with nonmaterial and nanocomposites, particularly, semiconductors. By using semiconductor materials have been made numerous types of devices, such as electronic devices and photon detectors, integrated circuits and thin film transistors, optoelectronic devices and others. Every time when the device is developed the problem of reproducibility of its work regimes and durability of their realization is appeared. The solution of this problem is very important for device manufacturing, and it depends on properties of used active semiconductor materials. The electrical properties often are not so transparent due to slightly nonlinear behavior of their characteristics. **Figure 1** shows the current-voltage and resistance-voltage characteristics for two silicon films prepared by plasma-enhanced chemical vapor deposition technique with gas mixture of silane diluted by hydrogen and silicon tetra fluoride gas: amorphous and nanocrystallized [11]. It is seen, than their current-voltage characteristics are similar in this voltage range, but resistance-voltage characteristics are strictly different. Such difference can be explained by the disorder of amorphous phase and generating of numerous point defects by applying external electrical field. It is clear, that the voltage is varied in the range from the −10 to 10 V. Hydrogenated amorphous silicon was widely used in last decades in electronics. In recent years the nanocrystalline silicon are studying for many

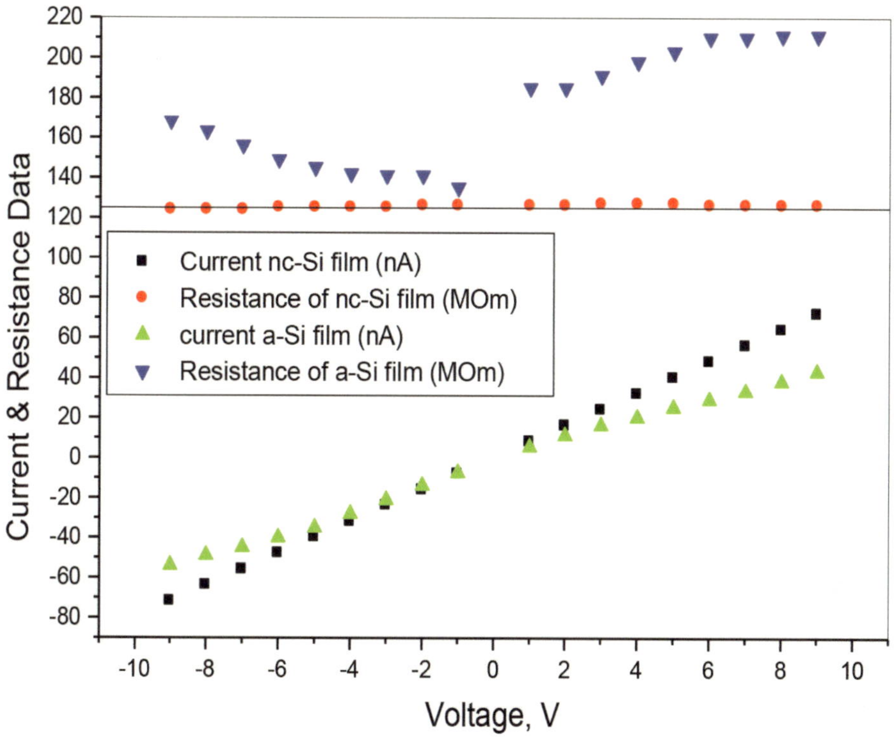


**Figure 1.** The current-voltage and ohm-voltage characteristics for two silicon films: amorphous and nanocrystallized.

technological applications. The structural transformation from crystal to disordered materials, however, is investigated very poor, mainly resulted in Staebler-Wronski photo-stimulated effect. However, the electric field applied to the nanostructured silicon thin film gives the new possibility to change structural order. Such kind of structural transformation is caused because of there are numerous defects inside the silicon film.

The anomalous characteristic of resistance-voltage can be explained by random distributed the point defects inside the amorphous film along with the hydrogen atoms, and existing the dipoles Si-O which turn to compensate the external electric field. But, for the nanocrystalline silicon film, the point defects are incorporated into silicon nanostructured net and cannot move freely, because there is a stabile electric characteristic for nanocrystalline silicon film, and anomalous for amorphous.

The other new area of scientific interests is crystal-amorphous phase transformations by applying electric fields and role of local fields in phase transition from order to disorder. Because, it is important to investigate the point defects which can be responsible for local electrical fields generation in polarized media, such as dielectric silicon oxide media or semiconductor thin film of silicon. The main role plays here the silicon-oxide bonding in side thin film of silicon. Si-O dipoles play a dramatic role in crystal phase destruction by applying electric fields. The induced dipole moment by applied electric field can be written in the following form [12]:

$d_{i\alpha} = \alpha_0 \left( E_{i\alpha}{}^{(0)} - \sum_{j=1}^{N} \left( \delta_{\alpha\beta} - 3\frac{(r_{ij})_\alpha (r_{ij})_\beta}{r_{ij}{}^2} \right) \frac{d_{j\beta}}{r_{ij}{}^3} \right)$; where $E_{i\alpha}{}^{(0)}$ is the field at the i-th monomer, $r_{ij}$ is a relative vector between i-th and j-th monomer molecules, $\alpha_0$ is the dipole polarizability. However, the Hamiltonian of semiconductor cluster can be surely expressed by using donor and acceptor states in bulk material [13]: $H = H_0 + H_D + H_A$, where $H_0$ is the Hamiltonian pure semiconductor for electron matrix elements of transitions between own conductive and valence bands, but $H_D$ and $H_A$ are the exchange energies matrix elements due to the donor and acceptor states, and for them it is clear be presented the following expressions:

$(H_D)_{ij} = \sum_k \frac{D_{ki}{}^2}{E-\varepsilon_k-\sigma_k} \delta_{ij}$; $(H_A)_{ij} = \sum_l \frac{A_{li}{}^2}{E-\varepsilon_l-\sigma_l} \delta_{ij}$; where $D_{ki}$and $A_{li}$ are coupling between k-th donor state and l-th acceptor state, but the value $\sigma_{k,l} = \frac{1}{2}\left( \theta_{k,l} - i\sqrt{4\gamma_{k,l}{}^2 - \theta_{k,l}{}^2} \right)$ are the self-energy corrections which depend on locality of sites, and $\theta_{k,l} = E - \varepsilon_{k,l}$, and $\gamma_{k,l}$ are the ionization energy for electrons in sites, and overlap integral between atomic orbital for the donor in k-site and acceptor in l-site, respectively. Therefore, different explanations of polarization properties of semiconductor media can describe only partially in their own borders the electric induced local fields' appearance and various theoretical predictions based on them are not satisfied. The macroscopic description and molecular nanoscopic model are poor for investigating the mechanism of crystal semiconductor structures destruction by applied electric fields because the their dielectric functions is not so transparent for these complicated media, for example $SiO_x$, from one side, and their ratio between covalent and ion fractions of inter atomic bond are not so homogeneous, from the other side. Because, there is a necessity of detail investigation of nanoscopic nature of local fields appearance and mechanisms of crystal-amorphous phases' transformations.

The present work is devoted to the nature of local field appearance in silicon nanoscopic material and role point defects in phase transformation of material.

## 2. Experimental researches of field-assisted destruction of silicon nanocrystals

Nonlinear polarization associated with the phonons can be written as $P = \frac{1}{V}\sum_{i=1}^{N} < \mu >_i = \frac{N}{V}\left(\frac{\partial\alpha}{\partial Q}\right)_0 < Q > E$. Using these equations it is possible to obtain the wave equation for field: $\Delta E + \frac{\eta^2}{c^2}\frac{\partial^2 E}{\partial t^2} = -\mu_0 \frac{\partial^2 P}{\partial t^2}$, where $c^2 = \frac{1}{\mu_0 \varepsilon_0}$, $\eta = \sqrt{\frac{\varepsilon}{\varepsilon_0}}$.

Raman effect is result from the interaction of an electromagnetic field and optical phonon mode. The vibration wave $< Q > = \frac{Q}{2}\exp(i(\omega_v t - k_v z) + c.c.$ The polarization at the Stokes frequency $P_S = \varepsilon_0 \chi_R(\omega_v)|E_i|^2 E_S$. Raman susceptibility

$$\chi_R(\omega_v) = \frac{N\left(\frac{\partial \alpha}{\partial Q}\right)_0^2}{4MV\varepsilon_0} \frac{1}{\Omega^2 - \omega_v{}^2 - i(\Delta\omega_L)\omega_v}. \quad (2)$$

The microcrystal wave function is a superposition of Eigen functions with $k$ vectors centered at $k_0$ . We suppose that $k_0 = 0$. If we are using the weighting function as Gaussian we will have

$$W(r, L) = \exp\left(\frac{-8\pi^2 r^2}{L^2}\right)|C(0, k)|^2 \cong \exp\left(-\frac{k^2 L^2}{16\pi^2}\right) \quad (3)$$

the normalized first-order correlation function $g^{(1)}(\tau) = \frac{<E^*(t)E(t+\tau)>}{<E^*(t)E(t)>}$. First-order coherence $g^{(1)}(\tau) = \exp\left(-i\omega(k)\tau - \frac{\Gamma_0}{2}\tau\right)$, where $\omega(k)$is the phonon dispersion curve, $\Gamma_0$ is the natural line width. The spectrum of natural broadened value is $F(\omega) \propto \frac{1}{(\omega-\omega(k))^2+\left(\frac{\Gamma_0}{2}\right)^2}$. The phonon probability $\rho = \Psi_0\Psi_0{}^* = u_0{}^2(r)\int d^3k|C(0, k)|^2$ . The first-order Raman spectrum

$$I(\omega) \cong \int \frac{d^3k|C_0(k)|^2}{(\omega - \omega(k))^2 + (\Gamma_0/2)^2}. \quad (4)$$

For a microcrystalline and nanocrystalline silicon with sizes of crystals L if the weight function is Gaussian the first-order Raman spectrum is following [14]: $I(\omega) = \int_0^1 \frac{\exp\left(k^2L^2/4a^2\right)}{[\omega-\omega(k)]^2+(\Gamma_0/2)^2} d^3k$, $k = \frac{2\pi}{a}k$, $a$is lattice constant, $k$is dimensionless value,$\Gamma_0$˜$3.6cm^{-1}$ line width of the Si LO phonons in c-Si. The dispersion of the LO phonon in c-Si $\omega^2(k) = A + B\cos(\pi k/2)$, where A = 1.714 × $10^5$ $cm^{-2}$ and B = $10^5$ $cm^{-2}$ [15]:

$$I_{nc-Si}^{a-Si,}(\omega) = 16\pi L(w, \rho) \int_{-\Delta q}^{\Delta q} \frac{|C(0, q)|^2 q^2}{4((\omega - \omega(q))^2 + \Gamma_T{}^2} dq, \quad (5)$$

where local field factor can be written as

$$L(\omega, \rho) = \frac{\rho}{4\pi} \frac{(\varepsilon_{c-Si}(\omega) - \varepsilon_{SiO_x}(\omega))}{1 + (\varepsilon_{c-Si}(\omega) - \varepsilon_{SiO_x}(\omega))(\Lambda - \beta\rho))}$$

where $q$ is a vector of inverted lattice, $2\pi/a$, a is a lattice constant. If the value $\lambda_0$ is the bond length Si-Si in bulk silicon, $\lambda_1$ is the weak bond resulted the silicon-oxygen surrounding. By the value of density of bonds is $N_{SiO}$ = $210^{21}$ $cm^{-3}$ and for crystalline silicon the density of silicon bonds $N_{Si-Si}$ = $510^{22}$ $cm^{-3}$ the lattice constants are following: $a_2 = 0.98\ a_0$ and $a_1 = 0.996\ a_0$.

For the silicon nanocrystalline and microcrystalline films the phonons can be generated in crystals by laser field or annealing. The wave of deformations can be generated by picoseconds laser pulse [16]. The acousto-electric effect was observed in n-type germanium [17]. The electric field which was appeared by ultrasound waves can be estimated by using the formula $F = eE = \frac{q^2 S}{c^2 KT} \frac{\omega^2 \tau}{\left[1+(\omega\tau)^2\right]}$, where $E$ is an acousto-electric field, τ is a relaxation time $\tau^{-1} = \left(\frac{4}{3\tau_\partial}\right) + k^2 D$, D is a diffusion coefficient, The deformation potential causes the appearance of effective acoustical charge (**Figure 2**).

The nanocrystalline film was made by me using CVD method of silane diluted by hydrogen (gaze flow rates ratio is 1:10) at low temperature of substrate (80°C). The RF power was 20 W. Working pressure was 0.2 Torr. The crystalline volume fraction was 66%. The crystal orientations for nanocrystals were determined by means of X-ray diffraction technique (111) and their average size was 24 nm. The thickness of silicon film was more than 300 nm. **Figure 5** shows the changes in Raman scattering spectral data by applying the external electric field. It is seen, that there is phase destruction by the relatively high voltage. It is assumed that the nanocrystals which have grain boundary with oxygen atoms incorporated into silicon were destroyed in their crystal structure by Si-O dipoles reorientations caused by applied field. The initial crystal orientation was (111). The incorporated oxygen atoms are adsorbed in determined places. Their position results the appearance of numerous dangling bonds which are multiplied by the electric field and create the deep cracks in crystals. The crystal order is damaged along the axis that is perpendicular to (111). According to the Raman data for $SiO_2$ [18] the Raman spectrum of $SiO_2$ has the variation modes D1 (at 490 $cm^{-1}$) with defects and activation energies 0.14 eV and pure mode $w_1$ of Si-O-Si bridge. The sum dipole moment consists of dipole moment that is created due to the ellipsoidal shape and because of surface charges are appeared by silicon net deformation due to the oxygen incorporation in silicon *SiO* or hydrogen termination of dangling bonds and creation the *SiH* bonds: $P_{SiM} = \frac{8}{3}\pi a b^2 N_{SiM} \delta_{SiM}(b-a)$, where $N_{SiM}$, $\delta_{SiM}$ are the density of the bonds and partial charge for SiM bond (O or H), respectively (**Table 1**).

It is necessary to note that the fractal structure of several kinds of nanocrystals may cause the dramatically changes (four orders of magnitude) in intensity of Raman scattering due to existing of plasmon resonance into the gaps between the fractals [19]. The Raman intensity by these conditions can be expressed as

$$I = \frac{|\alpha|^2}{|E_0|^2} \int |\sigma(r)|^2 |E(r)|^4 dr. \tag{6}$$

where $\alpha$ is a polarizability and $\sigma$ is a local conductivity of a fractal structure.

In addition, the light irradiation of amorphous silicon film causes the point defects generating and, mainly for amorphous hydrogenated silicon films, causes the appearance new dihydride configurations: (H-Si Si-H)2(H-Si Si-H) and $SiH_2$ [20]. The two atoms of hydrogen in the $SiH_2$ unit show an average proton separation of 2.39 Å. Because, for the hydrogenated silicon nanocrystalline films under influence of applying the external electric field the hydrogen diffusion increases and polysilane chains are created, surely.

For poly-Si films with nanocrystals the values of densities of SiO and SiH bonds varies in wide range from $10^{19}$ to $10^{21}$ $cm^{-3}$. The density of bonds were estimated for the poly-Si films

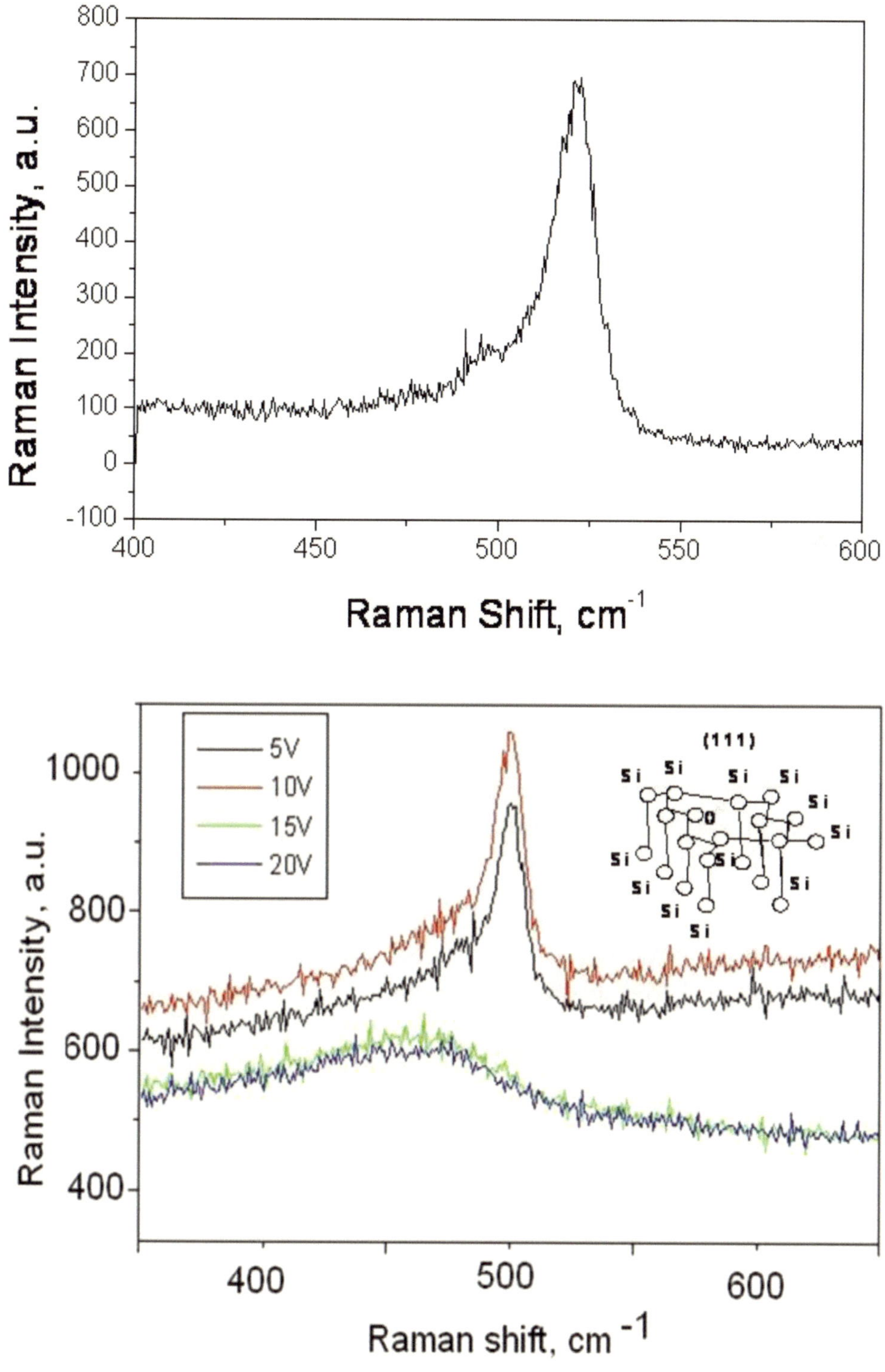


**Figure 2.** Raman spectra from nc-Si film with nanocrystal mean size 24 nm.

| $N_{DB}$,cm$^{-3}$ | $P_{ellipsoid}$, Debye | ΔE = μE$_{ext}$, μeV by the E$_{ext}$ = $10^6$ V/m | ΔE = μE$_{ext}$, μeV by the E$_{ext}$ = $10^7$ V/m |
|---|---|---|---|
| $10^{17}$ | 1.1 | 20 | 200 |
| $10^{18}$ | 10.8 | 200 | 2000 |

**Table 1.** The polarization of elliptic silicon grains evaluated by fixed $N_{DB}$ and stark energy shifts for levels in electronic structure by external electromagnetic field.

prepared by using PECVD as following: for Si-Si bonding the density of bonds is equals to $510^{22}$ cm$^{-3}$, but densities of SiO and SiH bonds are $10^{21}$ cm$^{-3}$. In these films there is an oxygen contamination on the 2% level. The S/V ratio is 1.25%. I suppose that all the oxygen is concentrated around crystals in their grains boundaries. By these values of densities the dipole moments causes by surface charges can be estimated as $P_{SiO}$ = 0.04 D and $P_{SiH}$ = 0.015 D. By applying external voltage it is clear that all Si-O dipoles move to compensate external field and destroy the crystal structure. Because, by applied electric field I observed the crystal phase destruction and $SiO_x$ creation by using the Raman scattering data which correspond to the results reported in work [21].

## 3. Matrix Hamiltonian by small perturbation of Si-Si-Si bridge

For very small nanocrystals with sufficient ration S/V the mechanism of three elements simultaneous interaction is important for precise calculations. The energy shift due to the stress appearance for crystal orientation (111) is less than 0.14 eV for the vacancy-oxygen (VO) complex by a stress 0.3 GPa. The Hamiltonian of such system of n atoms as for example,-Si-Si-Si- and, particularly the Hamiltonian of interaction between atoms with indexes k−1 and k + 1 can be explained in matrix form is given by using the operators of creation and elimination of boson particles, such as phonons: $H' = \alpha c_{k-1}^{+} c_{k+1} + e.c.$ The Hamiltonian matrix of interaction between three atoms (with upper indexes k−1, k and k + 1) in chain by a small perturbation $\alpha$ due to the VO appearance (for bonded $Si_1$ and $Si_3$ atoms without oxygen) can be written as $\hat{H} = \begin{Vmatrix} {A_{11}}^{k-1} & A_{12} & \alpha \\ A_{21} & {A_{22}}^{k} & A_{23} \\ \alpha & A_{32} & {A_{33}}^{k+1} \end{Vmatrix}$. This matrix can be transformed into next triangular form:

$$\hat{H}' = \begin{Vmatrix} A_{11} & A_{12} & \alpha \\ A_{21} & \hat{A}_{22} & \hat{A}_{23} \\ 0 & A_{32} & \hat{A}_{33}' \end{Vmatrix}. \tag{7}$$

Here, the matrix elements $\hat{A}_{22} = A_{22} - A_{21}A_{11}^{-1}A_{12}; \hat{A}_{23} = A_{23} - \alpha A_{11}^{-1}A_{21}; \hat{A}_{33}' = A_{33} - \alpha A_{11}^{-1}$ $\alpha - -\hat{A}_{32}\hat{A}_{22}^{-1}\hat{A}_{23}; \hat{A}_{32} = A_{32} - \alpha A_{11}^{-1}A_{12};$ and according to the Vilandt-Hoffman theorem of matrix perturbation theory [22] the following inequalities can be written.

$\left(\sum_{i=1}^{n}|(\lambda_{Ai}-\lambda_{Ai})|^2\right)^{1/2} \leq \|(A+E)-A\|_E$ or $\hat{\lambda}_{33}^2 \leq \lambda_{33}^2 + \frac{\alpha^2}{\lambda_{11}}$. Because, for the creation operator $\hat{c}_{2k+1}(t)$ is true the following expression $\hat{c}_{2k+1}(t) = c_{2k+1}(t)\exp\left(-i\int_0^t \frac{\alpha^2}{E^{\wedge}1\,d\tau}\right)$. Therefore, the dispersion curve is written as $\omega_{phonon} = \omega_0 \pm \xi$, where $\omega_0$ dispersion curve without applied field (for energy 0.14 eV the estimated additional frequency is less than $3.4 \times 10^{13}$ Hz, and additional frequency $\xi = \frac{\alpha^2}{\lambda_{11}}$ by stress due to the appearance of VO defects. According to the proposed model that is suitable for description Raman scattering phenomena caused by nano-sized cavities it is assumed that the results Raman frequency of radiation after scattering can be written as $\omega_S = \frac{\mu E}{\hbar}$; where $\mu = dQ$ is dipole moment of nanocrystal, $E$ is field, $d$ is the size of crystal, $Q$ is charge (**Figure 3**).

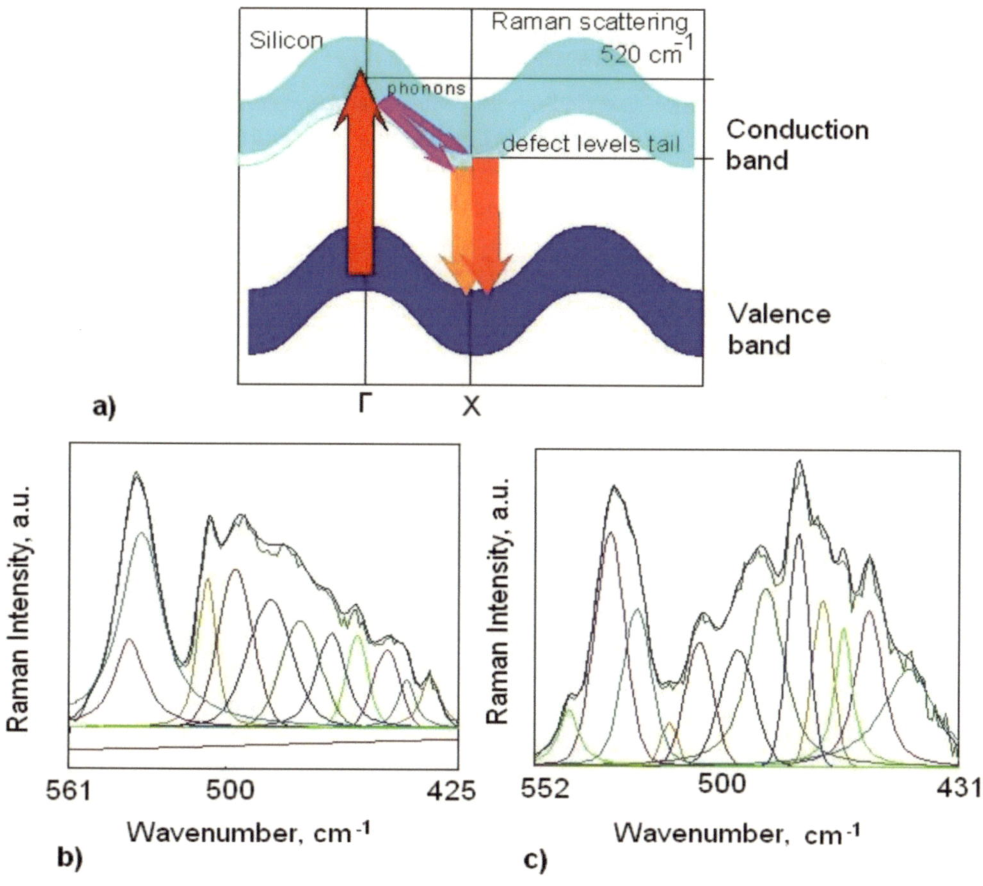


**Figure 3.** Scheme of Raman scattering of photons on phonons (a) and Raman spectra for silicon film without external field (b) and by applied external electric field with additional spectral peak at 540 cm$^{-1}$ which was created due to the defects generating by electric field (c).

## 4. Model of polaron state in silicon nanocrystals

The probability of changing the polarization state from one to another can be described by using Golden rule of Fermi

$$W_{ij} = \frac{2\pi}{\hbar} \sum_{k} \left| \langle \psi_i | H | \psi_j \rangle \right|^2 \delta(E_j - E_j - \hbar\omega_{polar}), \tag{8}$$

where $\omega_{polar} = \frac{PE}{\hbar}$ is a polarization energy that can be expressed in frequency units. Matrix element for such changing of polarization state from i to state j, according to theoretical work of V. Lakhno [23], can be written as

$$\langle \psi_i | H | \psi_j \rangle = hk \sqrt{\frac{hN_1}{6V\omega_{polar}\varepsilon} \frac{e}{m}} \left| \int \Psi_{iC} \frac{e^{ikr}}{\sqrt{V}} d^3 r \right|^2, \tag{9}$$

where V is a nanocrystal volume (**Figure 4**).

The second-harmonic generation is forbidden for center symmetric crystal such as bulk silicon because the sum dipole moment is zero, but is possible due to the surface breaking symmetry and quadruple terms contributions. The opposite situation is for nanostructured oxidized silicon film, the surface area for a great amount of nanocrystals is significant, the breaking symmetry is permanent and lateral isotropic. The oxygen atoms with concentration up to the values of $10^{20}$–$10^{21}$ $cm^{-3}$ show the sharp increase in SHG by increase in polarization properties of material, that have its properties as silicon nanocrystals, as silicon oxide inclusions. **Figures 5** and **6** illustrate the SHG spectra of radiation reflected from silicon films.

The reflected SHG response was measured by using the radiation of optical parametric oscillator/amplifier pumped by the third harmonic (355 nm) of a Q-switched Nd: YAG laser (Spectra-Physics, MOPO 730) at a 10 Hz repetition rate with spectral range between 440 and 1700 nm. The bandwidth of radiation is 0.3 $cm^{-1}$. The SHG response was detected by a photomultiplier tube

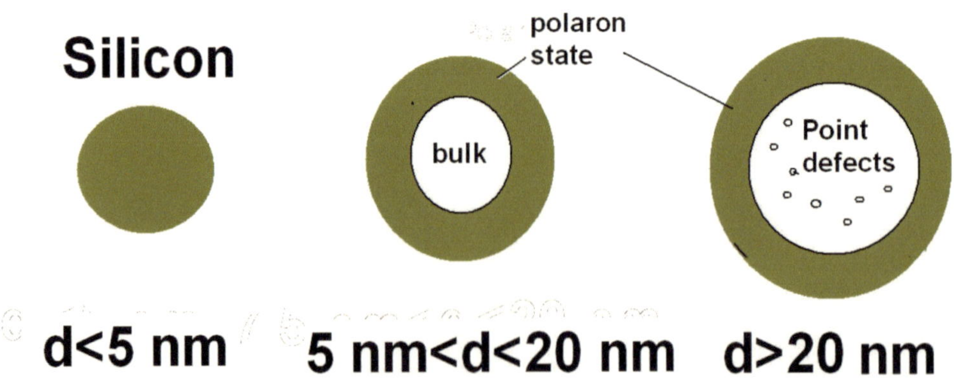


**Figure 4.** Silicon nanocrystals with different sizes by applying external electric field have various electronic structure: with polaron state in all volume of nanocrystal, and partially polarized nanocrystal according to relation between the values of size and "skin"-layer or depth of field penetration.

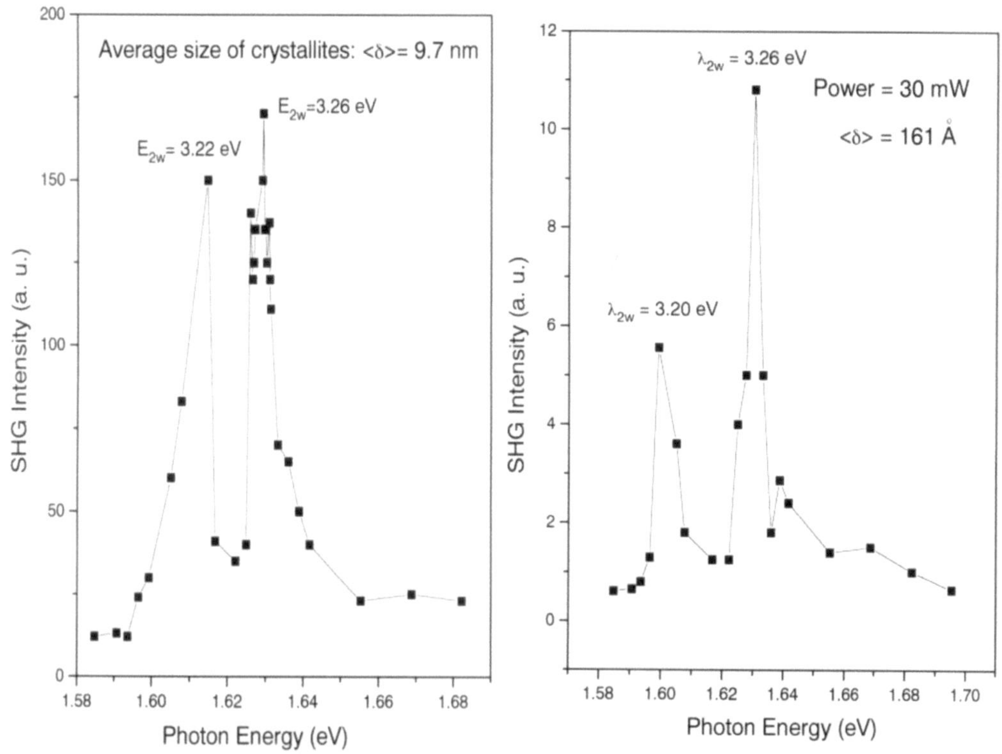


**Figure 5.** SHG spectra for $\chi^{(2)}{}_{xxx}$ component of susceptibility.

and gated electronics with an average of 100 pulses. The linear polarized radiation was focused on the surface of the sample at the angle 45° and detected SHG signal was observed at the angle 45°, too. Such optical scheme arrangement was useful for surface contributors' detection from the silicon surface (111). The diameter of irradiated spot was 0.5 mm. The energy of the primary laser beam was 4 mJ. The second-harmonic intensity can be written as

$$I(2w) = \frac{32\pi^3\omega^2\sec^2\theta_{2w}}{c^3\varepsilon(\omega)\varepsilon^{1/2}(2\omega)}\left|L(2\omega,\rho_{\min})L^2(\omega,\rho_{\min})\right|^2|\chi_2(\omega)|^2 I^2(\omega) \tag{10}$$

where the L(ω, ρ) value is a local field factor of film with crystalline volume fraction equals to $\rho = 70\%$.

The SHG intensity as a function of the average grain size in poly-Si films, with crystalline volume fraction 70%, is presented in **Figures 5** and **6** where

$$K = \frac{I_{\exp}}{I_{\exp}{}^{\min}}\frac{\left|L(2\omega,\rho_{\min})L^2(\omega,\rho_{\min})\right|}{\left|L(2\omega,\rho)L^2(\omega,\rho)\right|} \tag{11}$$

is normalized SHG signal, $L(2\omega,\rho)$ and $L(\omega,\rho)$ are the factors of the local field, where $\varepsilon_c$ and $\varepsilon_a$ are dielectric functions of crystalline and amorphous silicon, respectively. For sphere depolarization factor Λ is equal to 1/3, β is the Lorentz constant (for homogeneous spherical

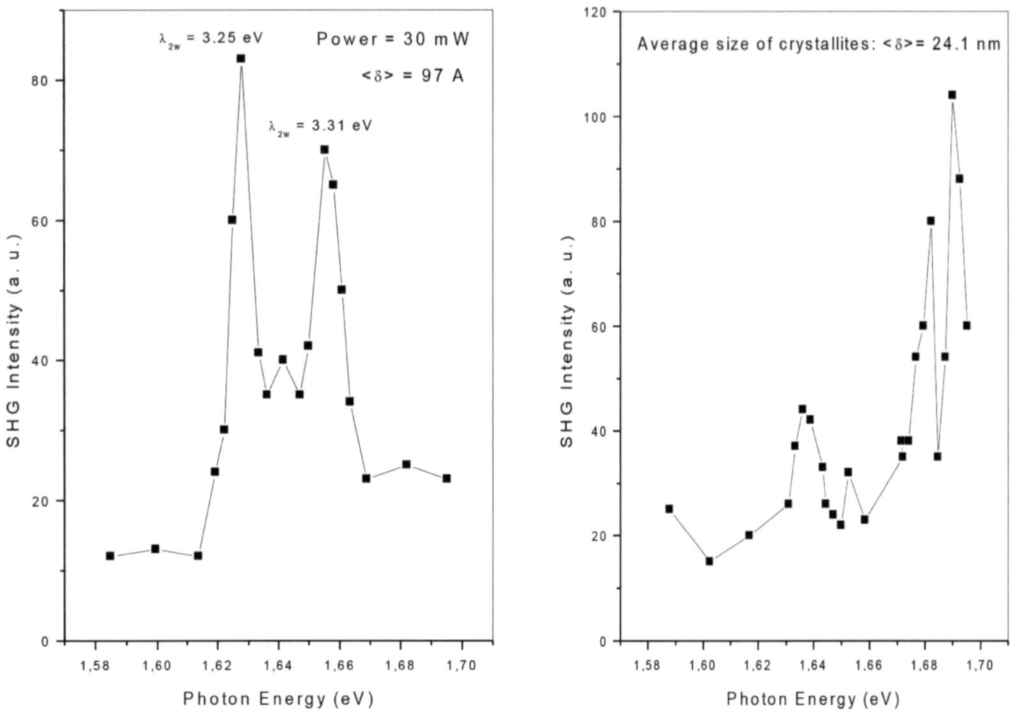


**Figure 6.** SHG spectra for $P_{in}\rightarrow P_{out}$ laser scheme and $\chi^{(2)}_{xxx}$ component of susceptibility.

surrounding β =1/3). For calculations the following estimated values were $\lambda_w$ = 1064 nm, $\lambda_{2w}$ = 532 nm, $\varepsilon_c'(w)$ = 13, $\varepsilon_c''(w)$ = 0.03, $\varepsilon_a'(w)$ = 12, $\varepsilon_a''(w)$ = 0.23, $\varepsilon_c'(2w)$ = 18, $\varepsilon_c''(2w)$ = 0.5, $\varepsilon_a'(2w)$ = 18, $\varepsilon_a''(2w)$ = 7.5. Resonance spectra consist of two sharp peaks. The peak at 1.6 eV is caused by SHG response due to $E_0' = \Gamma_{21} - \Gamma_{15}$ transition. We suppose that the second peak can be recognized as SHG response due to $E_1 = L_{2'} - L_1$ transition in silicon nanocrystallites.

$$L(\omega, \rho) = \frac{\rho}{4\pi} \frac{(\varepsilon_c(\omega) - \varepsilon_a(\omega))}{1 + (\varepsilon_c(\omega) - \varepsilon_a(\omega))(\Lambda - \beta\rho)}$$

The contribution of point defects as deviations of local fields and external applied electric field for a phonon generation in silicon nanocrystalline can be described by using the perturbation theory. The model Hamiltonian's matrix for two-level system including the point defect as small perturbation $\varepsilon$ that causes the violence of phonon energies of system $E_1$ (for unperturbed state) and $E_2$ (perturbed state):

$$\begin{pmatrix} E_1 & \varepsilon \\ \varepsilon & E_2 \end{pmatrix} \tag{12}$$

The changes in Eigen values from the $E_1$ and $E_2$ by the field $E = 0$ to the new values of energies are following

$$\lambda_{1,2} = \frac{E_1 + E_2}{2} \pm \sqrt{\frac{1}{4}(E_1 - E_2)^2 - E_1E_2 + \varepsilon^2}. \quad (13)$$

It is assumed, that the field value $\varepsilon$ determined as linear combinations of external applied field and all local deviations due to film structural disorder. We assume that there is no strict disorder media, but some small disorder is determined. $\varepsilon \approx (E_1 - E_2)$; $\Delta E \approx 2(E_1 - E_2) = 2\Delta$. Therefore, by a small perturbation of system $\varepsilon \leq (E_1 - E_2)$, the energetic gap between the two levels which are located closed each to other increases from the zero value to $2\Delta$. The value of perturbation of atomic orbital for dimer molecular like assembly Si-Si in a point defect as $(VO)^-$ can be evaluated by using the energy of their interaction of atoms with dipole Si-O, that has its polarized charge 0.2 e: $F = \frac{1}{4\pi\varepsilon_0}\frac{0.2e^2}{r^2}$; where r is a distance from the dimer Si-Si to dipole Si-O equals to 1–2 Å.

From the other side, for drift of particle, such as hydrogen atom, by driving forces in condensed matter can be expressed by using formula for force $F = \frac{H_M}{kT}k\nabla T$, where $H_M$ is an enthalpy or energy of transport by heat, $\nabla T$ is a gradient of temperature [24]. For elastic medium the force that treats the defect is given by using stress value $\sigma$ and displacement field of defect in its surrounding medium as $u$: $F = \int_\Sigma (u\nabla\sigma - \sigma\nabla u)ds$. David Emin studied the deformable lattice in 1972 [25] by using short-range electron-lattice interaction in dielectric or semiconductor materials which have weakly coupled electron-lattice interactions or small polaron states, the distortion related energy of which expressed in terms of distortion parameter x, $E = \frac{1}{2}M\omega^2x^2$, the energy reduction due to the linear electron-lattice interaction as $E = Ax$.

## 5. Modular group translation model for crystal phase destruction by applied electric field

I propose the modular group translation (MGT) model for crystal phase destruction by applied electric field for explanation the Raman data which are on **Figure 7** and show the dramatic

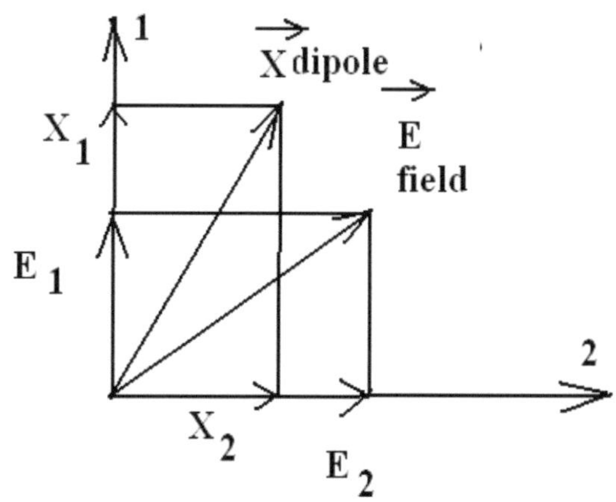


**Figure 7.** Scheme of dipole's and electric field vectors location on 2D plate.

changes in silicon crystal phase related spectral component at 520 $cm^{-1}$ due to the applying electric field.

It is assumed that the electric fields of external field and local polarized field can be written as $\vec{E_1}=\begin{pmatrix}E_{11}\\E_{12}\end{pmatrix}$; $\vec{E_2}=\begin{pmatrix}E_{21}\\E_{22}\end{pmatrix}$; $\vec{X_1}=\begin{pmatrix}X_{11}\\X_{12}\end{pmatrix}$; $\vec{X_2}=\begin{pmatrix}X_{21}\\X_{22}\end{pmatrix}$; for applied and local electric fields, and for dipoles vectors on 2D plate. For free energy it is easy to write the expression by using binding energy and energies of dipoles in external field and local field: $F=\frac{1}{2}\sum_N N_i \left(E_j - \delta Q_j \vec{E_j}\,\vec{x_j}\right)$, where $N_j$ is a number of neighbor atoms, $\vec{E_j}$ and $\vec{x_j}$ are the vectors of electric field and dipoles along the j direction. For j components of free energy along the axis 1 and 2 on 2D plate it is surely can be presented other form of such formula:

$$F_1 = E_1 - \delta Q_1 E_{11}x_{11} - \delta Q_2 E_{21}x_{21}$$

$$F_2 = E_2 - \delta Q_1 E_{12}x_{12} - \delta Q_2 E_{22}x_{22}.$$

For analysis of deformation by applying the external electrical field it is clear to use ratio between free energy components for different bonding and directions:

$\frac{F_1}{F_2} = \frac{E_1 - \delta Q_1 E_{11}x_{11} - \delta Q_2 E_{21}x_{21}}{E_2 - \delta Q_1 E_{12}x_{12} - \delta Q_2 E_{22}x_{22}}$; and, according to the conservation law for a system shown on **Figures 8**, and **9** charge neutrality for all elements of system influenced by electric field equals to nonzero, it can be possible to write the following formula:

$$\delta Q_1 E_{11}x_{11} + \delta Q_2 E_{21}x_{21} + \delta Q_1 E_{12}x_{12} + \delta Q_2 E_{22}x_{22} = 0.$$

Accordingly, the ratio between the polarization charges for two dipoles inside the electric field is given by

$\frac{\delta Q_1}{\delta Q_2} = \frac{E_{11}x_{11} + E_{12}x_{12}}{E_{22}x_{22} + E_{21}x_{21}}$; or $\delta Q_2 = -\delta Q_1 \frac{E_{22}x_{22} + E_{21}x_{21}}{E_{11}x_{11} + E_{12}x_{12}} = -A\delta Q_1$.

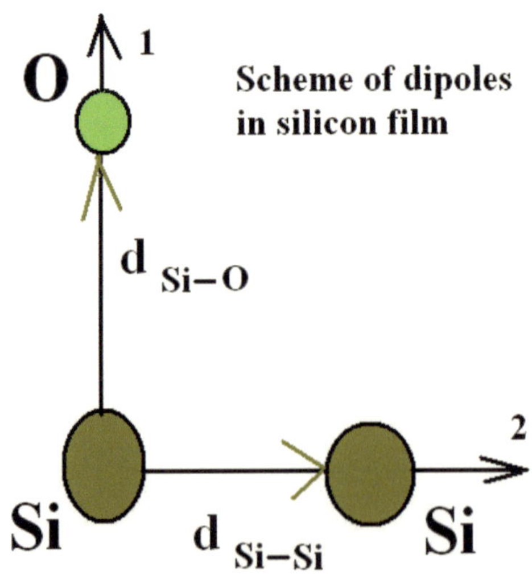


**Figure 8.** Scheme of arrangement of dipole vectors along the axis.

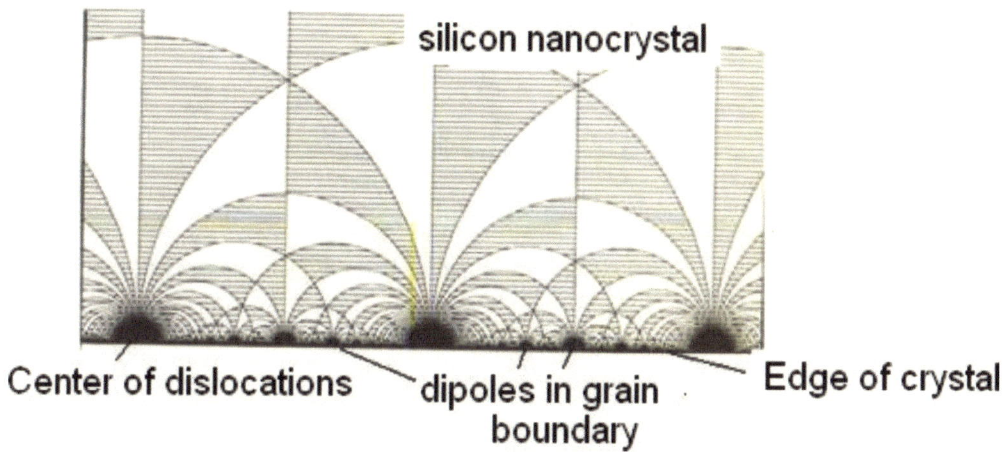


**Figure 9.** Scheme of nanocrystal's destruction by external electric field.

The first dipole is devoted to the description of electrical polarization properties of Si-Si bond, but the second – Si-O bond. By substituting the expression for Si-O polarization charge in expression for free energy we can easily to obtain the following expression

$$F_1 = E_1 - \delta Q_1 (E_{11} x_{11} - A E_{21} x_{21})$$
$$F_2 = E_2 - \delta Q_1 (E_{12} x_{12} - A E_{22} x_{22}).$$

The relation between the deformation values along the 1 and 2 axis can be used for analysis the translations consequences of modular group.

$\frac{W_1}{W_2} = \frac{(E_{11}x_{11} - AE_{21}x_{21})}{(E_{12}x_{12} - AE_{22}x_{22})}$; or $\frac{W_1}{W_2} = \frac{(p+r)}{(q+s)}$; where the first pair $p,q$ describe the translation result applied field, but the second pair $r$, $s$ causes the translation due to the local and external fields interaction with Si-Si dipoles. Crystal phase destruction can be generated by using significant values of electric field combining with sufficient density of Si-O dipoles inside the silicon film (**Figure 10**).

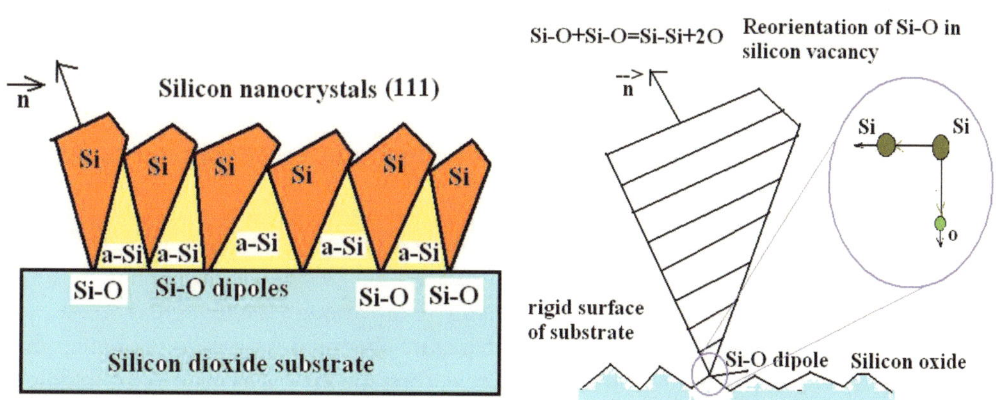


**Figure 10.** Scheme of silicon nanocrystal film with orientation (111) and precipitate of crystal growth on silicon dioxide substrate with nuclear silicon-oxide dipole in the top of pyramid.

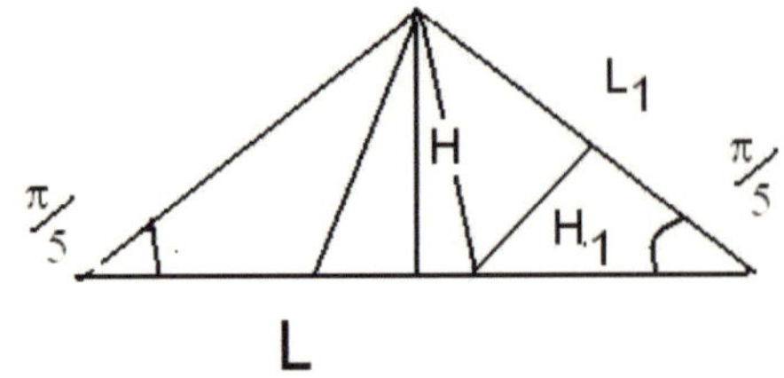


**Figure 11.** Scheme of geometrical transformation the primary triangle with L and H parameters into triangle 1 with parameters $L_1$ and $H_1$.

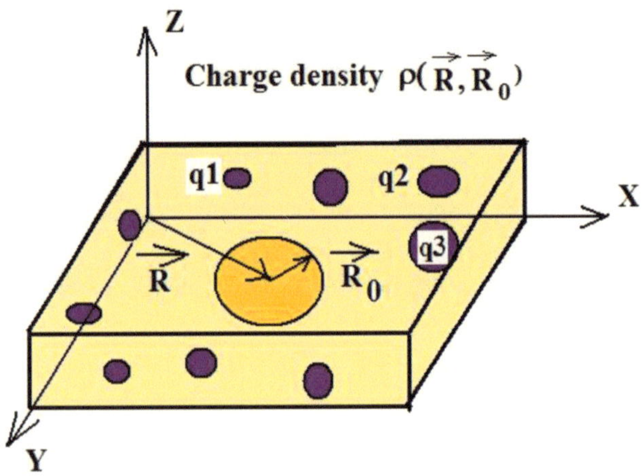


**Figure 12.** Scheme of charges density for anisotropic and non-homogeneous media.

Transformations which create the low dimensional model structure are illustrated on **Figures 11** and **12** and are explained as combination of modular group transformation of two-dimensional nanocrystal and Knop triangular transformations into Kantor dust set or fractal structure with low dimension. The surface and interface point defects and impurities cause the local electric fields which can generate by applying external field great values of field.

## 6. Possible scenario of nanocrystal destruction: from bulk silicon to Kantor dust

Model of phase destruction by modular group substitution [26] which consists of arc series and Knop transformation of two-dimension area under arc through the triangular decomposition [27] to one dimensional structure. The down picture illustrates the creation of Kantor dust by dividing the triangular angle on the top and neglecting the area of triangle in the middle of primary triangle area. Such nonlinear triangular transformations can be caused by a point defects and impurities which were included in bulk silicon net of nanocrystal.

$$S = \frac{1}{2}HL.,$$

$$S_1 = \frac{1}{4}{L_1}^2\, tg\,(\pi/5).$$

$$H = \frac{1}{2}L_1\, tg\,(\pi/5),$$

$$L_1 = \sqrt{H^2 + \frac{L^2}{4}}; L = \frac{2S}{H}; H^2 = Stg(\pi/5);$$

$$S_1 = \frac{1}{4}\left(H^2 + \frac{S^2}{H^2}\right) tg\,(\pi/5); \frac{S}{S_1} = 4\frac{1}{(\,tg^2(\pi/5) + 1)} = 2.63$$

For such transformation the equations for triangular quantities $X_N = 2X_{N+1}$ and for areas $S_{N+1} = 0.38{*}S_N$. The estimated value of Hausdorf dimension for such mathematical set is $d = \frac{\ln 2}{\ln 2.63} = 0.72$.

## 7. Classical and quantum mechanical models of charge and current densities

We have to propose the new model that is more suitable to explain electric properties of nanometrical scale media with strong anisotropic and non-homogeneous properties (see **Figure 12**). It will be necessary to describe the further possibility to design new nanoelectronic devices based on quantum conductivity properties and atomic scale sizes.

$$\begin{aligned} &Q(R) = 4\pi \int \rho(R, R_0){R_0}^2 dR_0; \\ &V = \frac{4}{3}\pi {R_0}^3; \\ &dV = 4\pi {R_0}^2 dR_0. \end{aligned} \tag{14}$$

For current density of homogeneous media with charge density $\rho$ in classical theory we usually use the formula

$$\begin{aligned} &\vec{j} = \frac{\vec{I}}{S} = \frac{N\,\vec{v}Q}{S} = \frac{\vec{v}\,\rho}{S}; \\ &\rho = NQ \end{aligned} \tag{15}$$

For $\rho$ value of non-homogeneous anisotropic media it is possible to use the expression:

$\begin{pmatrix} \rho_{XX} & \rho_{XY} & \rho_{XZ} \\ \rho_{YX} & \rho_{YY} & \rho_{YZ} \\ \rho_{ZX} & \rho_{ZY} & \rho_{ZZ} \end{pmatrix}$. For current density the following form is given:

$\vec{J}=\frac{\vec{e}}{S}\begin{pmatrix}\rho_{XX} & \rho_{XY} & \rho_{XZ}\\ \rho_{YX} & \rho_{YY} & \rho_{YZ}\\ \rho_{ZX} & \rho_{ZY} & \rho_{ZZ}\end{pmatrix}\begin{pmatrix}v_X\\ v_Y\\ v_Z\end{pmatrix}$; where the absolute value of vector $\left|\vec{e}\right| = 1$.To determine the charge density and current density on nanoscopic scale the quantum mechanical approach is applicable by the way

$$\rho = e\Psi^*\Psi;\ \ J_{ij} = \frac{i\hbar e}{2m}\left(\Psi_{3piSi}\frac{\partial\Psi_{3pjSi}{}^*}{\partial r_j} - \Psi^*{}_{3pjSi}\frac{\partial\Psi_{3piSi}}{\partial r_i}\right).$$

It is clear, that such approach is approximate and can be applicable to study the electric properties of point defects.

Nonlinear polarization concludes as linear as nonlinear terms: $P = \alpha E$, where $\alpha$ is a linear polarizability. Canonical equation for electro-magnetic fields can be written by using the Maxwell equations by a simplification of model and assuming that the first order of derivates are much more than others. The second supposal is that the solution can be explained as harmonic function. It is clear, that the equation for the field for second harmonic can be presented in the following form

$$\frac{\partial E}{\partial z} = -\frac{\sigma}{2}\sqrt{\frac{\mu}{\varepsilon}}E \tag{16}$$

It is supposed that currents which was created due to the electromagnetic field of second harmonic generation and induced in nanocrystals dominate in surface layers and grain boundaries. Because, such currents can be explained by the first term in equation and relate to absorption and emission of photons. Equation (17) can be written in suitable form:

$$\frac{\partial E}{\partial z} = -\frac{J}{2}\sqrt{\frac{\mu}{\varepsilon}} \tag{17}$$

Because, the surface current can play a significant role in nanostructured silicon film and the current density can be written as following J:

$$J = \sigma E = -\frac{2}{\sqrt{\mu}}\frac{\partial E}{\partial z}\sqrt{\varepsilon};\ \ E(z) = E(0)\exp\left(-\frac{\sigma\sqrt{\mu}}{2\sqrt{\varepsilon}}z\right). \tag{18}$$

By a symmetrical form of wave functions $\Psi \approx \exp(i(kx - wt))$ the current value is zero. By a bonding of p orbital of silicon atoms for bonding and antibonding cases the energy gap between them is estimated as 10 μeV and their energetic location is closed to a bottom of conductivity band $E_c$*-0.17* eV.

$$\Psi(t) = \left(a\Psi_{3piSiA}\exp\left(\frac{-iE_A}{\hbar}t\right) + b\Psi_{3pjSiB}\exp\left(\frac{-iE_B}{\hbar}t\right)\right);\Delta E = E_B - E_A = 10\ \mu eV. \tag{19}$$

For two energetic levels (A and B) which are situated closed to each other the expression for the current density is following:

$$J_{ij} = \frac{i\hbar e}{2m}\left[\begin{array}{l} a^2\left(\Psi_{3piSiA}\frac{\partial\Psi_{3pjSiA}{}^*}{\partial r_j} - \Psi^*{}_{3pjSiA}\frac{\partial\Psi_{3piSiA}}{\partial r_i}\right) + b^2\left(\Psi_{3piSiB}\frac{\partial\Psi_{3pjSiB}{}^*}{\partial r_j} - \Psi^*{}_{3pjSiB}\frac{\partial\Psi_{3piSiB}}{\partial r_i}\right) + \\ +ab\left(\Psi_{3piSiB}\frac{\partial\Psi_{3pjSiA}{}^*}{\partial r_j} - \Psi^*{}_{3pjSiA}\frac{\partial\Psi_{3piSiB}}{\partial r_i}\right)\exp\left(\frac{-(E_B - E_A)}{\hbar}t\right) + \\ +ab\left(\Psi_{3piSiA}\frac{\partial\Psi_{3pjSiB}{}^*}{\partial r_j} - \Psi^*{}_{3pjSiB}\frac{\partial\Psi_{3piSiA}}{\partial r_i}\right)\exp\left(\frac{(E_B - E_A)}{\hbar}t\right) \end{array}\right].$$

The values of currents for two energy states A and B are different due to the difference in their energies, and their occupations are also varied because they depends on Boltzmann distribution for unperturbed case, and for the laser excitation of carriers they distributed according to the Gauss distribution. Because, the current of charges depends strictly on energetic location of defects levels which are closed to the bottom of conductivity band of silicon. Therefore, the nonzero current is appeared because the field of SHG is applied in silicon nanocrystals. The current spectrum has a resonant energetic peak by the electron energy became equal to the energy of defect level:

$$J \approx \frac{2he}{m} ab\cos\left(\frac{\Delta E}{h}t\right). \tag{20}$$

The dipoles-field interaction causes the appearance of oscillations on frequencies $\Omega_1 = \frac{\mu_1 E}{h}$ and $\Omega_2 = \frac{\mu_2 E}{h}$, which by reemission result in radiation with various harmonics such as $2\omega \pm \Omega_1$, $2\omega \pm \Omega_2$, $2\omega \pm \Omega_1 \pm \Omega_2$. The estimates for the values $a$ and b as a levels' widths are following $a = \frac{\Delta E_a}{\Delta E_a + \Delta E_a} g_a$; $b = \frac{\Delta E_b}{\Delta E_a + \Delta E_a} g_b$, and by using the Erginsoy formula for semiconductors impurities it is possible to determine the sizes of local area for such kind of effect ~$R^3$, where R is a size of area of surface SHG assisted currents are generated. Therefore, there is a surface current in nanocrystals which are generated by applying the SHG fields. Such currents are caused by repolarization of pairs of atoms of silicon by their distribution of the surfaces of all nanocrystals irradiated by laser light. By this irradiation the times of repolarization are much more than period of oscillation of electromagnetic field of laser radiation, because there is absorption of radiation.

The free energy for nanocrystal with volume V can be expressed as following: $F = F_0 - \frac{\kappa V E^2}{2}$. By using determination of deformation of solid, it is possible to use the other definition for free energy by using the Lame coefficients η and υ [28]: $F = F_0 + \frac{\eta}{2} u_{il}{}^2 + \upsilon u_{ik}{}^2$, By equating of two right parts from expression ((11)) and (12), it is easy to obtain the result: $-\frac{\kappa V E^2}{2} = \frac{\eta}{2} u_{il}{}^2 + \upsilon u_{ik}{}^2$.

The estimated value of free energy to destroy the silicon crystal phase is following $F < 1.23 *10^{-4}$ J/cm$^2$. There are various preliminary states between stabile Si-Si bonds and broken bonds. The bond length distortion is varied from 0–15% of initial bond length. There are minima in potential energy for hydrogen which are varied from 1.3 eV to 2.3 eV. The Coulomb force driving the migration of charged impurity or defect can be written as $F = ZeE$; where the value of $Z$ is an effective charge of defect, $E$ is applied electric field. According to [28] the force due to the appeared interstitial atom is written as an Eshelby formula $F = \alpha \frac{4}{3}\pi r_0{}^3 \nabla Tr(\sigma)$, where $Tr(\sigma) = \sigma_{11} + \sigma_{22} + \sigma_{33}$, σ is a stress, $r_0$ is a radius of point defect. By this way it is clear that the electric field necessary to make stress can be evaluated as following: $E = \frac{4\alpha\pi r_0{}^3}{3Ze}(\sigma_{11} + \sigma_{22} + \sigma_{33})$.

## 8. MGT model and mechanism of crystal phase destruction by local fields

$$F_i = \frac{1}{2}\sum_j N_j\left(E_j - e\vec{E_j}\vec{x_j}\right)$$

Free energy is determined as energy $E_j$ per one bond, $N_j$ is a quantity of neighbor atoms, $\vec{E_j}$ is electric field, and vector $\vec{x_j}$ of dipole that is due to the inter atomic bond, such as Si-O, for example. Inter atomic bond length (for Si-O bond) can be calculated by using the following formula proposed by S. Hasegawa and co-workers [29] $d_{SiO} = d_{SiO}{}^{nonpolar} - B_{Si}\delta q_{Si} - B_O\delta q_O$, where $\delta q_{Si}$and $\delta q_O$ are partial charges for atoms, and $B_{Si}$,and$B_{Si}$ are empirical factor that has its positive value. Therefore, for different directions is

$$\frac{F_1}{F_2} = \sum_1 N_1\left(\frac{E_1 - \delta Q_1 E \rightarrow_1 x_1{}^{\rightarrow})}{\sum_2 N_2(E_2 - \delta Q_2 E \rightarrow_2 x_2{}^{\rightarrow});}\right. \tag{21}$$

For estimation the ratio between the energy of deformation and weak bond length which can be appeared by applying the external electric field we can use the Einstein relation for relation between the drift velocity and applying force $v_{drift} = \frac{D}{kT}F{:}\frac{\vec{\mu}}{\frac{E}{d}} = \frac{kT}{D}vN$; where $N$ is quantity of atoms that was locations were deformed, $d$ is bond length of polarized silicon. The $N$ value can be evaluated as following $N = \frac{D}{kTv}\frac{\vec{\mu}}{\frac{E}{d}}$. Here D is a diffusion coefficient, T is temperature, $\mu \rightarrow$ is a dipole moment of bond. The model of kinetic behavior of densities of dangling bonds and weak bonds which are generated by local electric field can be described by using the following

$$\begin{aligned}\frac{dN_{DB}}{dt} &= -W_1(N_{DB} - N_{WB}) - W_2(N_{DB} - N_H)\\ \frac{dN_{WB}}{dt} &= W_1(N_{DB} - N_{WB}) - W_3(N_{WB} - N_{FB});\end{aligned} \tag{22}$$

where $N_H$ is the density of silicon-hydrogen bonds, $N_{DB}$ is the density of dangling bonds, and $N_{WB}$ is a density of weak bonds, $W_1$ is a rate of dangling bonds elimination and weak bonds creation due to the annealing of silicon, mainly, but the $W_2$is a rate of generating of dangling bonds due to the hydrogen diffusion, $W_3$ is a rate of floating bonds generating due to the decreasing of density of weak bonds. It is assumed, that the values $W_2$and $W_3$are the same order of magnitude. For atomic diffusion coefficient the following formula is $D = D_0\exp\left(-\frac{E_A}{kT}\right)$; where $E_A$ is activation energy value and theoretical equation for $D_0$ can be written as

http://dx.doi.org/10.5772/intechopen.74310

follows $D_0 = n\alpha \exp\left(\frac{\Delta S}{kva^2}\right)$ [30]; where $n$ is a number of neighbor interstitial places of location, $v$ is frequency of vibrations in interstitial position, $a$ is a lattice constant, $\alpha$ is a coefficient that value depends on the interstitial position, $\Delta S$ is an entropy $\Delta S = -\frac{\partial \Delta F}{\partial T}$, $\Delta F$ difference in free energy that equals to the energy of activation. However, the total fraction of neutral interstitial is low compare with density of bulk atoms and according to W. Harrison studying [31], their number is approximately $10^{14}$ cm$^{-3}$ by the melting temperature. Therefore, the diffusion as interstitial as dopant diffusion plays mainly if we observe only the hydrogen diffusion with

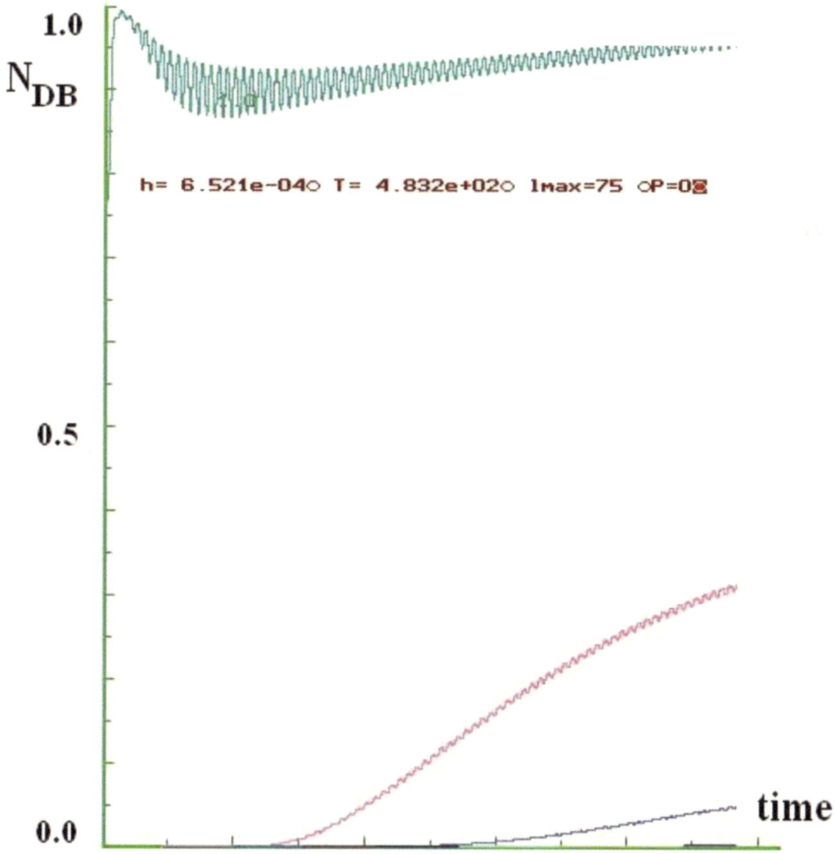


**Figure 13.** Evolution of density of dangling and weak bonds for different degrees of atomic sites order apart from the central of local electric field. It is assumed, that the values are $W_2 = W_3$. It is seen, that there is a reconstruction of the order by initial increasing of weak bond density, and redistribution of particles' local places according to the surrounding density values. It is assumed that this is a creation of order of amorphous phase. The third curve illustrates the evolution of density of perturbed bonds of next degree of order from the distance of a local electric field center. Oscillations of density of dangling bonds reflect the reconstruction of bonds and destruction by the annealing, hydrogen migration, according to the initial order of fixed atomic places of crystal structure and their new amorphous local placing. Over the period of time T the oscillations are suppressed by redestribution of initial impulse created by local field center and caused the increasing of weal bond density value. By this way the deformation penetrates in the surrounding of local area that will be larger than before.

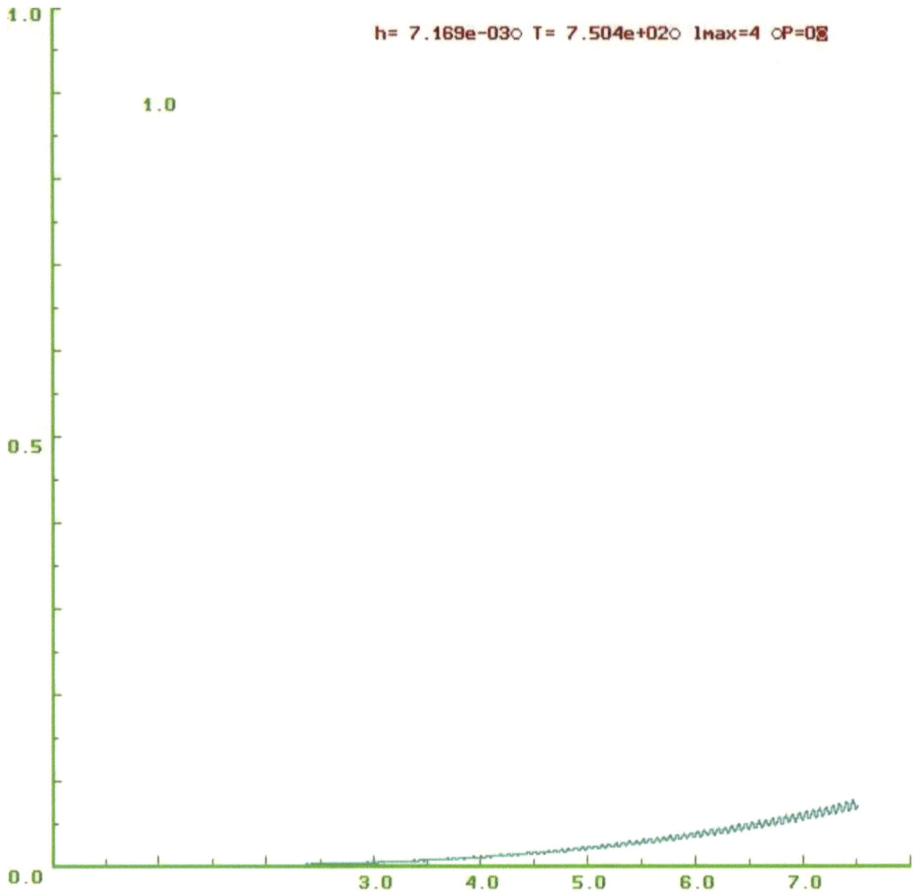


**Figure 14.** Evolution of density of dangling bonds for different degrees of atomic sites order apart from the central of local electric field, by the ratio between the values $W_2$ =0.001*$W_3$. This scenario can describe the poor hydrogen contamination of silicon network.

energy of its activation 1.5 eV [32] and investigate the hydrogen moving through the sites with different potential caused by variation in electric charges.

Analytical solution of system of differential equations results in the following expression for density of dangling bonds

$N_{DB} = N_{DB}{}^{(0)}\exp(-\frac{A}{2}t)\left[\exp\left(\sqrt{\frac{A^2}{4} - B}\right)t + \exp\left(-\sqrt{\frac{A^2}{4} - B}\right)t\right] - \frac{C}{B};$ where the values are given.

$A = W_2 + 2W_1 + W_3;\quad B = W_2(W_1 + W_3) - W_1W_3; C = (W_1 + W_3)W_1N_H + W_1W_3N_{FB}$. By the relation between values $W_2$ and $W_3$ as $W_2$ = 0.001* $W_3$, or $W_2 \rightarrow 0$, the coefficients A, B, C have view.

$A_{W2=0} = 2W_1 + W_3; B_{W2=0} = -W_1W_3; C_{W2=0} = (W_1 + W_3)W_1N_H + W_1W_3N_{FB}$ and the solution in this case can be written as.

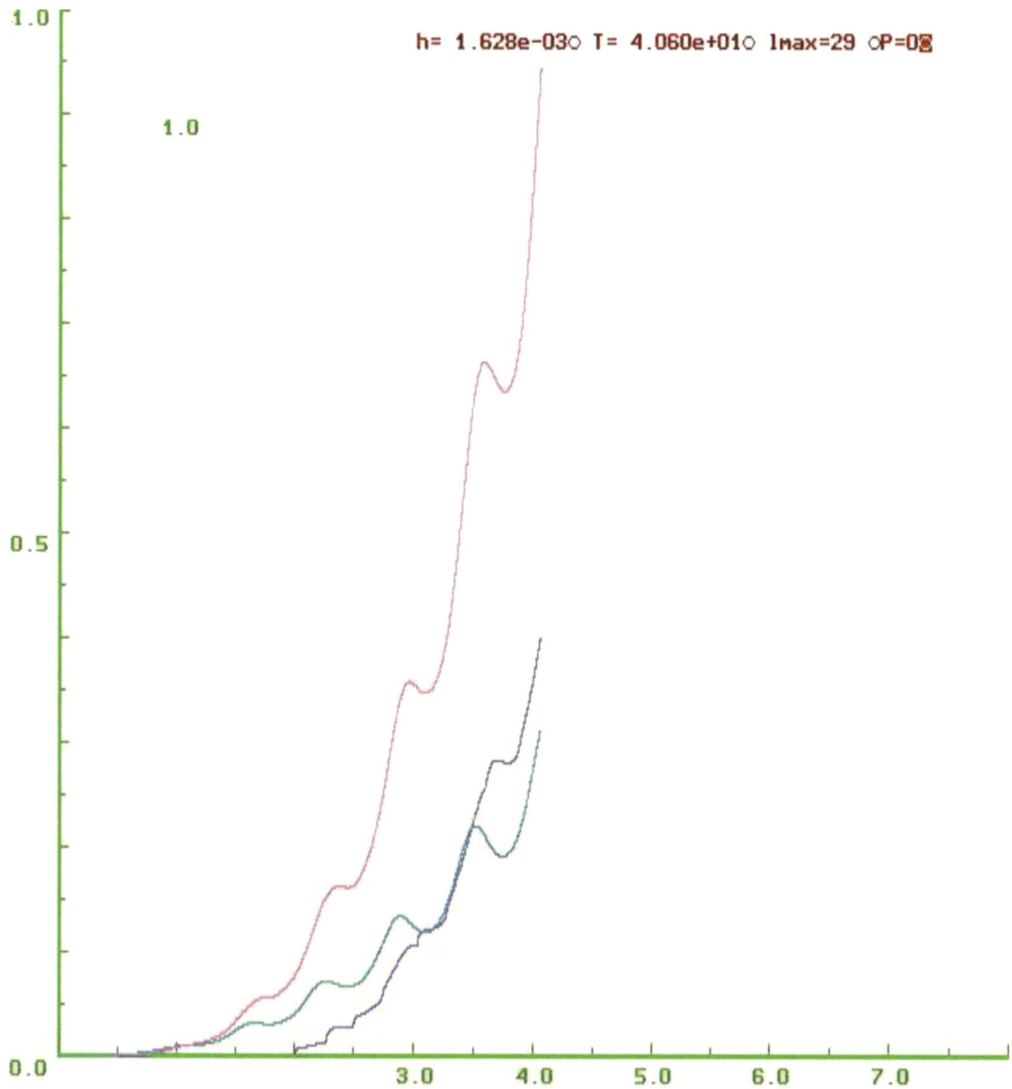


**Figure 15.** Evolution of density of dangling bonds for different degrees of atomic sites order apart from the central of local electric field, by the ratio between the values $W_2$ =0.01*$W_3$. It is clear, that there is amorphization of the solid because the atomic sites order was damaged. It is assumed that such scenario can be explained the difference in quantity of dangling bonds near the local field center $N_{DB}{}^{Local}$ and surrounding, $N_{WB}{}^{Surrounding}$, $N_{WB}{}^{Surrounding} >> N_{DB}{}^{Local}$. The conservation law can explain that the small number of particles which are placed near local field center cause the changes in impulses of numerous atoms in surrounding area.

$$N_{W2=0} = N_{W2=0}{}^{(0)}\exp\left[\sqrt{W_1{}^2 + \frac{W_3{}^2}{4}} - W_1 - \frac{1}{2}W_3\right]t^*\left[1 + \exp\left(-\sqrt{4W_1{}^2 + W_3{}^2}\right)t\right] + \frac{(W_1+W_3)}{W_3}$$

$N_H + N_{FB}$; where $W_1 > W_3/3$ should be realized for increasing the density of dangling bonds, and in this case the density of dangling bonds has a slow rising over the time (as it is shown in **Figure 13**). By using the Runge–Kutta method of 4-th order it is easy to calculate the evolution

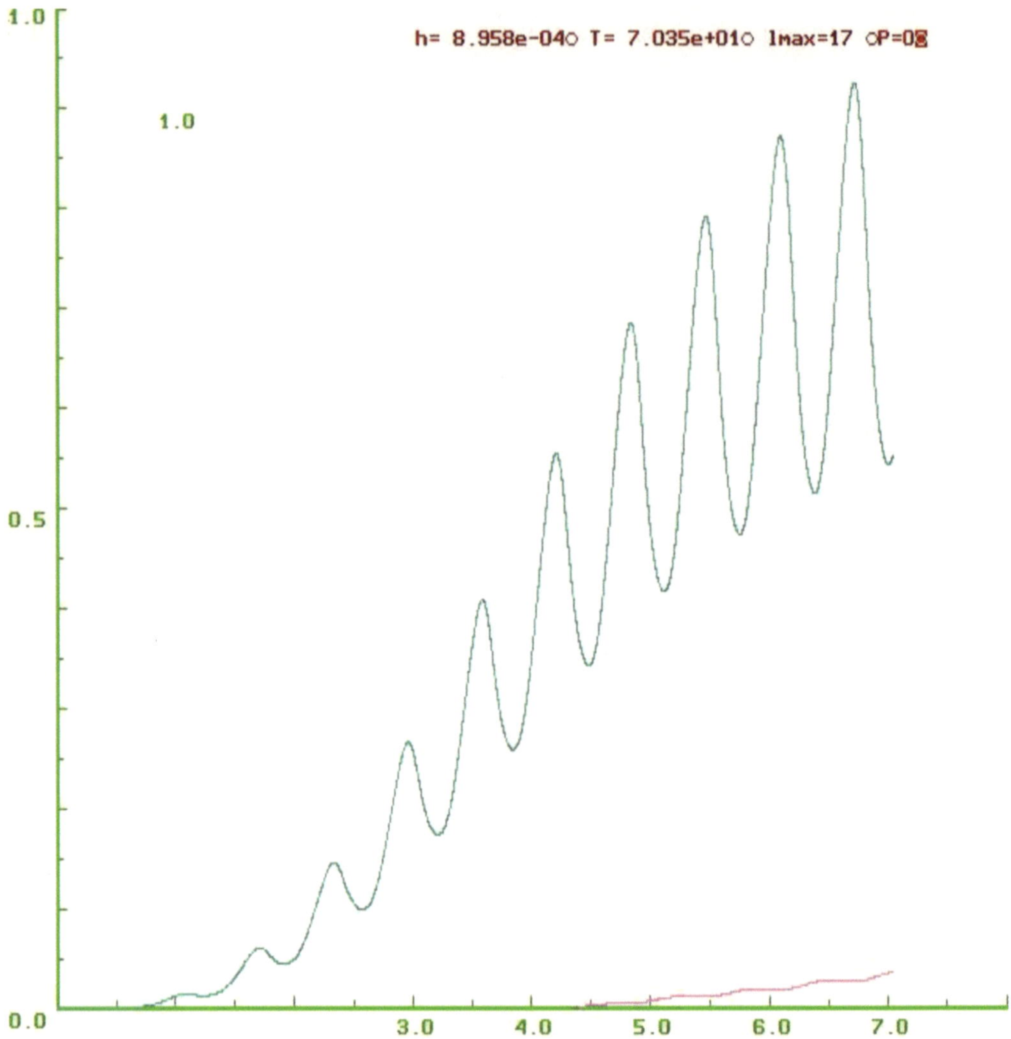


**Figure 16.** Evolution of density of dangling bonds for different degrees of atomic sites order apart from the central of local electric field, by the ratio between the values $W_2 = 0.1{*}W_3$.

of density of dangling bonds by the local electric field in several regimes for $N_{DB}(t)$, $N_{WB}(t)$ as normalized values of density of dangling and weak bonds.

It is known the model of defects generating by light irradiation in amorphous silicon which was proposed by the scientists of Ames Laboratory [33] which calculated the evolution of density of dangling bonds according to their proposed model. It is seen, that the evolution which was shown in **Figure 14** has the same increasing tendency as evolution stimulated by light irradiation. By the values $W_2$=0.001* $W_3$the evolution changes its sharp increasing to the slow behavior of density of dangling bonds function as it is seen in **Figure 14** .

By the ratio between rates $W_2$and $W_3$ equals to 100 the calculating evolution of density of dangling bonds can be presented as that was shown in **Figure 15**.

By the ratio between rates $W_2$and $W_3$ equals to 10 the calculating evolution of density of dangling bonds can be presented as that was shown in **Figure 16**.

## Author details

Dmitry E. Milovzorov

Address all correspondence to: dmilovzorov2002@yahoo.com

Fluens Technology Group, Ltd., Moscow, Russia

## References

[1] Maxwell-Garnett JC. Colors in metal glasses and metallic films. Philosophical Transactions. Royal Society of London. 1904;**203**:385. ISSN 1364-503X

[2] Lorentz HAW. Annalen. 1880;**9**:641. ISSN 0003-3804

[3] Brugeman DAG. Annalen der Phyziks. Leipzig. 1935;**24**:638

[4] Stoner E. Demagnetizing factor for ellipsoids. Philosophical Magazine. 1945;**36**:263. ISSN 1478-6435

[5] Osborn J. Demagnetizing factor of the general ellipsoid. Physical Review. 1945;**67**:351. ISSN 0163-1829

[6] Ishiguro M, Ueno M. Observation Studies of Interplanetary Dust, Lecture Notes in Physics. Nakamura, Mukai, Springer: Mann; February 2009

[7] Sankrit R, Blair W, Raymond J, Williams B. Dust destruction in the Cygnus Loop supernova remnant, Supernova Environmental Impact, Proceedings IAU Symposium No 296, 2013, eds. A. Ray and R. McGray

[8] Mann I, Czechovsky A. Dust destruction and ion formation in the inner solar system. The Astronomical Journal. 2005. ISSN 0004-6256

[9] Raymond P, Chavamian B, Williams W, Blair K, Borkovsky T, Gaetz R, Sankrit. Grain destruction in a supernova remnant shock wave. The Astrophysical Journal of AAS. 2013; **778**:161. 9 pp. ISSN 0004-6256

[10] Landi S, Meyer-Vernet N, Zaslavsky A. On the Unconstrained Expansion of a Spherical Plasma Cloud Turning Collisionless: Case of Cloud Generated by a Nanometer Dusty

Grain Impact on an Uncharged Target in Space, Plasma Physics and Controlled Fusion, April 2012, arXiv.1205.1718v.1, publication No 241779851. ISSN 0741-3335

[11] Milovzorov D. Point defects in amorphous and nanocrystalline fluorinated silicon. Journal of Materials Science and Engineering with Advanced Technology. 2010;**2**:41-59. ISSN 0976-1446. ISSN 0976-1446

[12] Stockman, M. Local fields' localization and Chaos and nonlinear-optical enhancement in clusters and composites, in Optics of Nanostructured Materials, ed. by V. Markel, T George, John Willey & Sons, 313–343 (2001)

[13] Zimbovskaya N, Gumbs G. Long-range electron transfer and electronic transport through the macromolecules. Applied Physics Letters. 2002;**81**:1518-1520. ISSN 0003-6951

[14] Compaan A, Trodahl HJ. Physical Review B. 1984;**29**:793. ISSN 0163-1829

[15] Richter H, Wang Z, Ley L. The one phonon Raman spectrum in microcrystalline silicon. Solid State Communications. 1981;**39**:625. ISSN 0038-1098

[16] Wright O. Thickness and sound velocity measurement in thin transparent films with laser picoseconds acoustics. Journal of Applied Physics. 1992;**71**:1617-1627. ISSN 0021-8979

[17] Weinreich G, Sanders T, White H. Acoustoelectric effect in n-type germanium. Physical Review. 1959;**114**:33-44. ISSN 0163-1829

[18] Geissberger AE, Galeener FL. Raman studies of vitreous $SiO_2$ versus fictive temperature. Physical Review B. 1983;**28**:3266-3271. ISSN 0163-1829

[19] Raldugin VI. Physico-chemistry of surface. Dolgoprudny. 2011;**568**. (on Russian). ISBN 978-5-91559-116-4

[20] Abtew TA, Drabold DA. Light-induced structural changes in hydrogenated amorphous silicon. Journal of Optoelectronics and Advanced Materials. 2006;**8**:1979-1988. ISSN 1454-4164

[21] Milovzorov D. Acoustoelectric effect in microcrystalline and nanocrystalline silicon films prepared by CVD at low and high deposition temperatures. Journal of Physics. 2012;**1**:38-49. ISSN 0953-4075

[22] Stewart GW, Sun J. Matrix Perturbation Theory. San Diego: Academic Press; 1990. p. 189 ISBN 0-12-670230-6

[23] Lakno V. Clusters in physics, chemistry, biology. Izhevsk. 2001;**256**. (on Russian)

[24] Britton D, Harting M. The influence of strain on point defect dynamics. Advanced Engineering Materials. 2002;**4**:629-633. ISSN 1438-1656

[25] Emin D. Energy spectrum of an electron in a periodic deformable lattice. Physical Review Letters. 1972;**28**:804-807. ISSN 0163-1829

[26] Milovzorov D. Crystalline phase destruction in silicon films by applied external electrical field and detected by using the laser spectroscopy. In: Huffaker DL, Eisele H, Dick KA,

editors. Quantum Dots and Nanostructures: Growth, Characterization, and Modeling XIII Edited. Vol. 9758. SPIE Proceedings; 2016. p. 11. DOI: 10.1117/12.2208270

[27] Hausdorff F. Grundzuge der Mengenlehre. Vol. 184. Berlin; 1914 3-540-42224-2

[28] Landau L, Lifshitz E. Elastic Theory. Moscow; 1987. pp. 51-56 (on Russian). ISBN 978-0-7506-2633-0

[29] Hasegawa S, He L, Amano Y, Inokuma I. Physical Review B. 1993;**48**:5315. ISSN 0163-1829

[30] Wert C, Zener C. Interstitial atomic diffusion coefficient. Physical Review. 1949;**76**:1169-1175. ISSN 0163-1829

[31] Harrison W. Diffusion and carrier recombination by interstitials in silicon. Physical Review B. 1998;**57**:9727-9735. ISSN 0163-1829

[32] Powel M, Dean S. Microscopic mechanism for creation and removal of metastable dangling bonds in hydrogenated amorphous silicon. Physical Review B. 2002;**66**:155212. ISSN 0163-1829

[33] Biswas R, Pan B, Ye Y. Metastability of amorphous silicon from silicon network rebonding. Physical Review Letters. 2002;**88**:205502. ISSN 0163-1829

Chapter 6

# Local Electric Fields in Dielectric and Semiconductors: Part II

Dmitry E. Milovzorov

Additional information is available at the end of the chapter

http://dx.doi.org/10.5772/intechopen.76660

**Abstract**

Local electric fields appeared in dielectric and semiconductors due to the destruction of symmetry, creating the vacancies, point defects, and chemical impurities in material. By increasing external electric field value, numerous structural changes will be generated. Point defects in silicon films were characterized by using electron-paramagnetic resonance spectroscopy and laser picosecond spectroscopy. EPR spectroscopy provides the detailed microscopic information about the point defects in silicon films, which have unpaired electrons. The switching effect for A or $(VO)^-$ defects was observed by applied bias voltage for nanocrystalline silicon films. The intensities of EPR signal change according to the switching of position of oxygen incorporated in silicon (111) from one pair of silicon atom to another pair of silicon atoms. In this case, the interaction between $p_x$ orbital is changed to $p_z$ orbital. The proposed mechanism of states' switching for the vacancy $(VO)^-$ is based on a model of coupled dangling bonds (DBs) of silicon atoms and interaction between atomic orbitals. The crystal phase destruction in nanocrystalline silicon film by applying external electric field was investigated by Raman spectroscopy. The possible mechanism of phase creation was proposed.

**Keywords:** local field, point defect, dangling bonds, Raman spectroscopy, nanocrystals, second-harmonic generation, silicon films

## 1. Introduction

Local electric fields appeared in dielectric and semiconductors due to the destruction of symmetry, creating the vacancies, point defects, and chemical impurities in material. By increasing external electric field value, numerous structural changes will be generated. Some of them will produce such great local fields that will destroy all material or change its physical properties. Studying the nature of local electric fields will open new tendencies in electronic device producing, from one side, and help to change materials' properties according to our needs, from another side. The quantum size effect for nanocrystalline silicon films with broad size distribution was observed by PL and SHG spectral measurements. The optical properties of films are likely to depend on the size distribution of nanocrystals, along with an effect of hydrogen termination on the surface of nanocrystals. The studying of physical properties of such nanoscale objects as nanocrystals, point defects and chemical bonds is very important for understanding the appearance of local fields in silicon films, as it is necessary to investigate them and to detect the physical phenomena which appear due to the local fields. For small particles such as electron, the wave features play a significant role because the de Broglie wave length is given by $\lambda = \frac{h}{p}$; where p is impulse and h is Planck constant. For electron rotating around nucleus of atom with circulation rate $3.54 \times 10^6$ m/s, $\lambda_{DB} = 2.5$ Å. For oxygen atom at room temperature and the rate of motion 250 m/s, the de Broglie wavelength is $\lambda_{DB} = 2.0$ Å. The probability of occupancy of each energy state is defined by wave function as a solution of the Schrodinger equation $H\psi = i\hbar\frac{\partial\psi}{\partial t}$; where H is full energy of system named as Hamiltonian. For example, for the quantum well with the size L = 100 Å, the solution of this equation is satisfied to a discrete energy according to the following expression $E_n = \frac{\pi^2\hbar^2}{2mL^2}n^2$, where numerical data are 3.37 meV for n = 1, 13.48 meV for n = 2, 30.33 meV for n = 3, and 53.92 meV for n = 4.

## 2. Quantum objects in silicon films

The probability to find the particle in each state can be calculated as $\rho_{nn} = |\psi_n(x)|^2$. Hamiltonian of system can be written in following form: $H = \sum_{i=1}^{N} H_i + \sum_{i=1}^{N}\sum_{j=1}^{N} V_{ij}$, where $H_i$ is diagonal matrix element and $V_{ij}$ is an interaction energy or non-diagonal matrix elements. For example, the quantum properties of interaction between the elements are explained by using resonance integrals ($H_{ab} = \int \psi_a H \psi_b d\zeta$), and overlap integrals (e.g., for two 3pSi-3pSi orbitals that are located on different Si atoms $\delta = \int_0^{2\pi}\int_0^{2\pi}\int_0^{\infty} \psi_{3pxSi}\psi_{3pxSi} r^2 \sin\varphi \cos\varphi \sin\varphi dr d\varphi d\varphi$) which can be calculated for a small distances: 0.675 for interaction of 3pSi-3pSi orbital and X = 2.35 Å, 0.837 for the same orbital and distance 2.21 Å and 0.0684 for X = 3.65 Å, 0.0484 for X = 3.83 Å; and by X = 9 Å it is $5.3 \times 10^{-7}$.For orbital interaction of $3p_x$Si-$2p_x$O, for X = 2.21 Å, the overlap integral equals to 2.8 $10^{-6}$, for X = 2.35 Å it is $1.2 \times 10^{-6}$, for X = 3.65 Å it is 3.1 $10^{-10}$, for X = 3.83 Å it is $9.55 \times 10^{-11}$ because the quantum phenomena can be surely observed on such scale as it is seen.

### 2.1. Nanocrystals in silicon films

The combination of laser picosecond spectroscopy with Raman scattering measurements is perspective for nondestructive analysis of composite thin films and silicon surface with complex relief. The method of laser picosecond spectroscopy is a powerful tool for surface analysis with high sensitivity. The pump-probe scheme is similar to the interferometric schemes with great possibility to measure the refractive index changes. For composite thin film measurements, such as poly-Si or porous silicon, the local factor field plays a great role because the crystalline volume fraction (or porosity) and dielectric functions' values of film's components are important. Also, due to the great amount of scattering centers (e.g., nanocrystallites or microcrystallites spatial distribution in film according to Gaussian distribution), the role of photo-acoustic distortion is negligible. However, the non-equilibrium perturbation of career density and its decay due to the diffusion or recombination (in case of great porosity or small value of $\rho$) are detected surely by using picosecond technique. In addition, it is necessary to note that the electron relaxation from surface and defect states can be reflected by optical response evolution, too. Also, by carrying these measurements, it is important to know the surface morphology. For elimination of the surface scattering background the correlation function value was measured.

#### 2.1.1. *Nanocrystal detection by Raman spectroscopy*

The Raman spectra for poly-Si films consist of a narrow line near 520 cm$^{-1}$ arising from a crystalline phase and a broad line around 480 cm$^{-1}$ from an amorphous phase. The $\rho$ value was estimated from the ratio of the Raman integrated intensity for the crystalline component to the total intensity, using the ratio of the integrated Raman cross-section for the crystalline phase to that of an amorphous phase. The penetration depth of the incident Ar-ion laser radiation ($\lambda$ = 488 nm) into silicon is within about 0.3 μm.

The crystalline volume fraction was estimated using Raman scattering measurements from the ratio [1]: $\rho = \frac{I_c}{I_c + I_a}$, where $I_c$ is the Raman integrated intensity for the crystalline component (sharp peak at 520 cm$^{-1}$) and Ia is for amorphous phase (smooth peak at 480 cm$^{-1}$). The random silicon network is Gauss distributed in their bond lengths with various deviations, and phonon wave numbers is spread. The regular bonding network has the Lorentz shape of its spectral line. Our possible interpretation of a-Si Raman spectra decomposition is to recognize the defects and impurities which cause the changes in electrical properties of films. We guess that the a-Si film with high density of defects such as silicon vacancies causes the spectral peak around 465 cm$^{-1}$, but after annealing there is a spectral shift in wave number 465 cm$^{-1}$→475 cm$^{-1}$. By hydrogen dilution of gas mixture by PECVD of a-Si the Raman peak position is changed from 475 cm$^{-1}$ to the 480 cm$^{-1}$ [2].

It was assumed that this change is because the bond angle variation value decreases by the starting crystallization process in a-Si film. We assume that the Raman data of hydrogenized amorphous silicon film resulting in the spectral peak around 445–447 cm$^{-1}$ corresponds the LO mode but it is 480 cm$^{-1}$ for TO mode. For the higher structural relaxed silicon thin film, by high level of hydrogen dilution, the TO mode reflects in 490 cm$^{-1}$ value of peak position. The spectral peak width changes from the value of 40 cm$^{-1}$ for a-Si to the 70 cm$^{-1}$ for a-Si:H.

**Figure 1** shows the various types of decompositions of Raman spectrum for microcrystalline silicon film with thickness around 100 nm. This film was deposited at 380°C. By using this procedure we use 3, 4, and even 9 peak approximations. The figure illustrates the three-peak

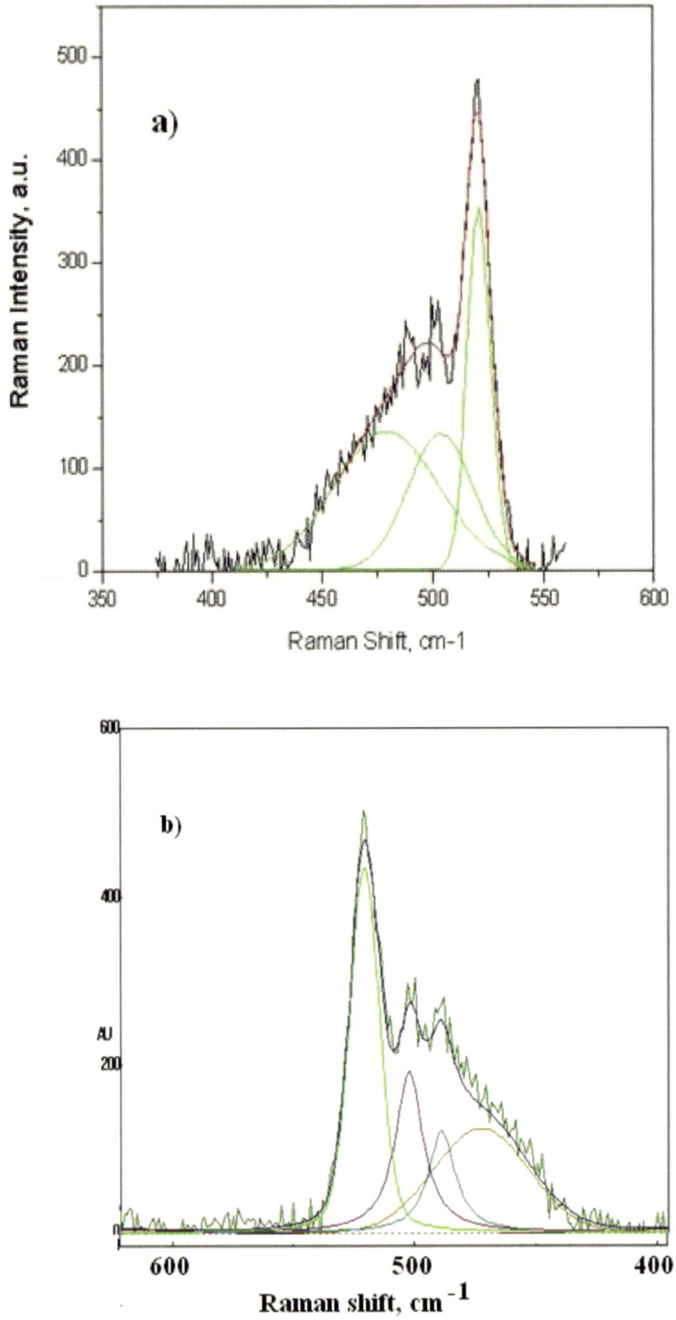


**Figure 1.** Raman spectra deconvolution by using (a) 3 peaks approximation: c-Si, for 520 $cm^{-1}$; a-Si for 480 $cm^{-1}$ and intermediate or nanocrystalline for 500–510 $cm^{-1}$; (b) 4 peaks approximation: for c-Si (520 $cm^{-1}$); a-Si (480 $cm^{-1}$); nanocrystalline (500–510 $cm^{-1}$) and hydrogenated a-Si:H for 490 $cm^{-1}$; approximation for different silicon networks for TO phonon modes: c-Si, nc-Si, a-Si, a-Si:H, a-Si high hydrogenated, a-Si with different defects; and for LO phonon modes in a-Si; a-Si:H.

decomposition with c-Si-related spectral line around 520 $cm^{-1}$, nanocrystalline or intermediate spectral line in the range 500–510 $cm^{-1}$ and for amorphous silicon it is 480 $cm^{-1}$. The next figure, **Figure 1b**, shows the Raman data with noise signal for comparison, and fourth spectral component (~490 $cm^{-1}$) is related to hydrogenated amorphous silicon network. The estimation of crystalline volume fraction was 27%. The hydrogen terminates all the dangling bonds (DBs). The spectral characteristics of amorphous-related peaks are changed by the hydrogen saturation of dangling bonds.

### 2.1.2. *Time-resolved picoseconds laser spectroscopy of silicon nanocrystalline films*

A mode-locked YAG:$Nd^{3+}$ laser radiation with wavelength 532 nm was used as an optical pump of media, but the second-harmonic radiation ($\lambda$ = 1064 nm) was used for probing the sample's **Figure 2**. The correlation between the reflected signal intensity and probe laser beam for a-Si and p-Si films surface is given. The pulse duration was 120 ps. The pulse repetition rate was 100 MHz and frequency of Q-switched modulation of second-harmonic radiation was 6.2 MHz.

The correlation function of reflected signal intensity of probe laser beam was detected by means of pump-probe laser scheme $G(t) =< I(t)I(t+\tau) >= \frac{1}{T}\int_0^T I(t)I(t+\tau)dt;$ where $G(t)$ is the averaged correlation function and $T$ is the time of detection. It is seen that $G(t) \sim exp.(-\Delta\omega t)$; $\Delta\omega$ is a width of level that is equals 8 μeV for porous silicon, and 16 μeV for a-Si:H. Figure illustrates the time evolution of correlation function for the amorphous silicon (a-Si)

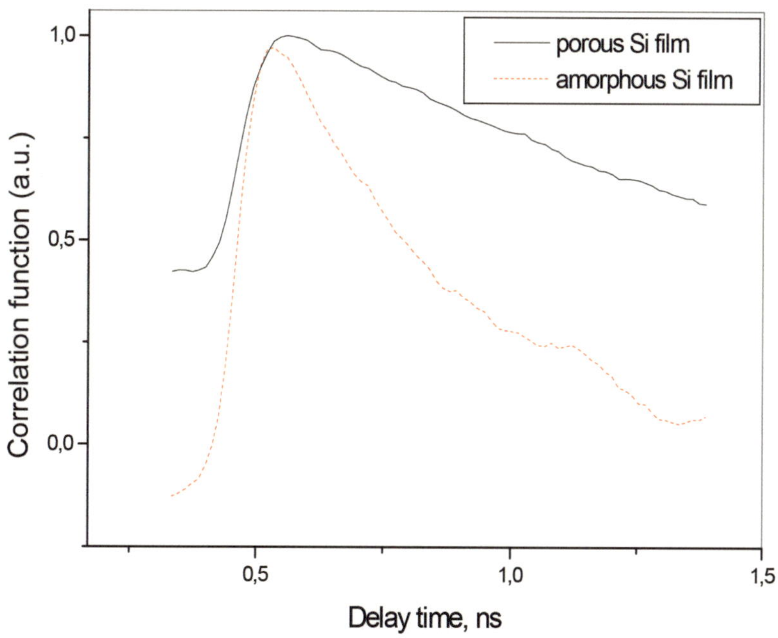


**Figure 2.** Evolution of correlation function for the amorphous silicon and porous silicon.

surface and porous silicon (por-Si) sample prepared by the value of current 5 mA/cm$^2$ in hydrofluoric acid (HF)/ethanol solution by the ratio 1:1. It is seen that the time decay of relative reflected intensity of signal for por-Si is slightly less due to the prepared nanostructured surface compared with a-Si. It is supposed that the time evolution corresponds to the surface electronic band structure. Also, we proposed the formula for estimation of recombination rate from the Fermi golden rule:

$$\tau^{-1} = \frac{16\pi^2}{3} n \frac{e^2}{h^2 m^2 c^3} E|< i|p|j >|^2 \tag{1}$$

where E is transition energy and < i| and |j > are the initial and final states of electron transition. The time evolution reflects the superposition of ensemble of different kinds of transitions such as band-to-band transition, transition through the surface or defect states. The width of the energy level depends on the chemical bond structure of silicon surface. The most important fact is appearance of quantum beats, which reflect the optical response from two-neighbor level. We suppose that the different fractions of oxygen incorporation in silicon film cause the production of neighbor levels inside the band gap with the spectral width: $\Delta = \frac{20\varepsilon_x h^3}{m^2 e^2} N$ a-Si reflects the presence of complex SiO configuration. The figure shows the correlation function as a function of delay time for oxidized Si (111) surface and poly-Si silicon film with <δ> = 9.7 nm. It is expected that the oxide defect levels and surface state level (for Si (111) surface) are responsible for the observed oscillation during the exponential decay of electron density. We suggest that such oscillations are the same as quantum beats of levels of occupation by the laser time-resolved spectroscopy of molecular levels with the period of ΔT. Also, it is seen that exponential decrement for poly-Si film is weaker than the other. It is assumed that the correspondence between the values of decrement and width of oxygen-related levels in the band gap of silicon is proved and the magnitude of correlation function oscillations illustrates oxide amount on the surface. Also, our assumption has strong evidence from the FTIR data for aged poly-Si films.

Intervalley scattering rate is given by [3]

$$\frac{1}{\tau_{intervalley}} \propto N_q (E_k + E_p)^{1/2} + (N_q + 1)(E_q - E_p)^{1/2} U(E_k - E_p); \tag{2}$$

where $E_p$ is the phonon energy and $U$ is the step function. For TA or LA phonon modes (with energies 10 meV) we can estimate the intervalley rate at around 2.5 $10^{12}$ Hz. Figure illustrates quantum beats observed in correlation function's evolution for oxidized surfaces: Si (111) surface and silicon film with <δ> = 9.7 nm. The period of quantum beats is estimated as ΔT=h/Δ. Width for level is Δω = 6 μeV for Si(111) surface but gap is Δ = 7.2 μeV, and Δω = 7.2 μeV and Δ = 12 μeV for silicon film with <δ> = 9.7 nm. It is seen in **Figure 3** that correlation function depends on the time according to exponential law with harmonic modulation of intensity because of quantum interference of levels $G(t) \sim exp.(-\Delta\omega t)(1 + C cos(\Delta t))$; Δω is a width of level and Δ is an energy gap between the two closed levels (see **Figure 4b**).

Dipole moment can be written in form [4]

$$d = \frac{er}{2} \frac{(\phi_2 - \phi_1)^2}{(\phi_2 - \phi_1)^2 + 4J^2(r)} [n_2(t) - n_1(t)]; \quad (3)$$

where $n_2(t)$ and $n_1(t)$ are levels populations, $J$ is resonance integral, and $\varphi_1$ and $\varphi_2$ are energies of electrons. Reflected intensity harmonic oscillations can be explained by using time-dependent dipole moment of coupled atoms of Si-Si bonding.

In conclusion, the new method of laser spectroscopy for testing nanocrystal silicon structures and measurements of band gap energy and energy position of defect levels inside the band gap is proposed. Also, the competition between the electron diffusion and recombination processes in poly-Si films by the transmitted probe-pump laser scheme is analyzed. It is clear that tunneling process between the silicon nanocrystals through the thin oxidized layer plays a great role and it is possible to design a new electronic device.

### 2.2. Point defects in silicon

For several decades, a-Si:H films are intensively studied by different spectroscopic and microscopic methods. Most of the a-Si films contain some amount of nanocrystals. This is because

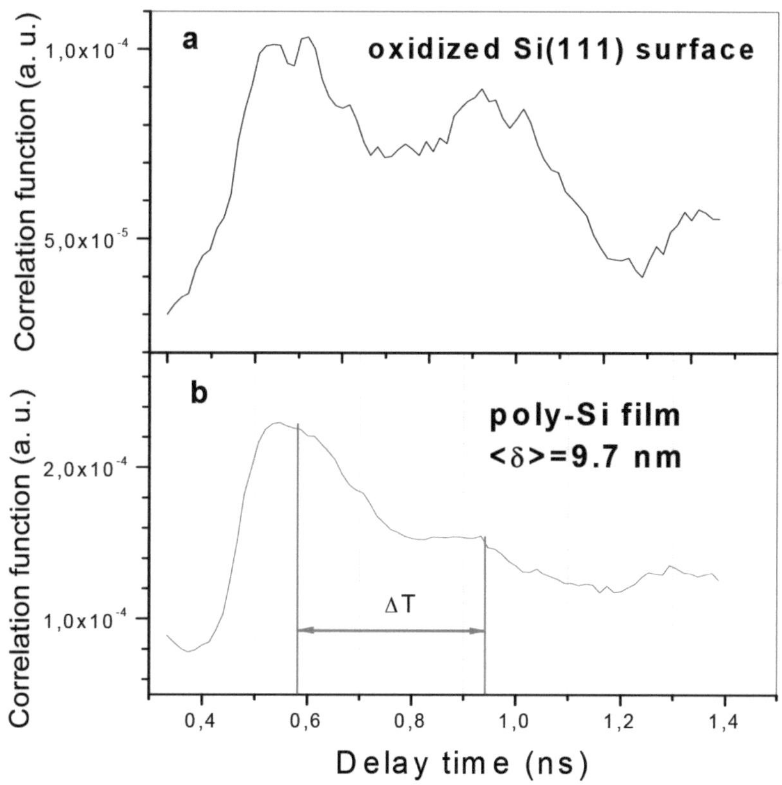


**Figure 3.** Quantum beats observed in correlation function's evolution for oxidized surfaces: Si (111) surface (a) and poly-Si film with <δ> = 9.7 nm (b). The period of quantum beats is ΔT=h/Δ.

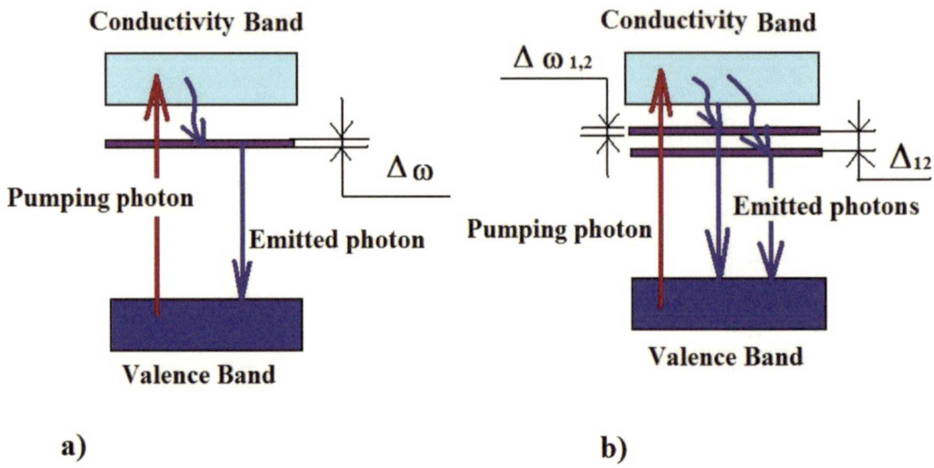


**Figure 4.** Energy diagrams and scheme of laser pump and light emission. The emission of light is realized by the carriers recombination through the one defect level (a) with spectral width $\Delta\omega$, and two defect levels which are located closed to each other (b) with energetic gap $\Delta_{12}$.

the PECVD method to prepare nc-Si films and a-Si films is the same but differs only in deposition conditions such as substrate temperature, gas mixture, working pressure, and RF power. **Figure 5** illustrates the CVD-prepared a-Si film with gas silane by 1 sccm flow rate. The RF power was 20 W and temperature of substrate was 25°C. It is clear that the sharp peaks relate to the various fractions of silane dissociation: $(SiH)_x$; $(SiH_2)_x$; $(SiH_3)_x$; $(SiH_4)_x$. It is

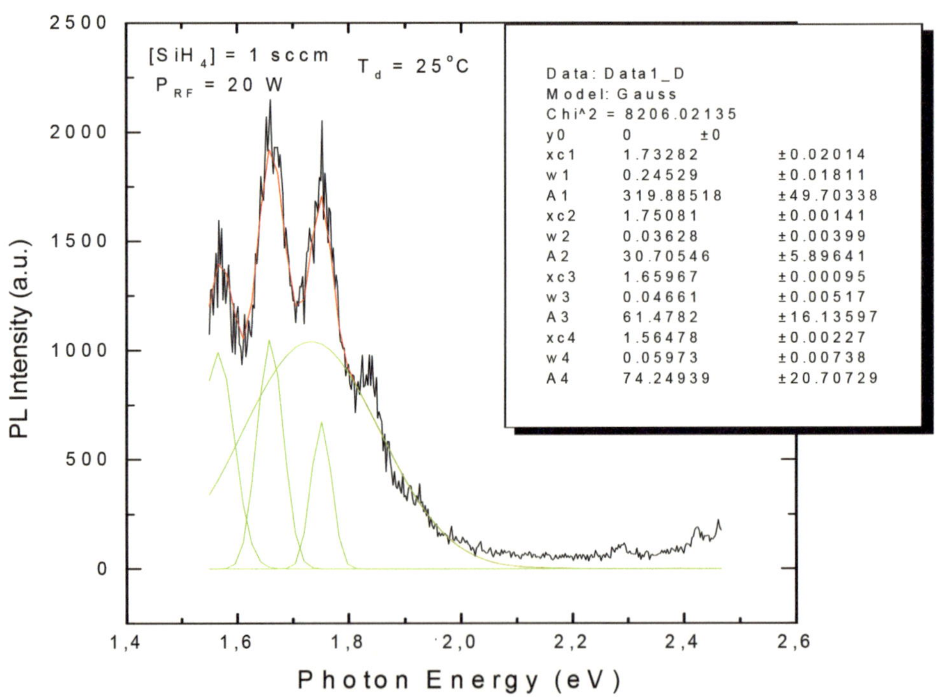


**Figure 5.** Photoluminescence spectrum of a-Si film.

http://dx.doi.org/10.5772/intechopen.76660

supposed that their responses are around spectral lines: 1.56, 1.65, 1.73, and 1.83 eV. Also, the broad spectral peak around 1.75 eV is related to the tiniest nanocrystals in a-Si film.

EPR spectroscopy provides the detailed microscopic information about the point defects in silicon films, which have unpaired electrons. It is seen in **Figure 6**, that there is a dangling bond signal on EPR spectrum. Therefore, the a-Si films contain dangling bond defects and are disordered due to the hydrogen termination of dangling bonds. The surface diffusion and desorption processes by growth are low at low temperatures of deposition.

### 2.2.1. *Dangling bonds*

For detail explanation of dangling bond (DB) defects several kinds of models were proposed such as defect-pool model of Powell and Dean [5]. The main principle of this model is the following. There is equilibrium between the quantity of dangling bonds and weak bonds. The chemical reactions that cause the appearance of dangling bonds are the following: WB ⇔ (2DB); SiH + WB ⇔ (DB + SiH) + DB; 2SiH + WB ⇔ (Si-H-H-Si) + 2DB. Thin film deposition at low temperatures of substrates results in the appearance of dangling bonds defects that can be terminated by hydrogen. The process of nucleation of crystal grains required the minimum quantity of silicon atoms.

These paramagnetic centers appear when there is an unpaired electron of the dangling bond of silicon atom that bonded with three silicon atoms (see **Figure 7**).

For coupled dangling bonds of silicon atoms on the surface (111) there is an interaction between atomic orbitals (as it is seen in **Figure 8**). The degeneracy of the level according to rules for wave functions of one silicon atom with free atom $3pSi_x$ orbital and the other silicon atom with one $3pSi_z$ orbital is $\begin{matrix} \Psi^+ = \Psi_{3pSix} + \Psi_{3pSiz}; \\ \Psi^- = \Psi_{3pSix} - \Psi_{3pSiz}. \end{matrix}$ Coupling two dangling bonds are transformed into A

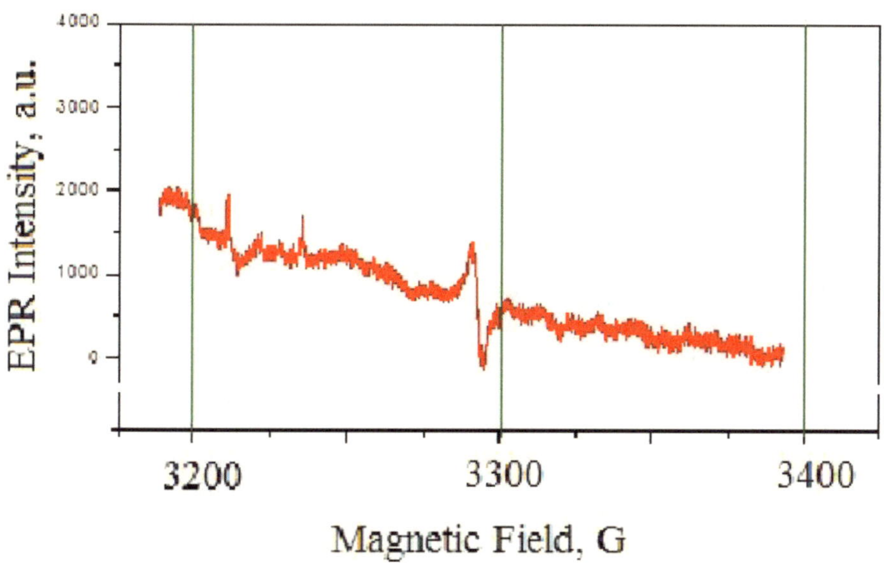


**Figure 6.** EPR data for a-Si film with dangling bonds defects.

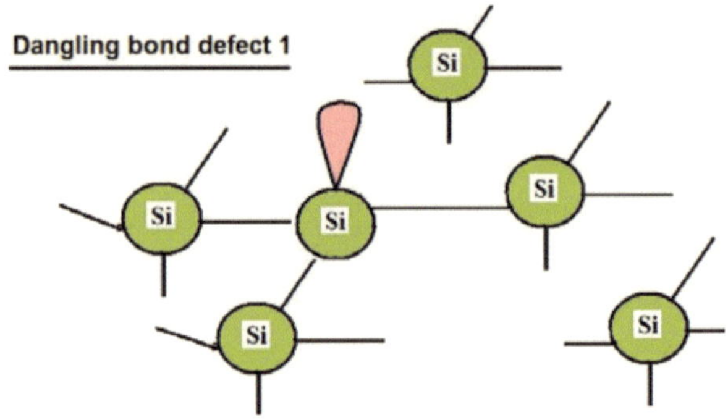


**Figure 7.** Dangling bond on the silicon (111) surface.

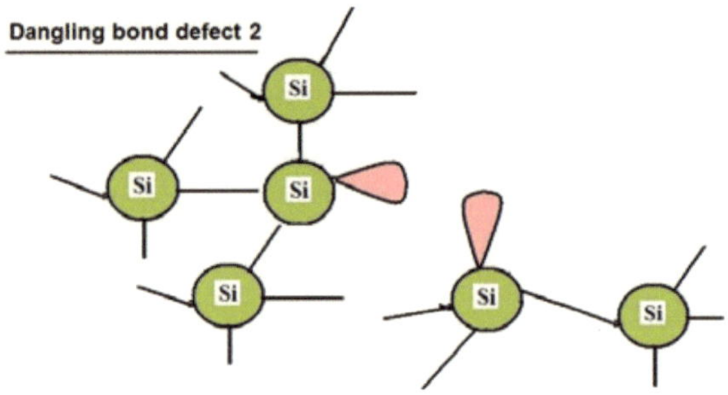


**Figure 8.** Dangling bonds defect in silicon crystalline (111) film.

defect with coupling bonds by the following way according to Elsner's theorem of matrix perturbation theory: $\begin{pmatrix} H_{11} & \varepsilon \\ \varepsilon & H_{22} \end{pmatrix} \rightarrow \begin{pmatrix} \hat{H}_{11} & 0 \\ 0 & \hat{H}_{22} \end{pmatrix}$, where $\lambda'_1 \le \left( \|H_{11}\| + \left\|\hat{H}_{11}\right\| \right)^{-1/2} \|\varepsilon\|^{1/2}$; $\lambda'_2 \le \left( \|H_{22}\| + \left\|\hat{H}_{22}\right\| \right)^{-1/2} \|\varepsilon\|^{1/2}$; are the eigenvalues for $( H^{\wedge 11}\, 00 H^{\wedge 22})$ system. Spectral characteristics of eigenvalues $\hat{\lambda}'_1, \hat{\lambda}'_2$ become broader by each bonding transformation or switching because there is a great difficulty to define the final state.

### 2.2.2. *Vacancies and oxygen incorporated in silicon (111)*

At first time the term A defect (or center) in silicon was used by Watkins and Corbett in 1961 [6]. They studied the irradiated by 1.5 MeV electron beam silicon with current density 2.5 μa/cm$^2$. The role of oxygen in A center creation is significant due to its definite bonding with silicon atoms. Spin resonance is caused by unpaired electrons which is trapped by splitting of the atomic orbital of pair-coupled silicon atoms. The figure illustrates the appearance of A defect paramagnetic centers in inter-grain area of nanocrystalline silicon films experimentally detected by means of the EPR spectrometer technique.

**Figures 9** and **10** show the switching effect for A centers by applied bias voltage for fluorinated nanocrystalline silicon films. The intensities of EPR signal change according to the switching of

http://dx.doi.org/10.5772/intechopen.76660

position of oxygen incorporated in silicon (111) from one pair of silicon atom to another pair of silicon atoms. In this case the interaction between $p_x$ orbital is changed to $p_z$ orbital. The frequencies of spectral component of EPR related to A defect on Si (100) surface disappeared by the annealing procedure because of surface thermo-diffusion.

**Figure 11** presents the various pictures of potential energy for oxygen atom bonded with couple Si atoms as a function of a distance by classical approach. It is clear that the distances that characterize the motion of oxygen atom are compared with its de Broglie wave length (2–4 Å) because there is a possibility to use quantum mechanical treatment of moving oxygen atom caused by the electric field from one silicon atom pair to another. **Figure 12** shows the scheme of Si-O-Si bridge with coupled Si-Si atoms as the quantum oscillator with eigenvalues of energy of the oxygen atom.

The oxygen atoms and dimers are incorporated in the silicon grain boundary and have weak covalent bonds with silicon. But the activation energy for molecular diffusion is low, 0.3 eV, in contrast with the activation energy value for atomic diffusion (1.3 eV). The energy of Si-O-Si bridge interaction can be written in form of Morse function

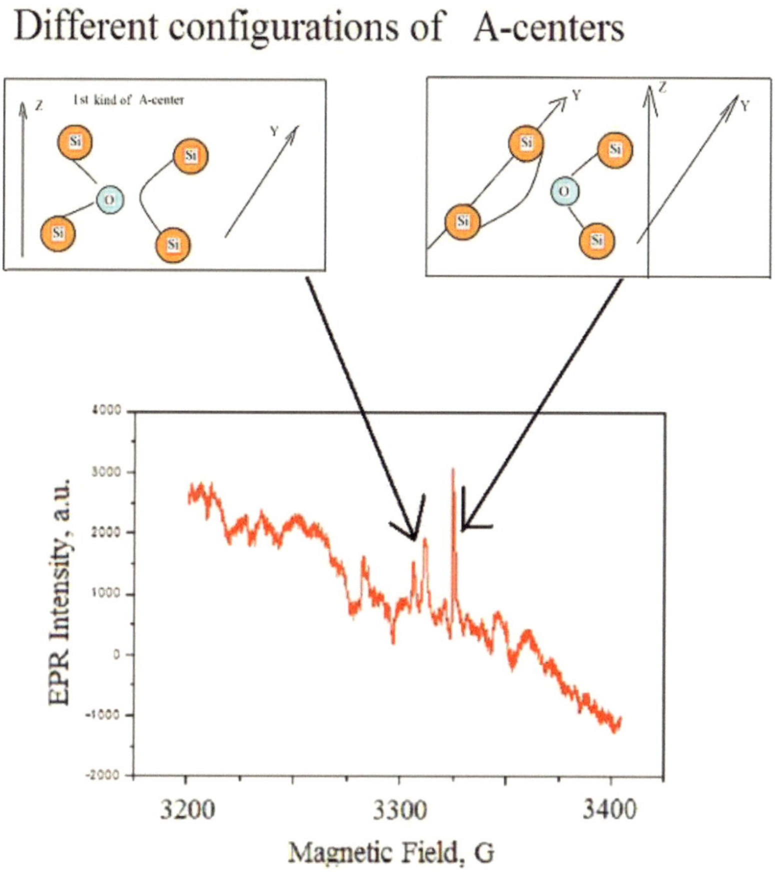


**Figure 9.** A defect paramagnetic center in inter-grains area of nanocrystalline silicon films detected by EPR spectrometer (Bruker Comp.).

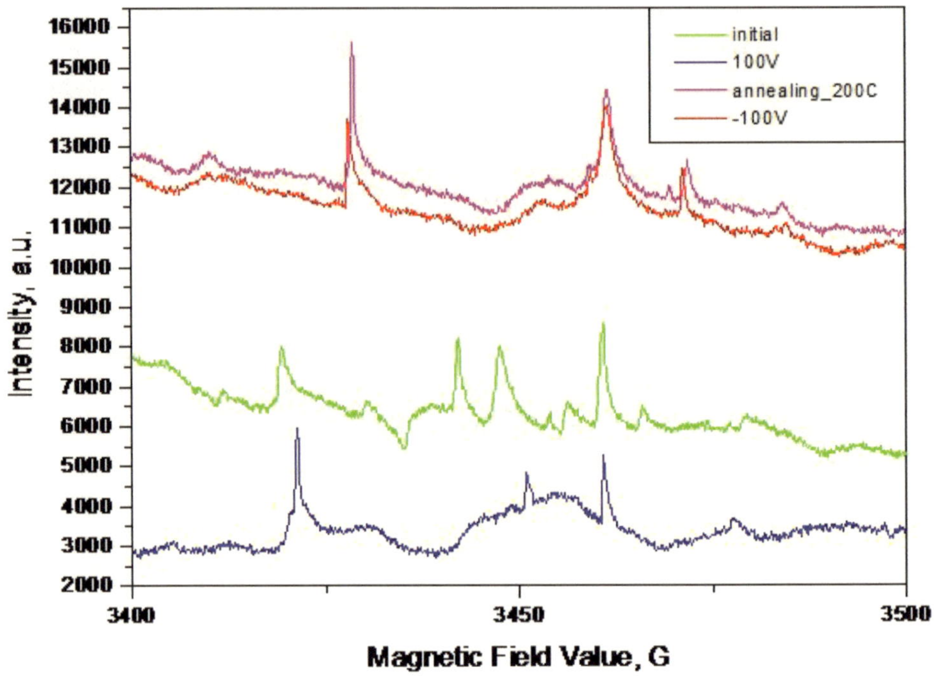


**Figure 10.** EPR data for fluorinated nanocrystalline silicon film by applied bias voltage and annealing.

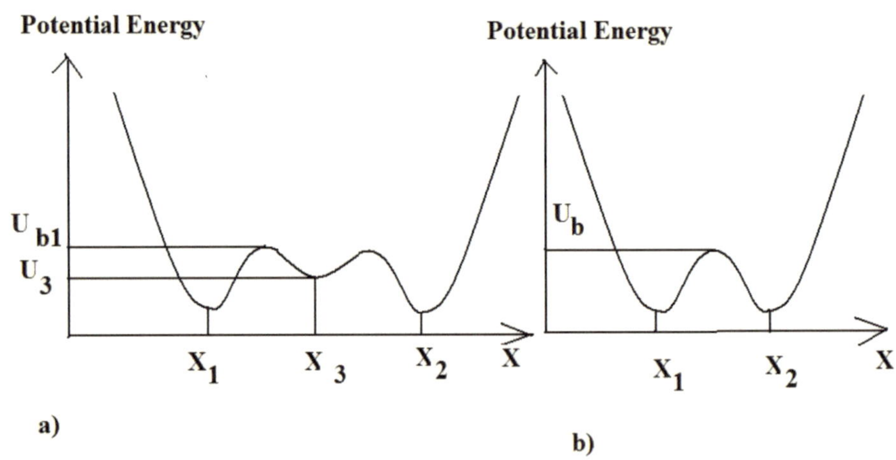


**Figure 11.** Potential energy of oxygen atom bonded with couple Si atoms.

$$U = U_0(1 - \exp(-\alpha(x - d)))^2 + U_0\left(1 - \exp(-\alpha(x + d)\right)^2 \tag{4}$$

where $U_0$ is the energy of Si-O (4.5 eV) and $\alpha$ is coefficient. The oxygen atom in A center oscillates between two points of stable positions. Transition of probability for oxygen atom through the potential barrier by applied external electrical field $U(r) = U_0\text{-}Er$ is given by

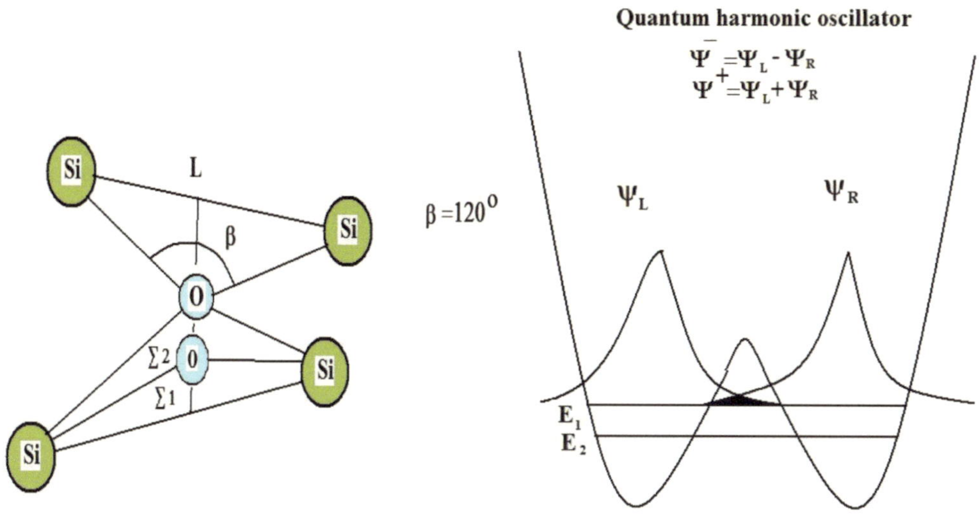


**Figure 12.** Scheme of Si-O-Si bridge with coupled Si-Si atoms.

$$T = \nu \exp\left(-\frac{2}{\hbar}\right)\int_{r_1}^{r_2} dr\sqrt{2M(U(r) - E_l)} \tag{5}$$

where ν is a frequency of oscillations of oxygen atom. Wave function of the oxygen atom is the following

$$\Psi = a_1|\phi_1 > \exp\left(-i\frac{E_1}{\hbar}t\right)\exp(-\Gamma_1 t) + a_2|\phi_2 > \exp\left(-i\frac{E_2}{\hbar}t\right)\exp(-\Gamma_2 t);$$

where $\varphi_1$and $\varphi_2$ are the wave functions of pure states 1 and 2 and $\Gamma_1$ and $\Gamma_2$ are the widths of the levels, respectively. We assume that all the Γ values are approximately equal to each other. Raman scattering data help to determine the dipole Si-Si orientation along the laser E field ax of incident radiation on the silicon surface by applied electric field switching of O atom spatial position. Applied electric field causes the tunnel of oxygen atom from one pair of coupled silicon atoms to another with perpendicular axis of dipole orientation. The annealing procedure can assist to restore spectral characteristics related to the primary-ordered atomic position due to the minimums in potential energy diagrams.

*2.2.3. Physical model of vacancy switching by applied electric field*

The role of applied electric field in order–disorder transition can be surely understood by using several mechanisms of electric dipole creation as it is seen in **Figure 10**. The first mechanism is the changing length of silicon-oxygen Si-O bond or silicon fluorine (Si-F) bonding and atomic transfer of oxygen inside the vacancy defect which causes the reorientation of VO or VF defects. Such bonds are strongest in silicon film because of high-electron affinity values for O and F atoms. The dipoles Si-O and Si-F have the great value in their polarized charges because the interaction between them and electromagnetic field is sufficient, and it results in their efficient reorientations because the dipole in external field changes its orientation to compensate it.

The second mechanism of order–disorder transition in the silicon film was initially proposed by Watkins and Corbett [7], which is devoted to hydrogen atom reorientation around the vacancy by the temperature above 200°C as the possible EPR data explanation. For nanostructured silicon thin film that contains a great amount of tiniest nanocrystals, the hydrogen atoms are distributed in grain boundary of nanocrystal silicon grains. But by electric field application to the thin film along the surface, the smallest nanocrystals disappear and, partially, the crystal structure is destroyed. By annealing, the opposite tendency is observed. The third assumption is the creation of excited hydrogen molecule inside the silicon film by annealing to the temperatures 150–200°C: Si-H + Si-H→Si-Si + $H^*_2$ [8]. It is clear that the silicon nanocrystal fraction decreases but the broad spectral line for amorphous phase significantly increases. The fourth mechanism is a deep valence hole and negatively charged OH- created by laser beam irradiation [9] (**Figure 13**).

The pump-probe laser picosecond spectroscopy data with phonon-stimulated stress pulses [10] confirm the existence of the quantum interference of closed neighbor levels resulting in quantum beats phenomena (see **Figure 3**). The estimation of the energy difference of level position is around 5 μeV. Picosecond acoustic longitudinal stress pulses cause the small displacement of interface media surrounding the silicon nanocrystals because the Si-O-Si bridges were slackened and the oxygen atom oscillates between two stable excess states of four coupled pairs of silicon atoms. The main principle of switching is a motion between two silicon coupled atoms with oxygen from one pair of silicon atoms to another. R. Biswas and coworkers proposed the flip model for hydrogen incorporation in amorphous silicon with two stable opposite positions for hydrogen atoms with axial symmetry relative to the bonding silicon atom [11].

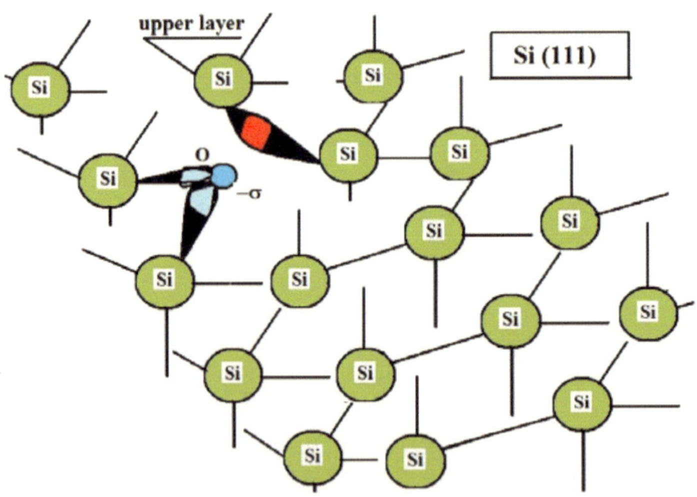


**Figure 13.** Scheme of AI defect location in silicon film with crystal orientation (111).

http://dx.doi.org/10.5772/intechopen.76660

The energy of Si-O-Si bridge interaction can be written in form of Morse function

$$U = U_0(1 - \exp(-\alpha(x - d)))^2 + U_0\left(1 - \exp(-\alpha(x + d)\right)^2 \tag{6}$$

where $U_0$ is the energy of Si-O (4.5 eV) and $\alpha$ is the coefficient. The oxygen atom in A center oscillates between two points of stable positions. The change of the spatial position of oxygen atom can be described by using simple model of oscillator

$$\ddot{x} + \gamma \dot{x} + \omega_0{}^2 x = f_0 \cos(\Omega t) \tag{7}$$

where the eigenfrequency of oscillating oxygen atom can be expressed as $\omega_0 = \sqrt{\frac{1}{M}\left(\frac{d^2U}{dx^2}\right)}$; $\gamma = \frac{kT}{h}$, k is Boltzmann constant. The solution of equation by the resonance conditions (by the $\omega_0/\gamma >> 1$) can be written as

$$x(t) = e^{-\gamma t}\left(\frac{f_0}{2\omega_0{}^2}\sin(\omega_0 t) + \frac{f_0}{2\gamma\omega_0}\cos(\omega_0 t)\right) - \frac{f_0}{2\gamma\omega_0}\cos(\omega_0 t) \approx -\frac{f_0}{2\gamma\omega_0}\left(1 - e^{-\gamma t}\right)\cos(\omega_0 t).$$

The eigen frequency value for Si-O bond was estimated by means of the FTIR spectroscopy technique. It equals approximately to $3 * 10^{13}$ Hz. The resonance will be by the equality between the phonon frequency $\Omega$ and $\omega_0$ value.

### 2.2.4. *Model of vacancy with oxygen as two coupled oscillators*

For the vacancy-oxygen atom (VO) defect center the model of two coupled oscillators is much more convenient:

$$\ddot{x}_1 + \gamma \dot{x}_1 + \omega_0{}^2 x_1 + a x_2 = f_0 \cos(\Omega t) \tag{8}$$

$x_2^{\bullet\bullet} + \gamma x_2 +^{\bullet} \omega_0{}^2 x_2 + b x_1 = f_0 \cos(\Omega t)$ and the energy of atom interaction can be expressed as the following for two pairs of silicon atoms which are oriented along the x and z axis:

$$\begin{aligned} U = {} & U_0(1 - \exp(-\alpha(x_1 - d)))^2 + U_0(1 - \exp(-\alpha(x_1 + d)))^2 + U_0(1 - \exp(-\alpha(x_2 - d)))^2 \\ & + U_0(1 - \exp(-\alpha(x_2 + d)))^2 \end{aligned}$$

where $x_1$ and $x_2$ are the displacement values for oxygen atom positions relatively, $Si\text{-}Si_A$ and $Si\text{-}Si_B$ pairs.

The solution of the system of two two-order differential equations contains two eigen frequencies $w_1$ and $w_2$. The oxygen atom oscillates in A center with two different frequencies corresponding two energy levels for oxygen atom. The quantum beats are observed in picosecond laser spectroscopy experiments shown on **Figure 19** caused by quantum interference of two energetic states of oxygen atom.

*2.2.5. Dipole moments of Si-O bond and Si-O-Si bridge*

By using the Slater atomic orbital for Si and O atoms which were written in my previous work [12] we can easily estimate dipole moment for S-O configuration. Dipole moment by the zero displacement is $\mu_{Si-O} = ex\int_V \psi_{3pSi}\psi_{2pO}d\nu = 4\pi ex\int_0^{2\pi}\int_0^{\xi}\int_0^{\pi}\Psi_{3p_xSi}\Psi_{2p_xO}r^2 d\phi dr d\theta$; where $d\nu = 4\pi r^2 dr$, $0 < r \le \xi$, $\zeta$ is covalent radius of oxygen atom, x is bond length. For 3pSi-2pO orbital bonding we can $\int_0^{\pi}\sin^2\theta d\theta = \frac{\pi}{2};\ \int_0^{2\pi}\cos^2\phi d\phi = \pi$;

$\mu(r) = 230ea_0\exp\left(\frac{-3.655r}{a_0}\right)\left[0.27\left(\frac{r}{a_0}\right)^5 + 0.37\left(\frac{r}{a_0}\right)^4 + 0.41\left(\frac{r}{a_0}\right)^3 + 0.37\left(\frac{r}{a_0}\right)^2 + 0.18\left(\frac{r}{a_0}\right) + 0.05\right]$ By r = 1.6 Å, $r/a_0 = 3$, we can estimate the dipole moment value as $\mu_{Si-O} = 1.12$ D.

The comparison of calculated overlapping integrals as a function of distance is shown in **Figure 14**. It is seen that the Si-Si orbital is more significant even at large distances than Si-O.

For Si-O-Si bridge we can obtain

$$\mu_{Si-O-Si} = 4\pi e\int_0^{\xi}\int_0^{2\pi}\int_0^{\pi}\Psi_{3pSi'}\Psi_{3pSi''}r\Psi_{2pO'}\Psi_{2pO''}r^2 dr d\phi d\theta. \quad (9)$$

We can evaluate the dipole moment by zero displacement $\mu^0_{Si-O-Si} = 0.17D$. For Si-O-Si bridge the angle (see **Figure 15**) between two bonds is approximately 90–180°. It is clear that by $\alpha \rightarrow 180°$ the vector value of dipole moment is zero, but by $\alpha \rightarrow 90°$ the dipole moment can be estimated as $\mu_{Si-O-Si} \approx 1.4\mu_{Si-O}$.

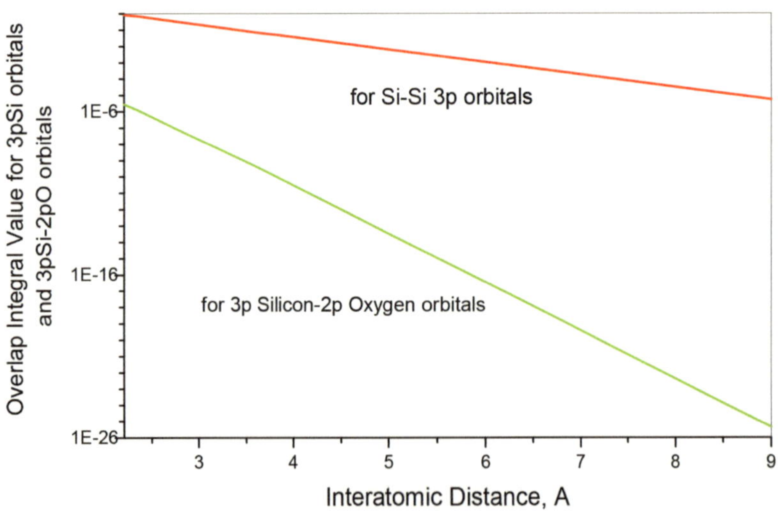


**Figure 14.** Overlap integrals for 3pSi-3pSi and 3pSi-2pO orbital as a function of inter atomic distance.

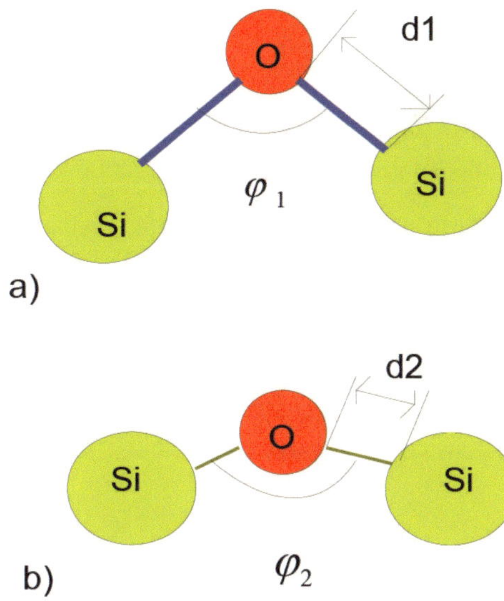


**Figure 15.** Si-O-Si bridge with (a) greater value of dipole moment; (b) smaller value of dipole moment, $d\varphi = \varphi_2 - \varphi_1$; $\Delta = d_1 - d_2$; $d_1$, $d_2$ are bond lengths.

By applied electrical field the dipole moment value for Si-O-Si bridge is defined by displacement evolution: $\mu = \mu^0 + \mu_x(t) \approx ex(t)$. In the case of $\mu < \mu_{critical}$ there is no localized bonded state. It is clear that for $\mu^0{}_{Si\text{-}O} > \mu_{critical}$ there is localized bonded state for electron, but for $\mu^0{}_{Si\text{-}O\text{-}Si} < \mu_{critical}$ such a state cannot exist because the Si-O bonding by zero displacement generates localized electron states on Si (111) surface which can cause troubles for writing and storing procedures of memory cell. By the annealing of Si (111) substrate to a temperature of more than 250°C we can eliminate the Si-O bonds and produce the strong siloxane bonding. Therefore, it is possible to prepare the oxidized Si (111) surface with strict siloxane Si-O-Si bridges without localized electron states on the surface. Applying the electrical or even acoustical field we can stimulate A center switching inside the film that causes the appearance of localized electron states. The polarization wave generated by laser radiation is given by

$$P = \tilde{\tilde{P}}(z)\exp(i(\Omega t - kz))\exp(-\gamma t) \tag{10}$$

We can estimate the yield of interference of two polarized waves with shift of phase's $\pi$.

$$\begin{aligned} &P_{sum} = P_+ + P_- = \tilde{\tilde{P}}_{sum}\cos(\Omega t + \phi); \tilde{\tilde{P}}\exp(i\phi) = \tilde{\tilde{P}}_-\exp(i\phi_1) + \tilde{\tilde{P}}_+\exp(i\phi_2); \\ &\tilde{\tilde{P}}^2 = \tilde{\tilde{P}}_-{}^2 + \tilde{\tilde{P}}_+{}^2 + 2\tilde{\tilde{P}}_-\tilde{\tilde{P}}_+\cos(\phi_2 - \phi_1) \end{aligned} \tag{11}$$

The spatial distribution of charges for polarized molecular bridges Si-O-Si on the clean Si (111) surface is used for electrical charge trapping. The potential energy barrier for electron trapping can be estimated as

$$W \approx 2\tilde{\tilde{P}}_{+}\tilde{\tilde{P}}_{-}\sin\left(\phi_2 - \phi_1\right)\Delta\tilde{\tilde{\phi}};\ \Delta\tilde{\tilde{\phi}} = \frac{\pi \Delta d}{\lambda} \tag{12}$$

where Δ d is the cell size and λ is the wavelength of polarization wave. By applied electrical field $10^7$ V/m W ≈ 100 μeV and by the value of field $10^8$ V/cm is 1 meV.

**Figure 16** illustrates the point defect on silicon (111) surface caused by a silicon atom vacancy and oxygen atom incorporation into the silicon network. There are three silicon bonds which are connected with oxygen atom through the $sp_3$ orbitals. Oxygen $2p_x$ and $2p_z$ orbital cause the splitting to the bonding and anti-bonding electron states. According to the perturbation theory the $H_{11}$, $H_{22}$, $H_{33}$ and $H_{44}$ are the energies for $p$ orbitals that interacted with themselves. It is clear that such configurations are produced, and switching of interacted orbitals between them can be expressed in diagonal and non-diagonal terms of matrix of Hamiltonian:

$$\begin{pmatrix} H_{11} & V_{12} & V_{13} & V_{14} \\ V_{21} & H_{22} & V_{23} & V_{24} \\ V_{31} & V_{32} & H_{33} & 0 \\ V_{41} & V_{42} & 0 & H_{44} \end{pmatrix} \rightarrow \begin{pmatrix} H_{11} & V_{12} & V_{13} & 0 \\ V_{21} & H_{22} & V_{23} & V_{24} \\ V_{31} & V_{32} & H_{33} & V_{34} \\ 0 & V_{42} & V_{43} & H_{44} \end{pmatrix}; \tag{13}$$

*2.2.5.1. $Si_3H^{-\sigma}$ vacancy defect*

The molecular-like spectrum of electron states for hydrogenated cluster $Si_3H^{-\sigma}$ has a series of electron states in the range from 0.7 to 1.0 eV. [13]. The eigen frequencies of oscillator $Si_3$-H are conjugated with the electron states for the same oscillators and results in appearance of broadening in the electron spectra. The defect can be described as tied three $Si_3$-H oscillators and one $Si_3$- oscillator. It is clear that the quantum properties of defect are reflected in their molecular orbital spectral positions. It should be observed surely for nanostructured silicon films with structured cell less than 1 nm or for amorphous hydrogenated silicon film. In such

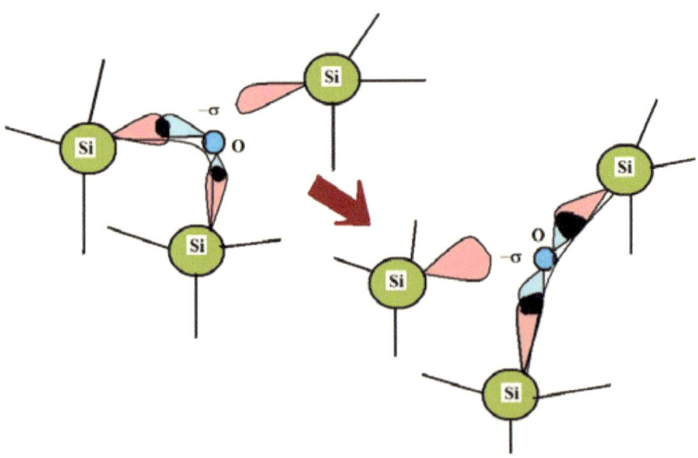


**Figure 16.** The point defect migration on the silicon surface (111) by applied electric field. Wave function is distributed among the 12 silicon atoms.

kind of media we surely can change the spectral positions of electron states by using bias voltage.

#### 2.2.5.2. *$Si_3$-F vacancy defect*

Silicon functional groups can be determined as the following: Si-Si, Si-H, Si-$H_2$, Si-$H_3$, $SiH_2$-F. The silicon bonding is arranged according to $sp^3$ hybridization: *3s* spin-paired electrons and *3p* electrons. The greatest energy of bonding is for Si-F bond equal to 135 kcal/mol, but for Si-H and Si-Si, it is 76 and 53 kcal/mol, respectively. The dipole moments due to the great value of electronegativity of fluorine atom will be significant. Accordingly, the charge distribution of electron density in the $SiH_3$-F molecular group is the following: for Si atom is it is +1.1, for H atom it is −0.15, and for F atom is −0.67 [14].

#### 2.2.5.3. *Si-$O_x$ clusters*

We assume that the defects are localized in Si-$SiO_2$ interface with complex composition of $SiO_x$. The oxygen atoms and dimers are incorporated in the silicon grain boundary and have weak covalent bonds with silicon. But the activation energy for molecular diffusion is low, 0.3 eV, in contrast with the activation energy value for atomic diffusion (1.3 eV). The increase in oxygen concentration results in increasing SiO bond length. For the sample with smaller grain size the hydrogen termination of dangling bonds is greater, and density of oxygen is low. But the film with large grains contains a great amount of oxygen in Si-O-Si and $SiO_2$ compositions. We assume that the exponential electron density decay is due to the exciton state decay. The short relaxation time can be described as possible relaxation process through the oxygen incorporation-related defect states, where density of states is the following: $n(E) = (n_{Si}(E) + xn_o(E))/(1 + x)$ [15].

The oxygen atom bonding with silicon atoms inside crystallites forms the dioxide composition along with defect production. **Figure 17** shows the two kinds of oxidized surfaces: Si (100) surface annealed by 1100 K and oxidized poly-Si film with a great value of $\langle\delta\rangle = 17.6$ nm. The oxide-silicon configuration for both surfaces is different. The Si (100) surface contains mainly the $SiO_2$ component with a high density of oxygen concentration but the poly-Si film is weaker oxidized with the presence of $SiO_{0.5}$ configuration.

**Figure 18** illustrates the oxygen incorporation into silicon grain boundary after transformation of the silanol groups into strong siloxane bonds.

## 3. Field-assisted crystal phase creation from amorphous

### 3.1. Fractal character of restored crystal phase

The first mechanism that can be used for explanation of crystal phase restoration phenomenon in silicon film is a difference in structural orders of silicon crystal structure and polysilane chains which are arranged into conglomerates after hydrogen migration, and further incorporation is

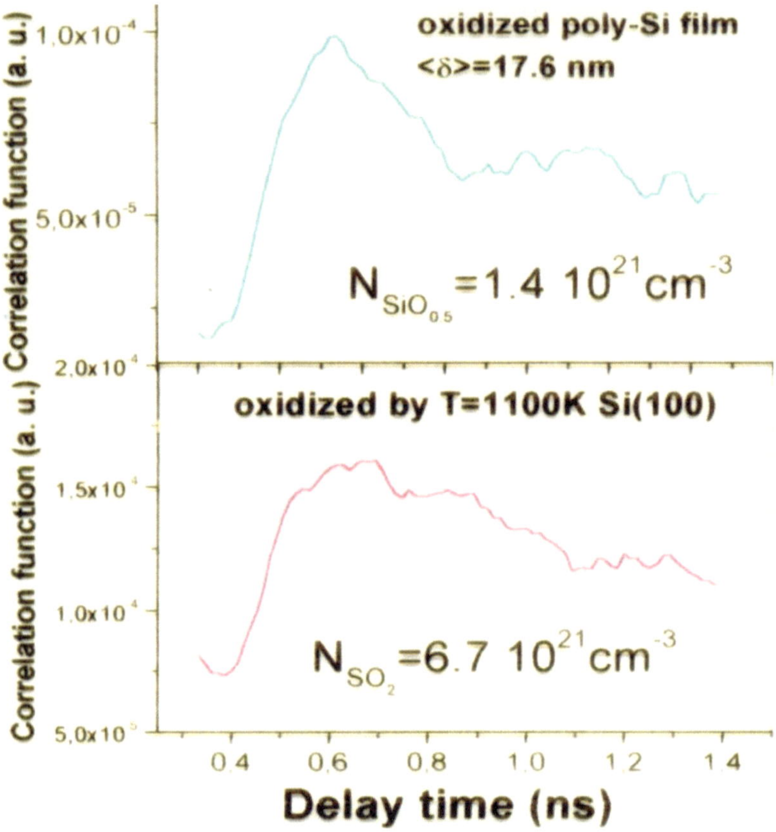


**Figure 17.** Correlation function g (τ) as a function of delay time between pump and probe pulses.

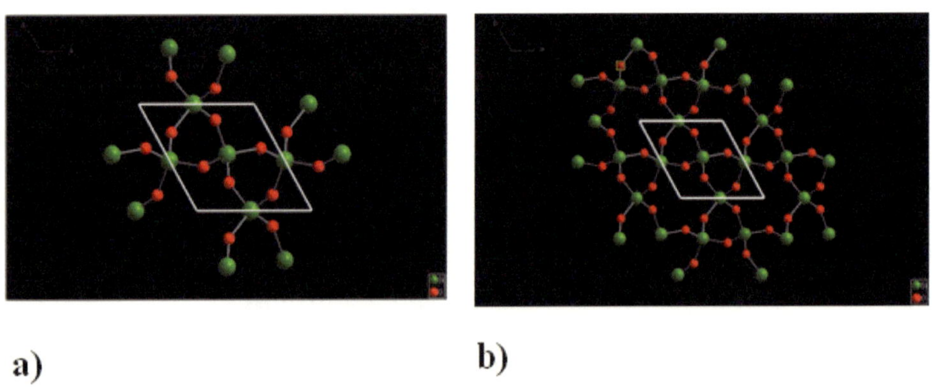


**Figure 18.** Silicon-oxygen incorporation in the grain boundary after transformation the silanol groups into strong siloxane bonds (a) 13 silicon atoms incorporated with 14 atoms of oxygen ($SiO_x$, $x \approx 1$)by the Si-O bond angle equal to 130° the width of interface is 8.86 Å; (b) 29 silicon atoms incorporated with 39 oxygen atoms ($x \approx 1.34$) create the intercrystal space with width equal to 14.77 Å.

transformed into the silicon net. The hydrogen surface diffusion by annealing also changes the phase fraction ratio in the silicon film. **Figure 19** shows the microphotography of silicon microcrystal that contains fractal structures of silicon bonding. By applying the external electrical field, the dipole orientations change to compensate the external field. Because the dipoles' reorientations cause numerous dislocations inside silicon nanocrystal, it is supposed that the order of crystalline structure inside the silicon film as grown usually is the same as order of fractals which is shown in **Figure 19**. The microphotography of the microcrystal of silicon on the glass substrate that is shown in **Figure 19a** demonstrates the hexagonal structure of the crystal along with fractal structures which are illustrated in **Figure 19b**. Because there is the necessity to analyze the mechanism of field-assisted phase transformation as destruction or phase restoration of native fractal nanoscale structures of silicon nanocrystals, for fractals, the dimension of Hausdorf can be surely evaluated as $d_C = -\lim \frac{\lg N(s)}{\lg(s)}$, where N(s) is the number of cubic volumes from which it is possible to substitute the rectangular nanocrystal with crystal orientation (111). Such a value of dimension equals 2.5. For determination of any structure the formula for Hausdorf metrics that is given by.

$\rho(x.y) = \sqrt{\sum_{i=1}^{n} (x_i - y_i)^2} = 2\varepsilon$ is suitable, where ε is radius of the open ball with centers in points X and Y. The space limit for the fractal structure can be evaluated using the silicon bonding size 2.5 Å. For small 3D objects such as nanocrystals, the density of substitution can be written as $N_V = 1/\varepsilon^3$. For a fractal set the set capacity $D_c$ can be written as the following: $N_V = 1/\varepsilon^{Dc}$: $D_C = \lim \frac{\log N(\varepsilon)}{\log \varepsilon^{-1}}$ by the changing of value ε by this way ε→0.

The second proposed mechanism of silicon crystal phase restoration can be illustrated by means of **Figure 19**. It is shown that the dislocations and dipoles in grain boundary of nanocrystal under the influence of external electric field cause the phase destruction into the arc series according to the modular group scenario. Such transformations cause the appearance of numerous arcs which will be destroyed after further triangular Knop scenario of transformation of the 2D silicon structure to 1 D silicon structure or polysilane chain. For the fractal structure the Hausdorf dimension can be varied and is less than 1. The down picture on **Figure 8** illustrates the creation of Kantor dust [16] by dividing the triangular angle on the top

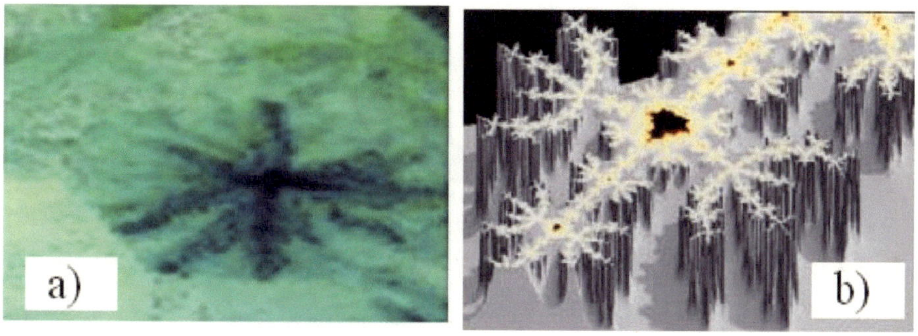


**Figure 19.** Photo of microscopic silicon crystal on glass substrate (a), and fractal structure of Mandelbrot set (b).

and neglecting the area of triangle in the middle of primary triangle area. For such transformation the equations for triangular quantities $X_N = 2X_{N+1}$ and for areas $S_N = 2.63^{d}{*}S_{N+1}$. Comparative analysis of last two equalities results in $2.63^d = 2$. The estimated value of Hausdorf dimension for such mathematical set is 0.72, because, there is a decrease in dimension by the destruction mechanism for media. It is obviously that the field-assisted transformation of silicon nanocrystals into objects with low dimension can be followed by the way of a polysilane chain creation $R_S = \frac{l\sqrt{n}}{\sqrt{6}}$;.

### 3.2. Raman spectra of silicon film with the created crystal phase

**Figure 20** illustrates the Raman spectra of amorphous silicon film which is deposited by magnetron sputtering atoms of Pt and further applying the external electric field. The scheme of experiments is shown in the figure. There is a crystal phase that was created by fixed value of applied electric field due to the charges accumulated by Pt clusters. Change in free energy by local field results in the appearance of silicon stripes of the crystalline phase in amorphous silicon.

Free energy functional can be expressed as follows

$F_{FE} = \int_V \left(\frac{1}{2}\varphi\left[\left(q_0{}^2 + \nabla^2\right)^2 + a\right]\varphi + \frac{1}{4}\varphi^4\right)dV$, where $q_0$ is a phenomenological parameter and $\varphi$ is a dimensionless average density. The solution of equation is

$$dF_{FE} = \frac{\partial F_{FE}}{\partial \varphi}\frac{\partial \varphi}{\partial t}dt + \frac{\partial F_{FE}}{\partial \varphi}\frac{\partial \varphi}{\partial x}dx + \frac{\partial F_{FE}}{\partial \varphi}\frac{\partial \varphi}{\partial y}dy + \frac{\partial F_{FE}}{\partial \varphi}\frac{\partial \varphi}{\partial z}dz = 0 \quad (14)$$

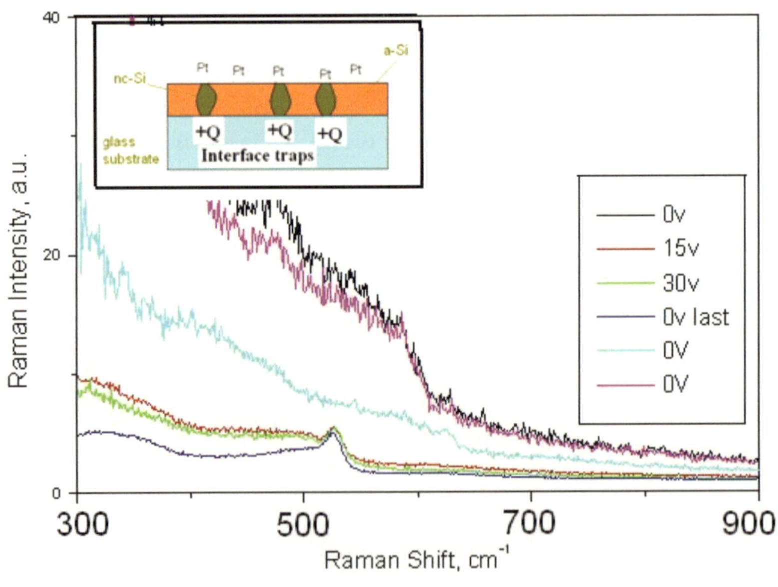


**Figure 20.** Raman spectra of amorphous silicon film deposited by magnetron sputtering atoms of Pt and further applying the external electric field.

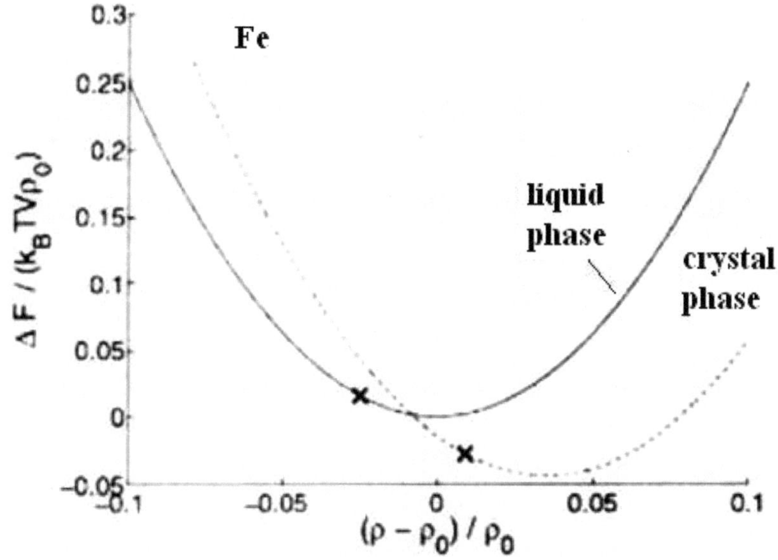


**Figure 21.** Free energy as a function of reduced particles density at the melting point of Fe according to the phase-field crystal model estimations [17].

or finding the minimum of the functional results in the investigations of crystal phase creation. For 1D in the uniform phase the expression $\psi(x^{-}) = \sum_j A_j(t)\exp\left(ik_j^{\to x\to}\right) + \psi^{-}$ can be used as a possible solution. **Figure 21** shows free energy as a function of reduced particle density at the melting point of Fe according to the phase-field crystal model estimations.

### 3.3. Renormalization group of translation for crystal phase creation

The resolving the equations for locality function by the free energy functional minimizing [17]: $\frac{\partial\psi}{\partial t} = \nabla^2 \frac{\delta F_i}{\delta\psi} + \eta;$ for $\psi$ locally conserved, and $\frac{\partial\psi}{\partial t} = -\frac{\delta F_i}{\delta\psi} + \eta;$ for $\psi$ locally non-conserved, from which we can easily find the equation for estimating the value of energy for order–disorder transition:

$\nabla^2 \frac{\delta F_i}{\delta\psi} + \frac{\delta F_i}{\delta\psi} = 0;$ also, it is assumed that the evaluation value of free energy (of heat) equals the energy of local field force to arrange in order all the atoms which should be placed on their sites: $\delta F \approx \mu E_{Local} - T\delta S$.

The solution of nonlinear Swift-Hohenberg equation also can be presented as a linear combination of solutions if we derive the terms of $\psi$ as partial solutions and further combine them. The $\psi$ local function can be transformed into a series according to perturbation theory as follows $\psi = \psi_0 + \varepsilon\psi_1 + \varepsilon^2\psi_2 + \varepsilon^3\psi_3 + \varepsilon^4\psi_4 + \ldots$, where the value $\psi_0$ is a steady-state solution, and next values are the next order of perturbed decomposing procedure. For such a purpose it can be used as a first order of the solution that can be expressed as: $\psi_1 = P_1(x,t)\exp(ik) + Q_1(x,t)\exp(2ik) + R_1(x,t)\exp(3ik) + c.c.$ and amplitude $A_j(t) = A_j(X,t)\left(1 + \sum_{j=1} \varepsilon^j Z_j\right)$, where $Z_j$ is a renormalization constant [18].

## Author details

Dmitry E. Milovzorov

Address all correspondence to: dmilovzorov2002@yahoo.com

Fluens Technology Group Ltd., Moscow, Russia

## References

[1] Tsu R, Gonzalez-Hernandez J, Chao SS, Lee SC, Tanaka K. Critical volume fraction of crystallinity for conductivity percolation in phosphorus-doped Si: F: H alloys. Applied Physics Letters. 1982;**40**:534. ISSN: 0003-6951

[2] Lucovsky G, Nemanich RJ, Knight JC. Structural interpretation of the vibration spectra of a-Si:H alloys. Physical Review B. 1979;**19**:2064-2073. ISSN: 0163-1829

[3] Yu P, Cardona M. Fundamentals of Semiconductors. Berlin: Springer; 1996. p. 617. ISBN: 3-540-61461-3

[4] Kahan VD. High frequency hopping transport in solids and dielectric transparency in compensated semiconductors. JETP. 2000;**117**:452-456. ISSN: 1063-7761

[5] Powell MJ, Deane SC. Improved defect-pool model for charged defects in amorphous silicon. Physical Review B. 1993;**48**:10815-10826. ISSN: 0163–1829

[6] Watkins G, Corbet JW. Defects in irradiated silicon I: Electron spin resonance of the Si-A center. Physical Review. 1961;**121**:1001-1014. ISSN: 0163-1829

[7] Watkins G, Corbett JW. Physics Review. 1965;**138**:A543. ISSN: 0163-1829

[8] Jones B, Coomer BJ, Goss JP, Hourahine B, Recende A. The interaction of hydrogen with deep level defects in silicon. Solid State Phenomena. 1999;**71**:173-248

[9] Bakos T. Defects in amorphous SiO2: reactions, dynamics and optical properties [PhD Thesis]. Nashville; 2003. p.70

[10] Milovzorov D. Electronic structure of nanocrystalline silicon and oxidized silicon surfaces. Electrochemical and Solid State Letters. 2001;**4**(7):G61-G63. ISSN: 0013-4651

[11] Biswas R, Li Q, Pan BC, Yoon Y. Mechanism for hydrogen diffusion in amorphous silicon. Physical Review B. 1998;**57**:2253-2256. ISSN: 0163-1829

[12] Milovzorov D. Memory cell with photoacoustic switching. In: Lai WY, Pau S, Daniel Lopez O, editors. Proceedings of the SPIE 5592. Nanofabrication: Technologies, Devices, and Applications. 2005. pp. 427–437. DOI: 10.1117/12.568290

[13] Xu C, Taylor T, Burton G, Neumark D. The Journal of Chemical Physics. 1998;**108**:7645. ISSN: 0021-9606

[14] Voronkov M. Topics in Current Chemistry. 1986;**131**:99. ISSN: 0340-1022

[15] Lanoo M, Allan G. A cluster plus effective tight-binding study of $SiO_x$ systems. Solid State Communications. 1978;**28**:733-739. ISSN: 0038-1098

[16] Hausdorff F. Grundzuge der Mengenlehre. Vol. 184. Berlin; 1914. ISBN: 3-540-42224-2

[17] Athreya B, Goldenfeld N, Dantzig J. Renormalization-group theory for the phase-field crystal equation. Physical Review E. 2006;**74**:011601. ISSN: 0163-1829

[18] Emmerich H, Granasy L, Lowen H. Selected issued of phase-field crystal simulations. The European Physics Journal Plus. 2011;**126**:102. ISSN: 2190-5444

Chapter 7

# Electric Field of a Medium Voltage Indoor Post Insulator

Mirza Sarajlić, Jože Pihler, Nermin Sarajlić and
Peter Kitak

Additional information is available at the end of the chapter

http://dx.doi.org/10.5772/intechopen.71871

**Abstract**

This chapter deals with the influence of the electric field on a Medium Voltage Indoor Post Insulator (MVIPI) with standard and modified external shapes. The goal of this chapter is to show the electric field behavior of the MVIPI with different external shapes and to introduce a numerical model with a favorable distribution of the electric field that relieves the dielectric from stress. The chapter describes an MVIPI with nominal voltage 20 kV AC and shows an existing MVIPI, an MVIPI with a different number of ribs, and an MVIPI with exceptional external shape. The MVIPI's new external shape does not have the typical shape of the ribs, but a new outline, using the Lagrange polynomial, that will acquire its optimal form through the optimization process. A Differential Evolution optimization algorithm is used for the optimal design of the insulator's external shape. The value of the Electric Field Strength (EFS) will be minimized within the permissible bounds during the optimization process. The important parameters during the minimization of the objective function are the value of the EFS in the interior and exterior of the insulator. EFS values are shown for every MVIPI example and are compared with the existing MVIPI. The obtained results are analyzed and discussed.

**Keywords:** electric field, medium voltage, indoor post insulator, shape, optimization

## 1. Introduction

Medium voltage indoor post insulators (MVIPIs) are the most common elements used in switchgear [1]. They are the significant components of electric power devices [2]. Electric insulation of conductive parts from the grounded parts is the primary task of the insulator [3].

Porcelain was used firstly as an insulator's material. Various epoxy resins appeared in the mid-60s, combined with various binders, glues, and fillers. Composite insulators are being produced increasingly recently for indoor and outdoor mounting. Araldite MVIPI (**Figure 1a**) is most used for internal mounting and has great insulation, mechanical, and thermic properties. With small dimensions, the MVIPI has a compound form and long lifespan [4].

The insulator model was built in MATLAB software, which has a dual task. Firstly, MATLAB does the task of a preprocessor, which is described in Section 2. MATLAB's second task is application of the optimization process. The selected optimization algorithm Differential Evolution (DE) [5, 6] is written and executed in MATLAB, which is described in Section 5, where the minimization of the objective function has been implemented. The essential parameter of the objective function is the value of EFS in the interior and exterior of the insulator. During the optimization process, the value of the EFS will be reduced within the admissible bounds and will be calculated using EleFAnT software [7].

The chapter is organized as follows. An existing MVIPI model is described in Section 2. Section 3 describes the MVIPI with a different number of ribs. Section 4 describes the novel-designed MVIPI using the Lagrange polynomial. The optimization calculation of the MVIPI's external shape and the best position of internal insulator components using DE algorithm is described in Section 5, with the Conclusion given in Section 6.

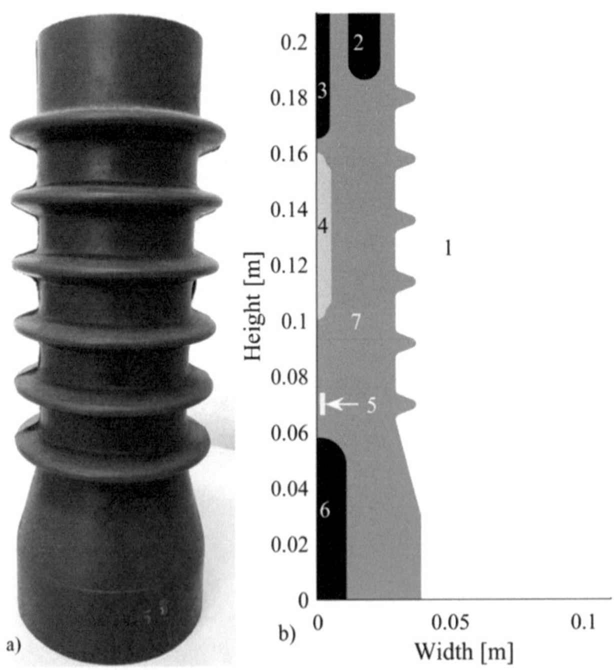


**Figure 1.** (a) Existing MVIPI and (b) MVIPI model in program tool MATLAB.

http://dx.doi.org/10.5772/intechopen.71871

## 2. MVIPI model

The existing MVIPI has six ribs (**Figure 1**). The MVIPI's parts are labeled with the numbers stated in **Table 1**, which also shows the materials and the potentials of each component required for the calculation of the electric field.

The complete MVIPI model geometry is generated using a mesh generator, which is part of the preprocessor. The mesh generator distributes the whole problem area into finite elements [8–10], which are isoparametric quadrangular elements in the case of axisymmetric model. The problem is separated into single geometric shapes that belong to respective types of material. There are seven geometric shapes: surrounding area (1), upper right connector (2), upper indicator electrode (3), capacitor (4), resistor (5), bottom electrode (6), and insulation material (7).

The upper right connector and upper indicator electrode are connected to high voltage. Their function is to connect the conductor (i.e., contact, busbar) on the insulator. The bottom electrode is grounded and functions to attach the insulator on grounded infrastructure (i.e., cells' housing).

The ceramic capacitor helps in achieving appropriate capacitance, defined by Standards IEC 61958 [11] and IEC 61243–5 [12]. The capacitance must be between 74 and 88 pF, according to [11, 12]. The electric field in the air and the outside surface of the MVIPI has to be less than 3 MV/m and inside the insulator less than 30 MV/m.

The MVIPI's designing requires input data, such as insulator geometry, materials, and boundary conditions. Potential's value passes from the source potential (upper indicator electrode) to the grounded part (bottom electrode). The Dirichlet boundary condition 0 V is on the right edge of the MVIPI's model. The material is described by means of the dielectric permittivity, which, for araldite, is $\varepsilon_{r,araldite} = 4.3$ and, for the air, $\varepsilon_{r,air} = 1$. The MVIPI model is written parametrically, which enables the rapid adjustment of geometry and full control of materials, potential, and boundary conditions. Finding the best parameters is executed with an iterative process of updating the MVIPI model with new input parameters.

| Number | Geometric shape | Material | Potential |
|---|---|---|---|
| 1 | Surrounding area | Air | Floating potential |
| 2 | Upper right connector | Metal | 125 kV DC[1] |
| 3 | Upper indicator electrode | Metal | 125 kV DC[1] |
| 4 | Capacitor | Ceramics | Floating potential |
| 5 | Resistor | Metal | Floating potential |
| 6 | Bottom electrode | Metal | 0 V |
| 7 | Insulation material | Araldite | Floating potential |

[1]20 kV MVIPI must endure 125 kV DC as the maximum test voltage (lightning impulse voltage).

**Table 1.** Explanation of geometric shapes from **Figure 1**.

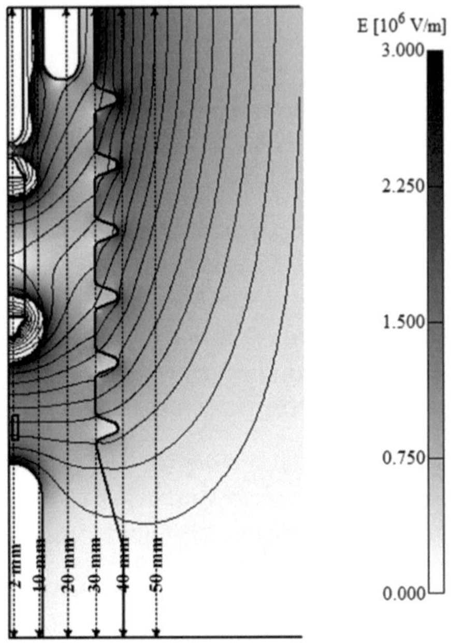


**Figure 2.** EFS of the existing MVIPI.

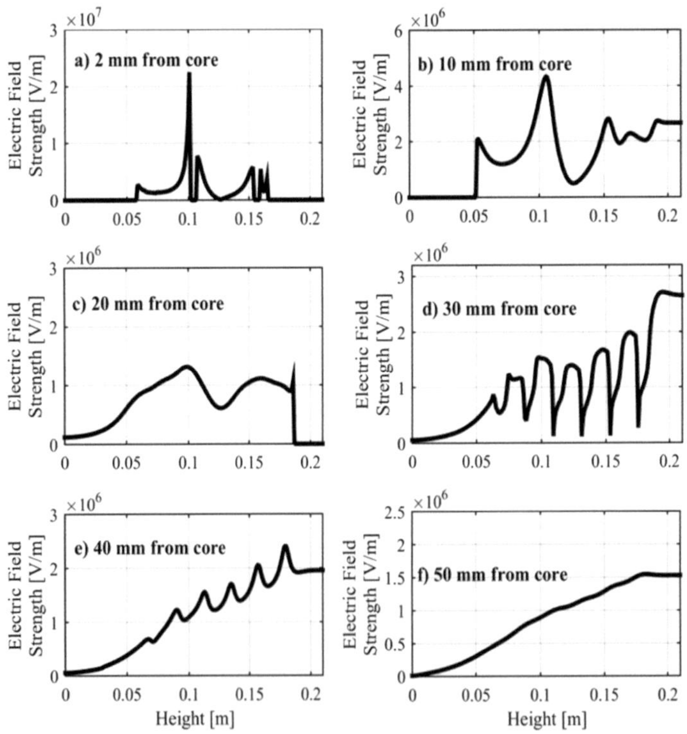


**Figure 3.** EFS of the existing MVIPI at different distances from the core of the MVIPI: (a) 2 mm for core, (b) 10 mm from core, (c) 20 mm from core, (d) 30 mm from core, (e) 40 mm from core and (f) 50 mm from core.

**Figure 2** shows a plot of the existing MVIPI's EFSs. **Figure 3** shows EFSs at different distances from the core of the MVIPI. As is apparent in **Figure 3**, the values of EFS are within the defined bounds (at 2, 10, and 20 mm), the dielectric strength is not exceeded (30 MV/m). Also, there was no excess of the EFS (3 MV/m) in the air (at 30, 40, and 50 mm).

## 3. MVIPI with a lower and higher number of ribs

Modeling of the MVIPI's exterior is presented in the following. Examples of an MVIPI with lower and higher number of ribs are shown and compared with the existing MVIPI. **Figure 4a** shows the EFS of the MVIPI with five ribs. **Figure 4b** shows the EFS of the MVIPI with nine ribs. **Figure 5** shows the comparison of the EFS between the existing MVIPI and MVIPIS with five and nine ribs at different distances from the core of the MVIPI.

The EFS of the MVIPI with five ribs is marked with a dotted line (**Figure 5**); a solid line marks the EFS with nine ribs, and a dashed line represents the EFS of the existing MVIPI. There is no significant change in the electric field in the interior of the insulator (**Figure 5a–5c**). The EFS values in **Figure 5d** and **5e** are approximately equal; the electric field is distributed better at the insulator with nine ribs. The differences between EFS values are negligibly small in **Figure 5f**.

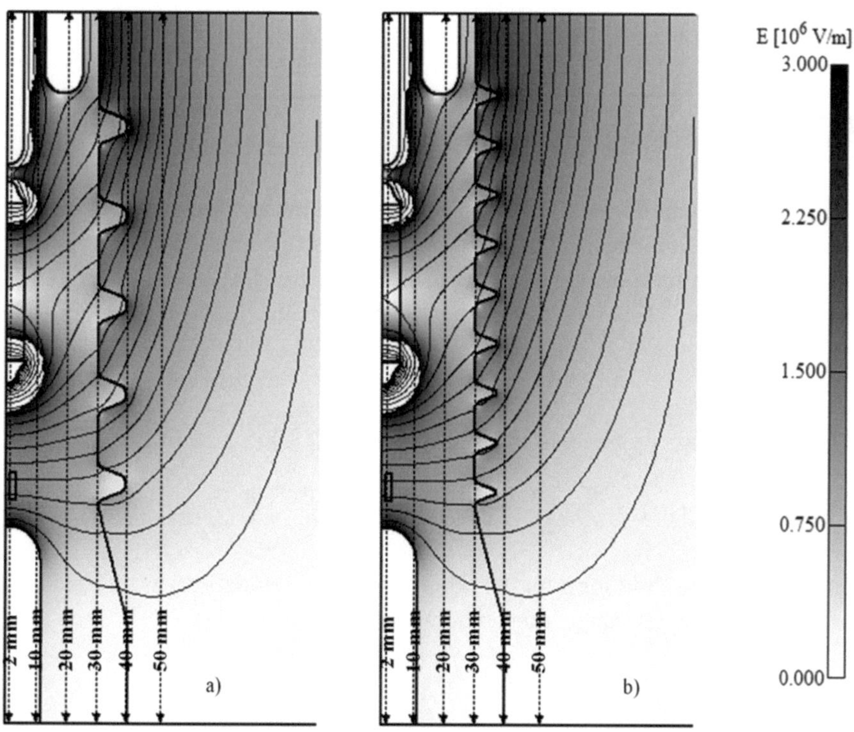


**Figure 4.** EFS of the MVIPI with: (a) five ribs and (b) nine ribs.

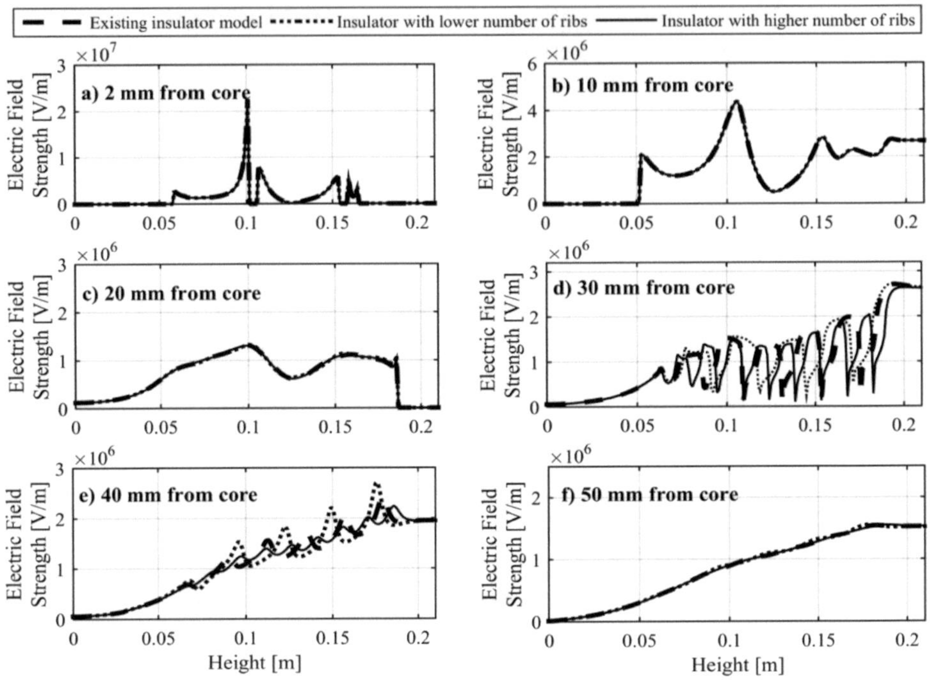


**Figure 5.** EFS of the MVIPI with five ribs (dotted line), nine ribs (solid line) and the existing MVIPI (dashed line) at different distances from the core of the MVIPI: (a) 2 mm for core, (b) 10 mm from core, (c) 20 mm from core, (d) 30 mm from core, (e) 40 mm from core and (f) 50 mm from core.

## 4. MVIPI with the exceptional external shape

The following presents the developing MVIPI's new external shape (**Figure 7**). The designed insulator exterior, as such, does not have the characteristically shaped ribs, since it has been acquired by using the Lagrange interpolating polynomial of sixth degree (**Figure 6a**) [2]. Lagrange's interpolating polynomial for a set of $n + 1$ given data points is written in the form (1) [2]:

$$M_n(r) = \sum_{i=0}^{n} G_i(r)\, z_i, \tag{1}$$

while $G_i(r)$ are Lagrange coefficients which is given by (2) [2].

$$G_i(r) = \frac{(r-r_0)(r-r_1)(r-r_2)\cdots(r-r_{i-1})}{(r_i-r_0)(r_i-r_1)(r_i-r_2)\cdots(r_i-r_{i-1})} \cdot \frac{(r-r_{i+1})\cdots(r-r_n)}{(r_i-r_{i+1})\cdots(r_i-r_n)}. \tag{2}$$

Taking into account the short form for the product of the differences (2) in (1) over (3), the final form is acquired for the Lagrange interpolation polynomial (4) [2]

$$\prod_{j=0}^{n}(r-r_j) = (r-r_0)(r-r_1)(r-r_2)\cdot\ldots\cdot(r-r_{j-1})(r-r_j)(r-r_{j+1})\cdots(r-r_n) \tag{3}$$

$$M_n(r) = \prod_{l=0}^{n}(r-r_l) \cdot \sum_{i=0}^{n} \frac{1}{(r-r_i)} \frac{f(r_i)}{\prod_{j=0, j\neq i}^{n}(r_i-r_j)}. \tag{4}$$

http://dx.doi.org/10.5772/intechopen.71871

For the sixth-degree polynomial, seven points are needed [2], which are labeled with $P_i(r_i, z_i)$, where $i = 0,1,\ldots,6$. Fixed points are $P_0(r_0, z_0)$ and $P_6(r_6, z_6)$: $r_0 := 0.185$ m, $z_0 := 0.029$ m and $r_6 := 0.064$ m, $z_6 := 0.029$ m. Coordinates $P_0$ and $P_6$ are marked in **Figure 6a** and **6b**.

The next step is to identify five middle points [2]. Such $r_i$ are selected, subject to $r_0 < r_i < r_6$, where $i = 1,2,\ldots,5$. According to step $s$ (5) [2], values $r_1$, $r_2$, $r_3$, $r_4$, and $r_5$ are selected:

$$s = \frac{(r_0 - r_6)}{6} = \frac{(0.185 - 0.064)}{6} = 0.0202 \text{ m}, \tag{5}$$

followed by the calculation of $r_1$, $r_2$, $r_3$, $r_4$, and $r_5$:

$$r_i = r_0 - s \cdot j; i = 1,\ldots,5; j = 1,\ldots,5. \tag{6}$$

After calculation (6), the next values are obtained: $r_1 = 0.165$ m, $r_2 = 0.145$ m, $r_3 = 0.125$ m, $r_4 = 0.105$ m, and $r_5 = 0.084$ m. Values $z_1$, $z_2$, $z_3$, $z_4$, and $z_5$ are determined by considering the creeping distance of the MVIPI, which must not be lower than the original MVIPI. The following state (7) [2] is defined for values $z_1$, $z_2$, $z_3$, $z_4$, and $z_5$:

$$z_1, z_2, \ldots, z_5 \geq \begin{cases} z_0 \\ z_6 \end{cases}; \tag{7}$$

therefore, the obtained values are $z_1 = 0.056$ m, $z_2 = 0.033$ m, $z_3 = 0.037$ m, $z_4 = 0.034$ m, and $z_5 = 0.036$ m.

Intermediate polynomials are used before the final polynomial form [2]. For $i = 0,1,2,\ldots,6$ is defined:

$$M_i(r) = \prod_{j=0, j\neq i}^{6} \frac{r - r_j}{r_i - r_j} \tag{8}$$

The sixth-degree polynomial [2] that will go through the points $P_0$ and $P_6$ (9) is given as:

$$m(r) = \sum_{i=0}^{6} M_i \cdot z_i = M_0 \cdot z_0 + M_1 \cdot z_1 + M_2 \cdot z_2 + M_3 \cdot z_3 + M_4 \cdot z_4 + M_5 \cdot z_5 + M_6 \cdot z_6. \tag{9}$$

The sixth-degree polynomial that replaces the shape of the MVIPI's ribs is shown in **Figure 6b**.

The MVIPI's novel external shape is shown in **Figure 7**, with the EFS value. **Figure 8** compares EFS values between the existing MVIPI (dashed line) and the MVIPI with the novel external shape (solid line). The values of EFS are within the defined bounds, with the exception of the boundary insulator to air (30 mm from the insulator core), where the EFS value of the MVIPI with the new external shape exceeded the permissible value slightly 3 MV/m (3.28 MV/m). The lowest possible value of EFS is wanted in that area; hence, the coefficients of the polynomial will be optimized with the goal of reducing the EFS at the boundary insulator to air and finding the best position of the internal components of the MVIPI.

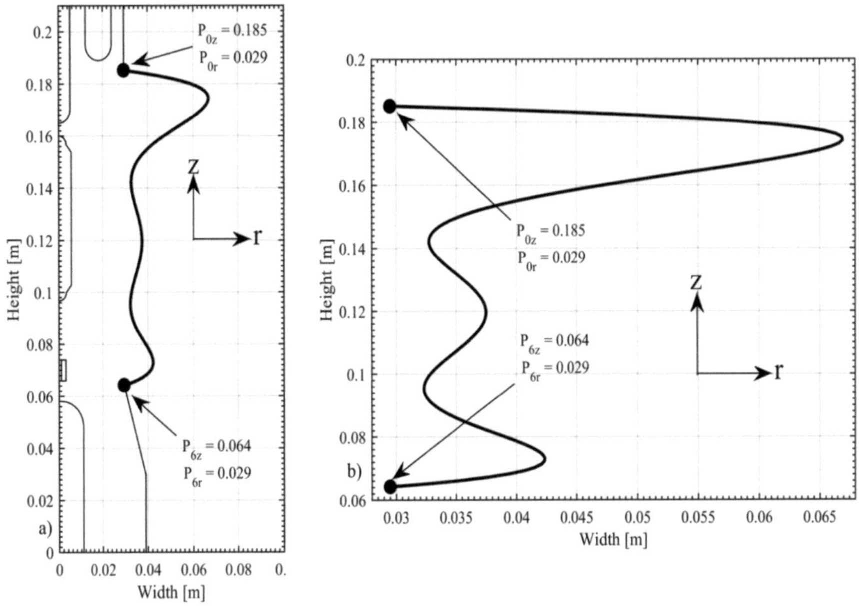


**Figure 6.** (a) Designed ribs using the sixth-degree polynomial. (b) Sixth-degree polynomial, which describes the exterior of the MVIPI.

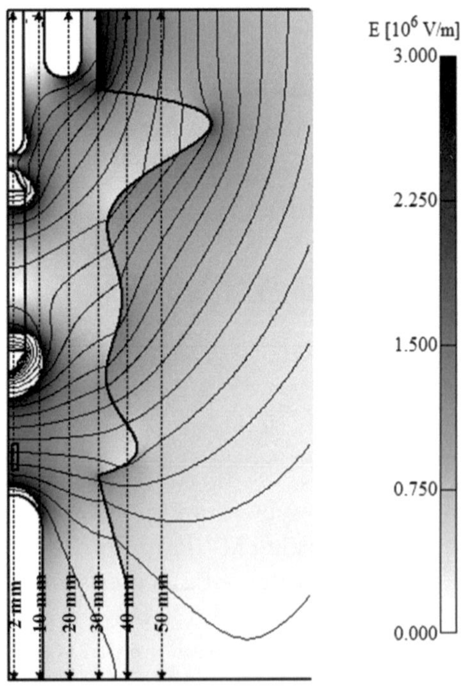


**Figure 7.** EFS of the MVIPI with the novel external shape.

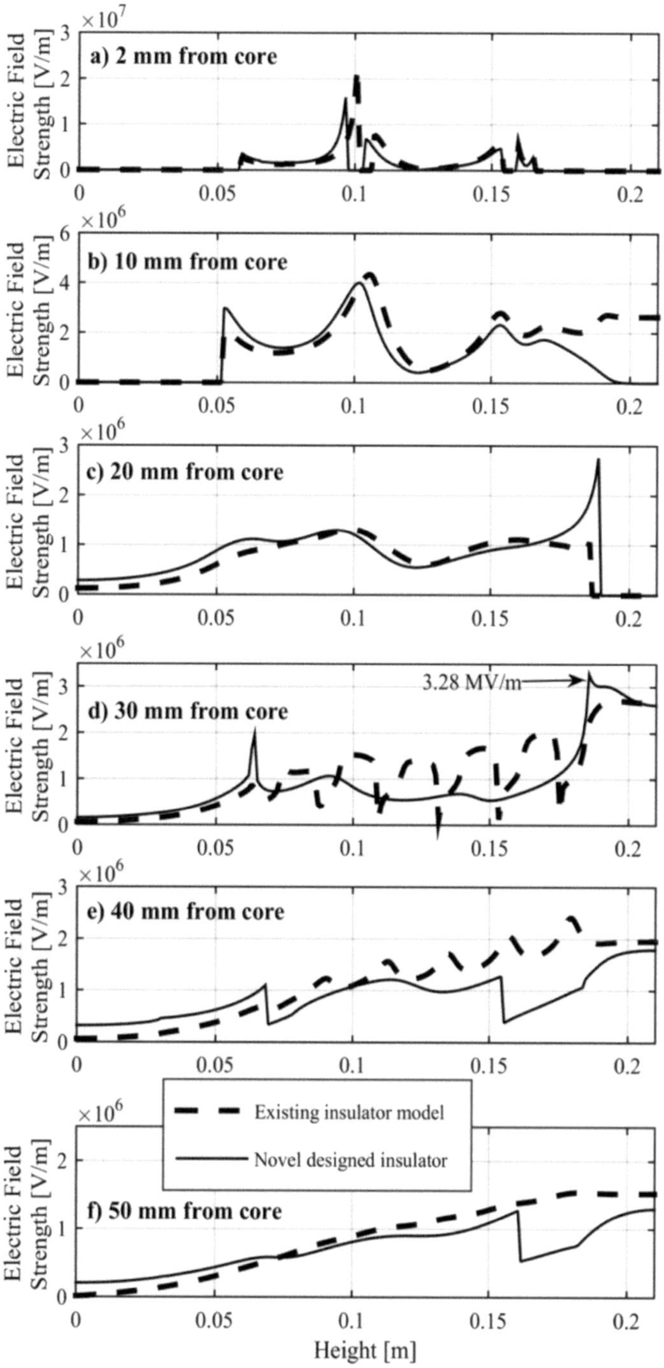


**Figure 8.** EFS of the MVIPI with the novel external shape (solid line) and the existing MVIPI (dashed line) at different distances from the core of the MVIPI: (a) 2 mm for core, (b) 10 mm from core, (c) 20 mm from core, (d) 30 mm from core, (e) 40 mm from core and (f) 50 mm from core.

## 5. Best external shape and the position of the internal MVIPI's components' calculation by means of the differential evolution algorithm

DE is a fast and robust population-based direct-search stochastic optimization algorithm that was first introduced by Storn and Price [5]. This algorithm is widespread among engineering audiences [2, 3, 13–17] due to its robustness in reaching global minima, suitable for solving nonlinear and constrained optimization problems. It requires only boundaries of expected solutions and has only a few control parameters to be defined. A detailed description of the DE algorithm is available from [5, 6].

The component to be optimized is written mathematically in the form of an objective function. In this chapter, the criterion of optimizing is the value of EFS in the most stressed points, which are settled in the air just above the insulator-to-air intersection [2], from the top of the insulator through ribs to the bottom of the insulator. The electric field, is in these points, defined as $E = 3$ MV/m, due to the dielectric strength of air, and the goal is to minimize this value to the utmost.

The wanted value $E_W$ of the electric field in the most stressed points has been defined while preparing the objective function, and its value is 2.2 MV/m. This value, 2.2 MV/m, is chosen to obtain a certain reserve with withstand test voltages. Thus, the objective function $q$ [2] is defined with the following statement (10):

$$q = \frac{\max(E_i, \ldots, E_n)}{E_W}; i \in (1, \ldots, n) \tag{10}$$

where $\max(E_i,\ldots,E_n)$ is the maximal value of the EFS in the most stressed points and $n$ is the number of those points.

Due to the optimization process of minimization of the EFS, the parts of the geometric areas of the MVIPI are written parametrically (polynomial coefficients, upper right connector, the distance between the resistor and the bottom electrode). By adjusting the parameters, the shape of the MVIPI will occur in which the value of the electric field is the lowest at critical points. Boundaries are determined by the parameters, within which they will be alternated [2, 3].

In total, there are six parameters that are labeled from $p_1$ to $p_6$. Parameters $p_1$, $p_2$, and $p_3$ change the MVIPI's external shape; $p_4$ changes the distance between the bottom electrode and resistor; and $p_5$ and $p_6$ change the height and width of the upper right connector. **Table 2** shows the values of searched parameters before and after optimization, as well as their boundary values (maximum and minimum). **Figure 9** shows searched insulator parameters before and after optimization. The optimization algorithm changed the MVIPI's initial design and searched for the suitable parameters' configuration for the most favorable electric field's distribution. The result of the optimization is a set of best parameters from $p_1$ to $p_6$ from **Table 2**. The resulting best values define the MVIPI's external shape and position of internal components, which provide the most favorable electric field's distribution. This result was obtained at the size of the population $NP = 30$ and the following control parameters of the DE algorithm $F = 0.6$; $CR = 0.8$ $strategy$ = DE/rand/1/bin.

| Parameter | Minimum value | Maximum value | Value before optimization | Value after optimization |
|---|---|---|---|---|
| $p_1$ | 0.041 | 0.056 | 0.056 | 0.044 |
| $p_2$ | 0.033 | 0.043 | 0.037 | 0.043 |
| $p_3$ | 0.034 | 0.050 | 0.036 | 0.042 |
| $p_4$ | 0.070 | 0.140 | 0.070 | 0.070 |
| $p_5$ | 0.021 | 0.038 | 0.015 | 0.038 |
| $p_6$ | 0.007 | 0.012 | 0.012 | 0.012 |

**Table 2.** Parameter values before and after optimization.

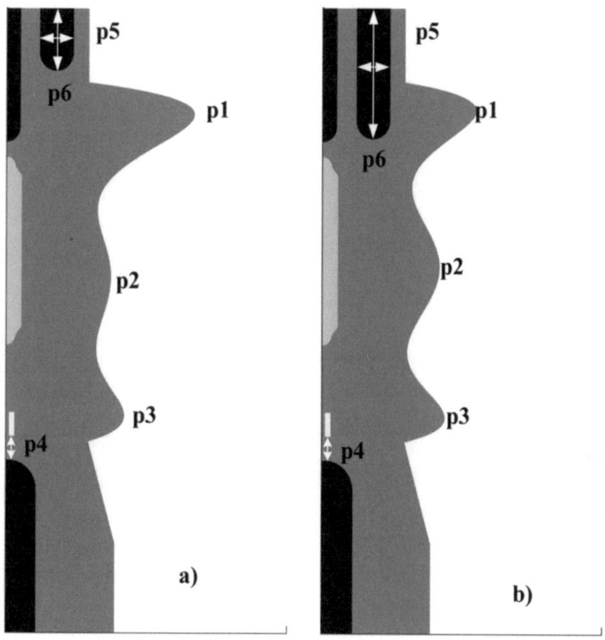


**Figure 9.** MVIPI parameters: (a) before optimization and (b) after optimization.

**Figure 10** shows the MVIPI's EFS distribution after the optimization. **Figure 11** shows EFS at different distances from the center of the MVIPI. The solid line marks the EFS of the novel-designed MVIPI after the optimization. The EFS of the novel-designed MVIPI before the optimization is marked with a dotted line. The EFS of the existing MVIPI is marked with a dashed line.

Comparing the results before the optimization, there was a decrease in the EFS inside the insulator and in the air after the optimization. 30 mm from the MVIPI's core (**Figure 11d**), the EFS does not exceed the value of 2 MV/m on the boundary insulator to air. Comparing EFS value between the novel-designed MVIPI before and after the optimization, there is a 35% decrease in the EFS at the same distance after the optimization. On the other hand, comparing EFS values between the existing MVIPI and novel-designed MVIPI after the optimization, there is a 22% decrease in the EFS at the same distance after the optimization.

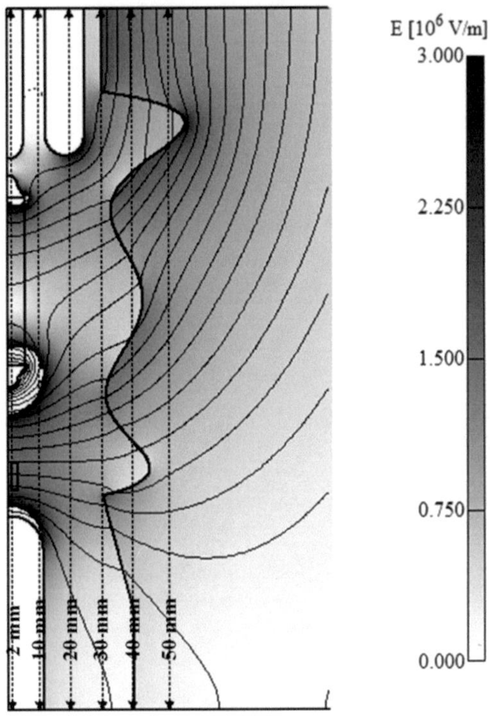


**Figure 10.** EFS of the MVIPI with the new external shape after the optimization.

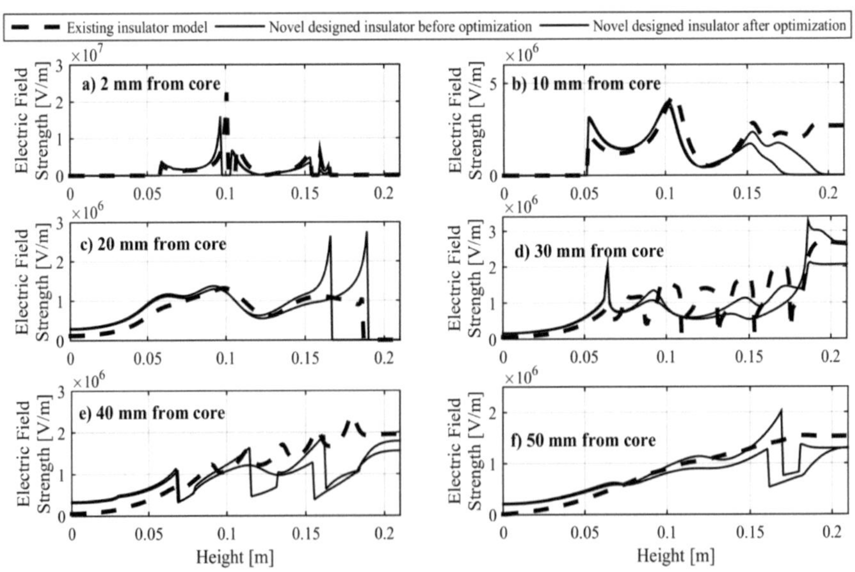


**Figure 11.** EFS of the novel-designed MVIPI after the optimization (solid line), before the optimization (dotted line) and the existing MVIPI (dashed line) at different distances from the core of the MVIPI: (a) 2 mm for core, (b) 10 mm from core, (c) 20 mm from core, (d) 30 mm from core, (e) 40 mm from core and (f) 50 mm from core.

http://dx.doi.org/10.5772/intechopen.71871

## 6. Conclusion

This chapter describes the influence of the electric field on the MVIPI with standard and modified external shapes. The electric field behavior of the MVIPI with different external shapes was shown, and the numerical model was introduced with a favorable distribution of the electric field. The MVIPI's external shape influence on the electric field is researched by means of changing the number of ribs and making a completely new external shape of the MVIPI, which does not have the typical shape of the ribs, but a new outline, using the Lagrange polynomial that has acquired its best form through the optimization process.

The application of the DE algorithm is shown in order to design the MVIPI's external shape and reach the best position of internal insulator elements. It was essential to satisfy the criteria for reaching a certain boundary of EFS inside and in the surrounding area of the MVIPI.

The improved novel-designed insulator, with its modified external shape, is returned mainly in the lower value of the electric field inside the insulator and in the air, which means less exertion of insulation material. Such an MVIPI can function over a long period of time. Producing such an MVIPI is also less challenging, due to the fewer elements inside the insulator and, thus, fewer risks for partial discharges. Another major achievement is scrap decrease, which is economically noteworthy. The tool for molding the MVIPI's external shape is less complex, which is favorable for production and that it will be less possible to form cavities in the casting.

## Nomenclature

| | |
|---|---|
| $CR$ | crossover rate |
| $DE$ | differential evolution |
| $E$ | electric field strength [V/m] |
| $E_W$ | wanted value of the electric field strength [V/m] |
| $F$ | mutation factor |
| $G_i(r)$ | Lagrange coefficients |
| $i, j$ | indices |
| $M_n(r); m(r)$ | Lagrange interpolating polynomial |
| $n$ | number of data points |
| $NP$ | size of the population |
| $p_1,\ldots, p_6$ | parameters of the insulator |

| | |
|---|---|
| $q$ | objective function |
| $(r, z)$ | coordinates [m] |
| $s$ | step [m] |
| $\varepsilon_r$ | dielectric permittivity |

## Author details

Mirza Sarajlić[1]*, Jože Pihler[1], Nermin Sarajlić[2] and Peter Kitak[1]

*Address all correspondence to: mirza.sarajlic@um.si

1 University of Maribor, Faculty of Electrical Engineering and Computer Science, Maribor, Slovenia

2 University of Tuzla, Faculty of Electrical Engineering, Tuzla, Bosnia and Herzegovina

## References

[1] Pihler J. Switchgear of Electric Power System. 2nd ed. Maribor: University of Maribor, Faculty of Electrical Engineering and Computer Science; 2003. pp. 1-273

[2] Sarajlic M, Kitak P, Pihler J. New design of a medium voltage indoor post insulator. IEEE Transactions on Dielectrics and Electrical Insulation. 2017;**24**(2):1162-1168. DOI: 10.1109/TDEI.2017.005947

[3] Sarajlić M, Pihler J. The influence of the number of ribs of a medium voltage post insulator on the electric field. In: 2016 International Symposium on Fundamentals of Electrical Engineering (ISFEE); Bucharest; June 30–July 2 2016. p. 1-5

[4] Tičar I, Kitak P, Pihler J. Design of new medium voltage indicator by means of electric field calculation. Journal of Microelectronics, Electronic Components and Materials. 2002;**32**(2):82-87

[5] Storn R, Price K. Differential evolution—A simple and efficient heuristic for global optimization over continuous spaces. Journal of Global Optimization. 1997;**11**(4):341-359. DOI: 10.1023/A:1008202821328

[6] Price KV, Storn RM, Lampinen J. A. Differential Evolution: A Practical Approach to Global Optimization. 1st ed. Berlin Heidelberg: Springer-Verlag; 2005. p. 539

[7] Program tools EleFAnT. Graz, Austria: Inst. Fundam. TU Graz Theory Electrc. Eng; 2000

[8] Sheikholeslami M, Hayat T, Alsaedi A, Abelman S. Numerical analysis of EHD nanofluid force convective heat transfer considering electric field dependent viscosity. International Journal of Heat and Mass Transfer. 2017 May;**108**:2558-2565

[9] Sheikholeslami M, Soleimani S, Ganji DD. Effect of electric field on hydrothermal behavior of nanofluid in a complex geometry. Journal of Molecular Liquids. 2016 Jan;**213**:153-161

[10] Sheikholeslami M, Ganji DD. Impact of electric field on nanofluid forced convection heat transfer with considering variable properties. Journal of Molecular Liquids. 2017 Mar;**229**:566-573

[11] IEC 61958:2000. High-voltage Prefabricated Switchgear and Controlgear Assemblies—Voltage Presence Indicating Systems. 1st ed.; 2000

[12] IEC 61243-5:1997. Live Working—Voltage Detectors—Part 5: Voltage Detecting Systems (VDS). 1st ed.; 1997

[13] Sarajlić M, Pocajt M, Kitak P, Pihler J. Bare conductor temperature coefficient calculation by means of improved differential evolution algorithms. In: Maribor: 13th Slovenian Power Engineering Conference CIGRE-CIRED; 22-24 May 2017. p. 1-10

[14] Sarajlić M, Pocajt M, Kitak P, Sarajlić N, Pihler J. Covered overhead conductor temperature coefficient calculation using differential evolution optimization algorithm. In: Neum: 13th BH K CIGRE Power Engineering Conference; 17-21 September 2017. p. 1-11

[15] Glotic A, Pihler J, Ribic J, Stumberger G. Determining a gas-discharge arrester model's parameters by measurements and optimization. IEEE Transactions on Power Delivery. 2010;**25**(2):747-754

[16] Glotić A, Glotić A, Kitak P, Pihler J, Tičar I. Optimization of hydro energy storage plants by using differential evolution algorithm. Energy. 2014 Dec;**77**:97-107

[17] Glotic A, Sarajlic N, Kasumovic M, Tesanovic M, Sarajlic M, Pihler J. Identification of thermal parameters for transformer FEM model by differential evolution optimization algorithm. In: 2016 International Conference Multidisciplinary Engineering Design Optimization (MEDO). 2016. pp. 1-6

Chapter 8

# The Application of Electric Fields in Biology and Medicine

Francis X. Hart and John R. Palisano

Additional information is available at the end of the chapter

http://dx.doi.org/10.5772/intechopen.71683

**Abstract**

We discuss a wide range of applications of electric fields in biology and medicine. For example, physiological strength (<500 V/m) fields are used to improve the healing of wounds, the stimulation of neurons, and the positioning and activation of cells on scaffolds for tissue engineering purposes. The brief, strong pulses used in electroporation are used to improve the insertion of drugs into tumors and DNA into cell nuclei. The references direct readers to detailed reviews of these applications. The mechanism by which cells detect physiological strength fields is not well understood. We also describe a field-transduction mechanism that shares features common to the detection of fluid shear by cells. We then provide some experimental evidence that supports our model.

**Keywords:** electric field, wound healing, electroporation, galvanotaxis, glycocalyx

## 1. Introduction

The applications of electric fields in biology and medicine are many and varied. Physiological strength (~100 V/m), direct current (DC) electric fields are important in the development, maintenance and control of cells and tissues. Their role in wound healing, embryonic development, and tissue regeneration is described in detail in the reviews of Pullar [1], McCaig et al. [2], and Robinson and Messerli [3]. Endogenous DC electric fields are also important in embryonic patterning [4]. In tissue engineering cell proliferation on scaffolds can be controlled by the application of such fields [5]. At the tissue level electric fields are used for the measurement of body composition [6] and the promotion of wound healing [7]. An important new development has been the use of strong electric fields for drug delivery. Very high, pulsed fields can promote the passage of drugs through membranes [8] or the insertion of DNA into the nucleus for genetic engineering applications [9].

There are still unresolved questions regarding the mechanisms by which the fields achieve their effects. For physiological strength fields a wide variety of biochemical pathways within the cell following the initial detection of the field have been studied [10], but the initial transduction mechanism is not well understood. For high voltage, pulsed fields the details of membrane pore formation remain unclear.

·Many papers have been published in this very broad area for many years. It is not possible to cover the many applications of electric fields in biology. We first describe some representative applications of electric fields in biology and medicine. Next we discuss mechanisms responsible for these effects and explore some remaining gaps in our understanding of these processes. References are provided to reviews so that readers who desire more detailed discussions of these applications can find them. For low field effects we also describe the experimental procedures used to accurately measure galvanotaxis, a process involved in several of the applications described here, and present some typical results.

## 2. The application of physiological strength DC and low frequency AC fields to cells and tissue

### 2.1. Wound healing

When tissue is damaged, an electric field is produced at the wound site. **Figure 1** illustrates how such fields are produced when epithelial tissue, such as skin, suffers a wound. The upper three layers of skin are the stratum corneum, the epidermis and the dermis. The figure illustrates three typical cells in a row in the epidermis. At the top or apical membrane of a cell sodium ions enter the cell via sodium-specific channels. At the bottom or basolateral membrane potassium ions leave the cell. These ionic transfers are associated with a biochemical reaction inside the cell involving ATP, which is a molecule with bonds for high-energy transfer in the cell. **Figure 1a** shows this transfer process prior to the development of a wound for the three cells. The cells are connected by tight junctions (TJ) that do not ordinarily permit the passage of charge. The flow of charge through the cell produced by this biochemical process has a current I associated with it. The layer above the apical membrane has given up positive charges to the sodium channels and thus has a relative negative charge. The layer below the basolateral membrane has acquired a relative positive charge. This charge imbalance is relieved by its return flow along a paracellular pathway i.e. through more distant junctions that are not so tightly bound. Because there is an electrical resistance R associated with this pathway, a potential difference $\Delta V = IR$ is produced across the entire epithelial layer. Depending on the particular tissue and pathway details, $\Delta V$ ranges between 15 to 60 mV [11].

**Figure 1b** shows the situation when the middle cell has been damaged. There is now an easy passage for the return flow to the apical layer. Positive charge from both directions will flow along the basolateral membrane layer toward the wound. There is thus an associated electric field, E, directed toward the wound from both sides. The field strength for this endogenous field is about 100–200 V/m at the wound site. For a more detailed explanation of the biochemical origin of the field see the review by Nuccitelli [11]. For examples of how widespread

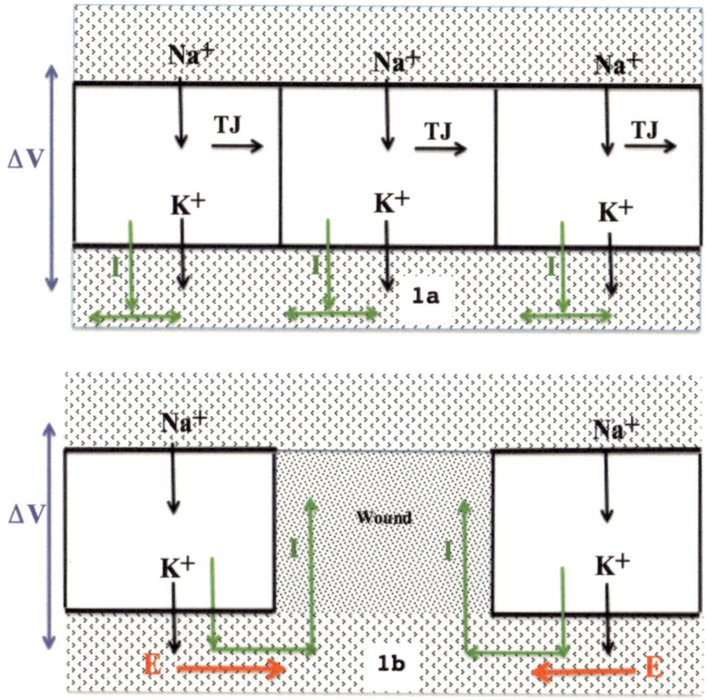


**Figure 1.** Schematic diagram of the electric field produced at a skin wound site.

wound-generated fields are in biological systems (e.g., plants), see the review by Robinson and Messerli [3].

Just as cells migrate in response to a chemical gradient (chemotaxis), most types of cell also migrate in response to an electric field (galvanotaxis or electrotaxis). In response to a nearby wound, cell types such as keratinocytes, fibroblasts, macrophages and lymphocytes migrate toward the wound under the guidance of the field [12, 13]. Most cells migrate in the direction of the field toward the negative pole. The situation is complicated, however. Macrophages move toward the positive pole although their precursors, monocytes, migrate toward the negative pole [14]. Moreover, when an endogenous field is not produced at the wound site, the wound does not heal properly. Application of an external electric field might then improve the healing of wounds that have not healed on their own by directing the migration of cells to the wound site and also by inducing those cells to produce biochemicals that promote healing [13]. For example, a negative electrode could be placed on the wound and one or more electrodes connected to a positive pole placed nearby. Kloth [12] has reviewed a number of clinical studies that have investigated whether the application of such fields improves the treatment of wounds compared to standard wound care. The applied fields in such studies are either constant direct current (DC) or pulsed direct current. Analyzing a wide variety of devices, Kloth [12] found that the most effective treatments used a current ranging between 250 and 500 μA. Some of these devices have been approved by regulatory agencies for wound healing in the EU and for antibacterial activity on wounds in the US.

### 2.2. Electrical stimulation of the nervous system

Pulsed and low-frequency alternating current (AC) fields, applied with either implanted or surface electrodes, are being used to either stimulate or suppress neural activity. In Deep Brain Stimulation (DBS) [15] electrodes can be surgically implanted into specific areas of the brain to apply pulse signals that suppress endogenous signals that produce Parkinson's disease tremors or epileptic seizures. The neurostimulator, connected to the electrodes is usually implanted under the collar bone. DBS is generally applied as a supplement to regular medications and only after the medications no longer provide relief of symptoms. Over time the electrodes can become coated and may need to be replaced.

Electrical stimulation (ES) [16] has been successfully used to restore muscular functionality in patients who have suffered a major spinal cord injury (SCI). Although the spinal cord has been damaged, the muscular systems that it ordinarily stimulates may remain undamaged. If electrical signals can be transmitted to those muscles, they may respond as usual. ES can be applied to the peripheral nervous system or directly to the spinal cord. For peripheral stimulation electrodes are preferably implanted into the tissue close to the nerves. Otherwise, surface electrodes are used. Bipolar pulses are used in order to prevent electrochemical damage to the tissue near the electrodes. Complex movements, such as grasping by the hands, require the stimulation by multiple nerves in a particular temporal pattern. For this reason the power supply must generate pre-programmed, separate signals to the individual neurons. For a detailed description of ES systems and their application in restoring functionality for standing, bladder control, pressure ulcer prevention and other conditions see the review by Ho et al. [16].

The spinal cord can generate its own complex neural stimulation patterns if it receives the proper electrical stimulation. SCI can damage the cord above the region where it ordinarily generates signals for standing and walking so that these activities cannot be performed. Rejc et al. [17] report that application of pulses from an array of electrodes inserted into the epidural spaces of the spinal cord allowed several paraplegic individuals to stand with minimal assistance. The pulse patterns must be tailored to the individual. Solopova et al. [18] used skin electrodes to apply biphasic rectangular pulses to the lower spines of children suffering from cerebral palsy. This treatment, combined with treadmill training, improved locomotor function compared to locomotor training only.

Transcutaneous Electrical Nerve Stimulation (TENS) is a treatment by which low frequency pulses are applied by skin electrodes to reduce pain. Although the electrodes are generally placed at the site of the pain (e.g., lower back), the primary effects appear to be due to stimulation of the central nervous system. The lower frequency (<10 Hz) pulses are applied with a relatively high intensity to produce non-painful motor contraction whereas the high frequency (>50 Hz) are applied with a relatively low intensity and do not produce contractions. Both types of stimulus activate opioid receptors in the spinal cord and brain that reduce pain, but the type of opioids produced differs for the two stimuli. For more details see the review by DeSantana et al. [19].

### 2.3. Induced electric fields and bone healing and brain stimulation

#### *2.3.1. Changing magnetic fields produce electric fields*

The electric fields just described are produced by an electrical potential difference that is generated by batteries or power supplies. The electric field lines flow from positive to negative charges. Such sources are well suited for surface applications such as wound healing, corneal repair or even brain and spinal stimulation with closely-separated, inserted electrodes. However, they are not suited for the broad application of electric fields to deeper tissue because of the tissue's electrical impedance. Another method for producing electric fields uses Faraday's Law that describes how a time-varying magnetic field produces an induced electric field with field lines that are closed loops. Magnetic fields penetrate tissue unimpeded so that significant induced fields can be produced well below the body surface.

Consider a spatially uniform magnetic field, B, that varies with time, t. Suppose that the field is applied perpendicular to the surface of a circular metallic disc. The electric field, E, induced at a distance r from the center of the disc is a circular loop of radius r with a magnitude

$$E(r) = \frac{r}{2}\frac{dB}{dt} \tag{1}$$

It should be noted that the same field would be produced even in the absence of the metal disc. The field itself is independent of the material. The induced electric current, however, does depend on the material through its conductivity. For systems more complicated than a circular disc the calculation of the induced field is complicated and requires numerical methods. For a simple example of a square dish with insulating inclusions (e.g., cells) see Hart et al. [20].

#### *2.3.2. Two examples illustrate the clinical use of time varying magnetic fields*

For many years electric fields have been used to promote the healing of bone fractures that have not responded to the usual treatments. Because the fractures are located deep within tissue, implanted electrodes are not generally used. There are, however, two other ways by which external delivery systems can produce the required internal fields. Consider, for example, a fractured tibia. Pad electrodes can be placed on either side of the patient's calf. A high frequency signal applied to these electrodes will be capacitively coupled to the internal tissues, including the fracture site, to produce an electric field there. Alternatively, a coil system could be wrapped around the calf. A time-varying current in the coils produces a time-varying magnetic field in the calf that generates, in turn, a pulsed electromagnetic field (PEMF) at the fracture site. The PEMF stimulates osteoblasts so that the bone grows and proliferates to heal the fracture. Two meta-analyses [21, 22] of clinical studies of electrically stimulated bone healing show some improvements in the patients, but more well-defined studies are needed.

In transcranial magnetic stimulation (TMS) coils of various designs are placed around a patient's head. As with the bone fracture treatment the coils produce a time-varying magnetic field that inductively generates an electric field at the desired site in the brain. The coil system design is chosen to produce a localized electric field at the desired site in the brain. This method is used, for example, to treat depression or to assess whether a particular motor area of the brain is stimulating the associated muscle system properly. For more details and examples of biomagnetic stimulation, see the review by Ueno and Sekino [23]).

### 2.4. Electric fields for diagnostic purposes

The preceding sections have described how the application of electric fields to cells and tissue can produce beneficial effects in those targets. Electric fields, however, can also be applied to monitor the physiological state of tissues for diagnostic purposes. All materials, including tissue, have an associated resistance, R, and capacitance, C [24]. If an AC voltage is applied to tissue, current that depends on both R and C will flow through the tissue. Capacitances restrict the flow of low frequency current, but allow high frequency currents to pass. The value of C determines the transition frequency range for passage and blockage of the AC current. For tissues that frequency is around 100 kHz–1 MHz.

**Figure 2** depicts the Fricke circuit model commonly used for tissue. Re and Ri represent, respectively, the resistances of the extracellular and intracellular resistances. Both resistances are determined primarily by water content. C represents the capacitance of the cell membrane. Low frequency currents are blocked by C from the cell interior and thus serve as a measure of extracellular water. C allows high frequency currents to enter the cell interior and thus measure the combination of Re and Ri. Use of two frequencies allows the separation of Re and Ri and thus the determination of intracellular and extracellular water content.

These principles can be used for the measurement of body composition [6]. In healthy individuals there is a well-defined relationship between total body fat-free mass and the total body water mass. Knowledge of the total body fat-free mass and the total body mass yields the body's fat mass and thus the percentage body fat. Depending on the positions of the electrodes, this method can be used for measurements on the whole body or on specific limbs. Determination of changes in the ratio of Re to Ri can yield information regarding the possibility of fluid leakage (edema) into a limb.

A different application of the Fricke model is useful in the monitoring of breathing. The air in the lungs does not permit the flow of current and can be modeled as a time-varying C. Measurement

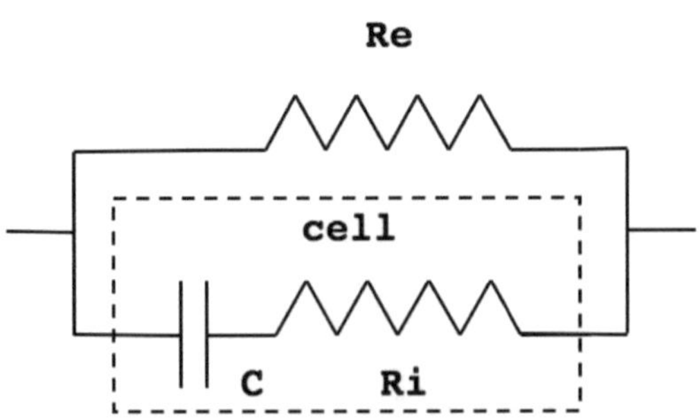


**Figure 2.** Fricke circuit model.

of the current over time yields the breathing rate and thus serves as a monitor of sleep apnea. How R and C vary over a wide range of frequencies could be used to differentiate healthy tissue from diseased tissue. The sensitivity of devices that identify breast tumors is not yet sufficient to receive governmental approval. For a detailed review of the various diagnostic uses of electric fields see the review by Hart [6].

### 2.5. Tissue engineering

The application of electric fields to cellular systems can produce a wide variety of physical effects in cells in addition to galvanotaxis. Such effects are described in detail in the CRC review monograph edited by Pullar [1]. This combination of electric field effects is proving useful in tissue engineering in which cells are deposited on a substrate (scaffold) and stimulated to grow into more complicated tissues. For example, the application of pulsed DC fields to cardiac myocytes cultured on a collagen sponge scaffold stimulated their alignment, coupling and synchronous contractions [25]. Such an assemblage would serve as a cardiac patch for heart attack victims. Pulsed fields promoted the outgrowth and orientation of neurites on a polypyrrole/collagen scaffold [26]. The goal of such research is to develop the ability to replace damaged nerves. A DC field applied to osteoblasts, the bone-forming cells, deposited on a titanium substrate increased their adhesion to the substrate and their proliferation [27]. The nature and structure of the substrate is important because it can provide mechanical stimulation that also affects cell behavior on a scaffold. Optimizing both contact guidance and the applied electric field produces results that are better than those obtained for either modality by itself for fibroblasts and cardiomyocytes [28] as well as corneal and lens epithelial cells [29].

Tissue engineering has typically used electric fields that were applied externally to the scaffold. Recently Arinzeh et al. [30] reported that the scaffold itself could provide the electric field used to control and stimulate the cells. They designed a scaffold that was part ceramic and part plastic. One of the plastic materials they used was piezoelectric; that is, it generated an electric field when mechanically stressed. With the proper piezoelectric-scaffold mechanical texturing, they were able to enhance axonal regeneration in neurites with the goal of promoting spinal cord repair. Moreover, using the proper piezoelectric materials and scaffold textures, they were able to induce stem cells to form either bone or cartilage. For a more detailed discussion of the use of electric fields in tissue engineering see the review by Hronik-Tupaj and Kaplan [31].

## 3. Cell dosimetry

In order to understand how cells produce the observed effects, it is first necessary to understand how the applied field is distributed within the cell as a function of time. Suppose that a spherical cell of radius R is placed within a culture medium far from parallel plate electrodes. At time $t = 0$ an electric field E is applied between the plates. The Fricke model shown in **Figure 2** can be used to understand the resulting charging of the cell membrane and induced transmembrane voltage TMVi. The cell behaves as a series RC circuit. When the field is applied, charge flows through the resistor to charge the capacitor. Once the capacitor is fully charged to establish TMVi, the flow of charge through R ceases. The time constant for the process is τ. For times significantly less than τ there is significant charge flow and an associated electric field within the cell interior (cytoplasm). The voltage across the capacitor, the TMVi, gradually increases. For times significantly greater than τ charge flow has ceased and the associated electric field within the cytoplasm is negligible. The TMVi has reached its maximum value. Without the applied field, there is a natural transmembrane voltage, TMVo, that is produced by a system of ion pumps and channels in the membrane. TMVo is on the order of 70 mV for most cell types with the cell interior being more negative than the exterior. The total transmembrane voltage, TMVt, is then the sum of TMVo and TMVi. The transmembrane voltage varies with time [32] according to:

$$\mathrm{TMVt} = 1.5\mathrm{ER}\cos\Theta(1 - \exp(-t/t)) + \mathrm{TMVo} \tag{2}$$

where Θ is the angle between the electric field and the normal to the cell surface. The 1.5 factor is related to the spherical shape of the cell and has higher values for more asymmetrically shaped cells. For irregularly shaped cells numerical modeling is required, as discussed in Section 6.

**Figure 3** is a stylized diagram that illustrates a cell placed in an initially uniform electric field, Eo. When the field is first applied (t < <t), charge has not had sufficient time to flow to charge the cell membrane and there is a significant field inside the cell. After the field has been applied for a long time (t> > τ), the membrane is fully charged and the field in the cytoplasm is negligible. For typical cells τ is on the order of 1 microsecond. TMVi can be quickly estimated for an applied field of 100 V/m. The diameter of a small cell is on the order of 10 μm so that the potential difference across the cell is on the order of 1 mV. At equilibrium there is negligible potential difference throughout the cell interior. The 1 mV is shared by the membrane at its ends to give an

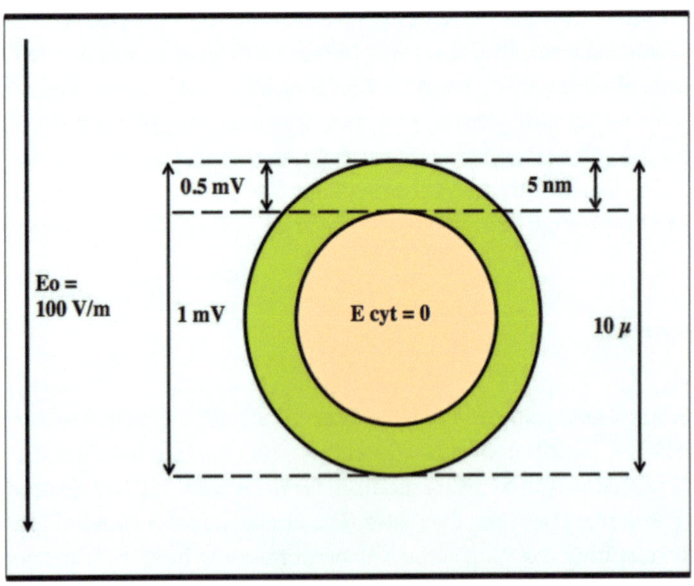


**Figure 3.** Dependence of the transmembrane voltage on the applied field.

estimated TMVi of 0.5 mV. The sphericity of the cell adds an additional factor of 1.5 to yield an actual TMVi of 0.75 mV. Note that this value is much less than TMVo. For a cell with a diameter of 50 mm the TMVi would be 3.75 mV, still small compared to the TMVo. For an applied field of 10 kV/m TMVi would be 75 mV for the 10 μm diameter cell and 375 mV for a 50 μm diameter cell. In these cases TMVi is comparable to or greater than Vo, respectively.

## 4. Electroporation

### 4.1. Basic principles

The applications described to this point are for applied fields of physiological strength; that is, less than about 300 V/m. The frequencies of these fields are also less than 1 MHz with corresponding time scales greater than 1 microsecond. As noted above, for such time scales the resulting cytoplasmic electric field is negligible, so any direct interaction must take place at the cell surface. In this section we discuss applications produced by fields that are on the order of 10 kV/m or greater with component frequencies often greater than 1 MHz. Such fields are strong enough to permeabilize the cell membrane; that is, to permit the passage of atoms and molecules to the interior of the cell. The principle model used to describe this process involves the opening of small (a few nm in radius) openings or pores. For this reason the process is referred to as electroporation although the term electropermeabilization is sometimes used. In this way previously administered drugs that ordinarily cannot pass through the membrane are able to enter the cell interior (electrochemotherapy, ECT) or DNA/RNA can be inserted into the cell nucleus (electrogenetherapy, EGT).

The applied field takes the form of square pulses with electric field amplitude, E, number, N, pulse duration, T, and pulse repetition rate, F. Typically, as any of these parameters increases, the probability of drug or DNA insertion increases, but so does the probability of cell damage. There is, thus, a trade-off between insertion success and cell survival. Typically, E is on the order of 10 kV/m to 100 kV/m and T is on the order of milliseconds to microseconds. Currently, nanosecond pulses with E on the order of 1 MV/m are being investigated for potential clinical applications. Below a certain threshold that depends on cell properties (e.g., type, radius, location in tissue, etc.) electroporation does not occur. For millisecond to microsecond pulses reversible electroporation begins above about 10 kV/m. Under these conditions transport to the cell interior occurs and the cell survives the process. At higher fields too much damage occurs and the cells do not survive. There are two categories of this irreversible electroporation. Cells can die because the membrane is so severely damaged that it does not reseal adequately. In such non-thermal irreversible electroporation (NTIRE) the target cells can be destroyed without damaging the surrounding veins and extracellular matrix [8]. At higher fields the cells and some of their surroundings are destroyed by heating.

Electroporation works by changing the potential difference across the membrane. As noted above, applied fields on the order of 10 to 100 kV/m can produce TMVi comparable to or greater than TMVo. Because the exterior cell surface is positive relative to the interior, TMVt will be greatest at the end of the cell facing the positive electrode (anode). At the opposite end facing the negative electrode (cathode), TMVo essentially subtracts from TMVi so that TMVt is large, but somewhat smaller than at the positive end. Eq. (2) above is useful in understanding parameters affecting ordinary electroporation. For pulses in the microsecond to millisecond range $t >> \tau$ so that the term in parentheses reduces simply to 1. At the midline where $\cos \Theta = 0$, TMVt = TMVo. Because the greatest TMVts are at the two ends, those are the places where maximum insertion is ordinarily expected. Minimum insertion is expected at the midline. As E increases, insertion will occur at places further and further from the poles. To achieve a particular TMVi, larger fields are required for smaller cells. Once the pore has opened, the molecules in the surrounding fluid pass through it to the cell interior. Larger molecules require longer pulse durations to widen the pores and to maintain the electrical driving force for a longer period of time. The area of the cell surface permeabilized depends on E whereas the level of permeability of the surface depends on the number and duration of the pulses.

A currently unresolved problem in electroporation theory is that the cell membrane does not completely reseal itself after the removal of the field. Electroporated simple lipid membranes reseal themselves on a time scale of microseconds whereas cell membranes remain partially permeabilized for time scales on the order of minutes. Two models have been proposed for the delay in complete resealing [32]. Both involve lipids, one of the major constituents of the cell membrane. One involves a change in the conformation of the lipids that results in a metastable membrane structure that requires considerable time to decay. The other is the oxidation of the lipids which changes a variety of the membrane properties.

### 4.2. Clinical applications of electroporation

Important anti-cancer drugs such as cisplatin and bleomycin cannot easily cross the cell membrane. Application of electroporation pulses opens pores in the membrane through which the drugs can pass. Because drug insertion into the cell is now more efficient, the dose given to the patient can be reduced significantly and have fewer side effects. ECT is being used at numerous centers in Europe for a wide variety of tumors [8]. For deep-seated tumors NTIRE is used to destroy the cancer cells while preserving the surrounding tissue. Deep-seated tumors require complex procedures. Typically an MRI is used to provide a detailed mapping of the tissue types surrounding the tumor. Knowledge of the dielectric properties of the various tissue types permits the calculation of the field that would be applied at the tumor site for various possible electrode placements. Once the optimum electrode placement has been decided the MRI can be used to guide the surgeon's placement of the electrodes [33].

The administration of drugs through skin patches (transdermal drug delivery) requires that the drugs successfully pass through the skin to the tissue below. Such passage is typically blocked by the stratum corneum, which is the outermost layer of the skin. The stratum corneum is an approximately 15 μm thick, dry layer of dead cells. Just as this layer protects the body by restricting the loss of fluids from the body to the surroundings, it also prevents the passage of fluids from the surface to the interior. In traditional electrically-assisted drug delivery systems (iontophoresis) electrodes are placed on the surface and a field is applied. Two factors are involved in the assisted drug transport [34]. First, the electric field will push positively-charged drugs from the anode through the skin to the cathode. Negatively-charged drugs are pushed from the cathode. Second, the charged ions pull some of the fluid along with them (electro-osmosis) and thus enhance the transport. Neutral drug molecules can be transported by this method.

Electroporation can offer an improved method of transdermal drug delivery [8]. Because the stratum corneum has a much greater electrical resistance than the lower skin layers, the electric field of the applied pulses is concentrated in it. The pulses increase its permeability to the drugs by opening pathways through it. In addition the field can apply a strong electrophoretic force to charged drug molecules to push them through the stratum corneum.

Because DNA is significantly larger than the anti-cancer drugs, it is more difficult to insert into the cell. The DNA is inserted in the form of plasmids, which are small, often circular segments that contain genetic information. Moreover, DNA is negatively charged as is the outer cell surface so that a natural repulsion of DNA from the membrane occurs. For these reasons two pulses are used to insert DNA (or RNA) into a cell. In the first stage a typical short, high voltage pulse is applied in order to open pores. In the second stage a lower voltage pulse is applied for a longer time to drive the negatively charged DNA plasmid into the cell membrane. This electrophoretic push will be effective only at the side of the cell facing the negative electrode. Use of bipolar pulses allows the entry of the DNA into both sides of the cell. At the end of the pulses the electric field part of the process is over with the DNA still in the membrane. At this point a process called endocytosis occurs in which the cell forms a vesicle around the plasmid and then draws it into the cell interior. The plasmid does not then passively diffuse into the cell nucleus but is actively transported by intracellular mechanisms. The endocytosis and active transport processes require hours after the end of the pulse for the plasmid to enter the nucleus and be expressed. The details of these processes are not well understood [9].

EGT is used clinically for vaccination or cancer treatment [9]. In typical vaccinations a molecule (antigen) is injected into muscle to produce a specific immune response by the body. In EGT vaccinations a plasmid that encodes the antigen is injected. Cell nuclei incorporate the plasmid and then produce the antigens that provoke the immune response. Clinical trials have been successfully carried out for conditions such as HIV and hepatitis B and C viruses. For cancer treatment the plasmids are injected into the tumors. Several strategies for the gene expression can be used to produce products that promote, for example, the stimulation of cell suicide, the activation of an immune response, or the suppression of blood vessel growth. EGT can be combined with ECT to provide a two-stage attack on the tumors.

### 4.3. Nanoporation

Recall from Section 3 that for times much less that the membrane time constant, $\tau$, the electric field penetrates the cell interior. It may then be possible to porate the cell nucleus and other subcellular structures. Because the size of those structures is much smaller than the cell itself, according to Eq. (2) correspondingly larger electric fields must be applied to do so. In nanoporation electric fields on the order of 1–10 MV/m are applied for times on the order of nanoseconds (ns). The actual pulse duration is important in determining where the nanopulses have their greatest effect. According to Eq. (2) with $\tau \sim 1$ μs, pulse durations of 10 ns or less will produce negligible changes in TMVi. Thus, the electric field is large in the cell interior and can porate interior structures. Conversely, pulse durations greater than 100 ns will produce a large membrane polarization and a reduced internal field. The application of very short pulses and their analyses are complicated so that reproducible results using such pulses are difficult to achieve. Pulses with duration in the range 11 to 100 ns are the most effective in producing reproducible results in the cell interior such as apoptosis (cell suicide) and release of calcium from intracellular stores [35].

## 5. Mechanisms and experiments

We present here an abbreviated description of the research we have conducted [36, 37] to determine the mechanism by which cells initially detect physiological strength electric fields. Identifying and understanding this mechanism is important for further developing the various applications described above. We also present an abbreviated description of some experiments that support this identification.

### 5.1. How cells detect electric fields

As described above, DC and low-frequency electric fields produce a wide variety of biological effects at the cellular and tissue levels. Although the emphasis here is on cell migration, wound healing and neural stimulation, electric fields produce a wide variety of other effects on cells as described in the CRC review [1]. Once a cell initially detects the field that information is transmitted throughout the cell to produce a wide variety of biochemical effects. A major question is—what is the initial transduction mechanism by which the cell detects the field?

As noted above, for DC fields and AC frequencies below about 1 MHz, applied electric fields cannot penetrate the plasma membrane because there is sufficient time for charge redistribution in the cytoplasm to essentially cancel the applied field. For DC fields, then, the initial transduction must take place at the plasma membrane or just beyond it. The transduction process must also transmit this information to the cell interior. Three mechanisms have been proposed for this initial transduction process: (1) electrodiffusion/osmosis [38, 39], (2) opening of voltage-gated channels [40], and (3) electromechanical torque exerted on the glycocalyx [41]. The electrodiffusion/osmosis model requires fields that are stronger than typical physiological strength fields applied for relatively long time intervals, and the changes in transmembrane potential produced by such fields are insufficient to open voltage-gated channels [37].

The fundamental principle of the electromechanical model is that electric fields and fluid shear forces share a common transduction mechanism—the production of torques on transmembrane glycoproteins. In the fluid force model [42] fluid shears on the glycocalyx core are transmitted to the cytoskeleton as forces that are then communicated as mechanical signals throughout the cell interior to activate biochemical signaling pathways. The glycocalyx is the carbohydrate rich zone on the cell surface [43]. It covers the surface of all eukaryotic cells, including cancer and stem cells [44]. Many of the glycoproteins comprising the glycocalyx are negatively charged, particularly those containing sialic acid [45]. Moreover, when placed in an electric field applied parallel to the cell surface, the negatively-charged glycocalyx experiences an electrical force tangential to the surface, analogous to that applied by a fluid shear. We applied this concept to the structural model of Weinbaum et al. [42] in which forces exerted on the glycocalyx brush structure are transmitted to the cytoskeleton and thus throughout the cell. Specifically, these forces produce an electromechanical torque about the cytoskeletal junction point and, thus, a force on the cytoskeleton itself [41, 46]. The magnitude of this cytoskeletal force is comparable to those of mechanical forces known to produce physiological effects [41]. A more complete description of this model is provided below.

The predictions of the electromechanical model are consistent with the galvanotaxis of keratinocytes [36] and amoeba [37]. Galvanotaxis serves as a convenient effect for comparing the three proposed mechanisms because it is a process that is readily reproduced and is easily measured in real time. One can record the effect of the field as it occurs rather than wait for the results of a biochemical analysis. The same initial transduction process should be present for a variety of other field effects. The model successfully predicted that the superposition of a 40 V/m, 1.6 Hz sinusoidal signal on a 100 V/m DC field would reduce directionality whereas the superposition of a 40 V/m, 160 Hz field would not reduce directionality compared to a pure 100 V/m DC field. Further confirmation of the electromechanical model, which identifies the glycocalyx as the transduction site, has been provided by Finkelstein et al. [47] who showed that removal of the major negatively charged molecules in the glycocalyx in 3 T3 and HeLa cells eliminated galvanotaxis. Moreover, the effect of the superimposed fields on cell motility (speed) is also explained by the model [37]. An increase in cell motility often, but not always, accompanies the increase in directionality toward the field. The applied field produces a bending torque on the negatively-charged glycocalyx between the cell and the substrate. The glycocalyx is bent away from the substrate, increasing the glycocalyx-substrate separation and increasing cell adhesion. As a result, cell motility can increase. However, if the cell adhesion is already optimal, the field does not produce a significant additional motility increase. The addition of positively-charged calcium ions that bind to the negatively-charged glycocalyx reduced the increase in directionality and motility produced by the field on amoeba—providing further confirmation of the electromechanical model [37]. The details are presented below.

Because similar behavior is exhibited by such different cell types, we surmise that this field-detection mechanism must have been present in early cells. We suggest [37] that this glycocalyx-bending mechanism was initially used to detect fluid flow changes. Because that detection system happened to be negatively charged, it could also detect electric fields. With the evolution of multicellular organisms, this ability might have become useful in controlling some developmental processes [4]. Other field sensing processes then followed.

### 5.2. Details of the mechanical model

#### *5.2.1. Directionality increase*

The first mechanism by which an electric field is detected by a cell involves the production of an electromechanical torque on the glycocalyx by the field that is analogous to the torque produced by fluid shear. Although a glycocalyx covers the surface of all eukaryotic cells [44], detailed modeling of its functionality has been carried out primarily for endothelial cells. For this reason the parameters used in the modeling rely on parameters for the glycocalyx of endothelial cells and their components. In their review of the structure of the endothelial glycocalyx, Curry and Adamson [48] describe an inner layer with a thickness of 100–150 nm that possesses a quasi-periodic structure. Beyond that layer there is a region out to 400 nm that is required for fluid shear detection. The fluid shear stress applied toward the top of that region is transmitted to the more rigid inner layer and then transferred to the membrane as a solid mechanical force [44].

**Figure 4** illustrates the basic electromechanical model [36]. It shows the forces acting on a charged, cylindrical glycoprotein that is connected at its base to the cytoskeleton. $F_{elec} = QE(t)$ is the electrical force exerted on the effective charge, Q, of the glycoprotein by the applied electric field, E(t), at time t. Because the glycocalyx is negatively charged, the force, F, exerted on it by the field, E, is opposite in direction to E. $F_{drag} = -cd\theta(t)/dt$ is the viscous force exerted on the rod by the surrounding extracellular fluid. The angular displacement of the rod, θ (t), is determined by torque balance [46] and c is the frictional drag coefficient. For a rod $c = 2\pi L'^2/$

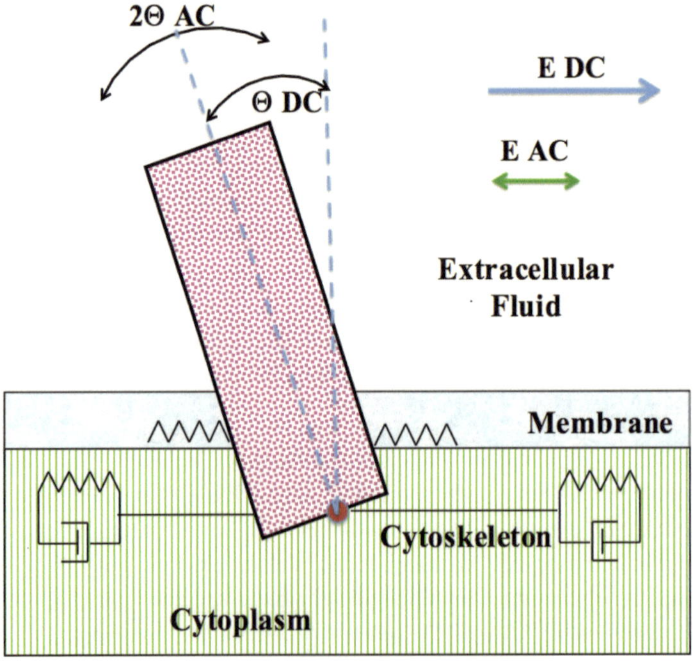


**Figure 4.** Schematic diagram of the forces acting on a glycocalyx rod under the application of an electric field.

(ln[L′/r] − 0.447) where L′ is the length of the rod and r is its radius [41]. $F_{memb} = -kh\theta$ (t) is the harmonic restoring force exerted on the oscillating rod by the cell membrane where k is the force constant and h is the distance from the center of the membrane to the base connection point with the cytoskeleton [41]. $F_{base}$ is the force exerted by the cytoskeleton on the rod at the connection point. According to Newton's Third Law, the force exerted on the cytoskeleton by the rod is $-F_{base}$. Newton's Second Law states the net force, $F_{net}$, acting on the rod is

$$F_{net} = Ma = F_{elec} + F_{drag} + F_{memb} + F_{base} \tag{3}$$

where M is the mass of the rod and a is the acceleration of its center of mass. Here a = (L′/2) $d^2\theta/dt^2$. Solving, one has

$$F_{base} = (ML/2)\, d^2\,\theta(t)/dt^2 - QE(t) + cd\theta(t)/dt + kh\theta(t) \tag{4}$$

The spring/dashpot combinations connected to the base in **Figure 4** indicate the transmission of the resulting longitudinal mechanical signals along the cytoskeleton to the rest of the cell. The angular displacement illustrates the situation in which the applied field has both DC and AC components. For a purely DC field θ AC and E AC are zero.

Application of the above model to the array of 27 glycocalyx rods in Weinbaum's glycocalyx model [42] yields a force applied to the cytoskeleton on the order of 0.5 pN for an applied electric field of 100 V/m. Such a force is comparable to some intracellular mechanical forces [41]. Hence, it is plausible that forces predicted by the electromechanical model would have physiological effects.

The viscous drag force exerted on the rod by the surrounding extracellular fluid increases linearly with the frequency of the applied field. The importance of this frequency dependence becomes apparent when the AC field is superimposed on a pure DC field. The drag exerted on the rods for the 1.6 Hz field is very small so that the total force (AC + DC) on the cytoskeleton is strongly modulated. The drag force for the 160 Hz field is 100 times larger so that the resulting modulation is negligible, and the total force is essentially identical to that produced by a pure DC field [41]. We suggest that the strong modulation at 1.6 Hz introduces significant mechanical signaling noise in the cytoskeleton that reduces the overall transduction relative to a pure DC field. No such reduction would be produced by the superimposed 160 Hz field. The electromechanical model is thus consistent with the results presented for keratinocytes [36] and for amoeba [37].

### *5.2.2. Motility increase*

The mechanism described above cannot explain the increase in speed produced by the applied fields, DC and/or AC. Cell motility is determined in part by the adhesion of the cell to the substrate [49]. Cells that are either too tightly or too loosely bound will have lower motility than those with an intermediate adhesion. The adhesion is reduced by the negatively charged glycocalyx [50, 51]. Increasing the distance between the glycocalyx and the substrate should then increase adhesion and increase motility.

The model described above, regarding directionality involves the interaction of the field with the relatively rigid inner glycocalyx. The model described here, regarding motility increase, involves the interaction with the relatively flexible outer glycocalyx. Recent measurements have shown that the entire glycocalyx for endothelial cells may extend out to 11,000 nm with a loose fibril mesh of elongated elements [52].

Amoeba have a negatively-charged glycocalyx [53], the full extent of which is not well known. Electron micrographs by Topf and Stockem [54] showed that the glycocalyx consists of a compact base layer that is approximately 70 nm thick with an outer, filamentous layer that is approximately 400 nm thick. As noted above, recent advances in microscopic techniques indicate that a typical glycocalyx thickness is on the order of several thousand nanometers or more. Grebecki et al. [55] showed that *Amoeba proteus* adheres to the substrate using discretely-spaced "minipodia" that are about 500 nm thick and up to 8000 nm long. Hence the outer glycocalyx between the amoeba's cell body and the substrate is not compressed but should be relatively free to reorient itself in an applied field.

Cruz-Chu et al. [56] used a molecular-dynamics simulation to investigate the effects of fluid shear on a model glycocalyx. They found that an applied shear of 0.47 MPa produced dramatic bending of the glycocalyx (cf. their Figure 5b). Stresses of about 3 MPa and greater significantly disrupted the glycocalyx. Although their model was applied primarily to the inner glycocalyx, similar bending should also be produced in the less rigid outer glycocalyx that interacts with the substrate.

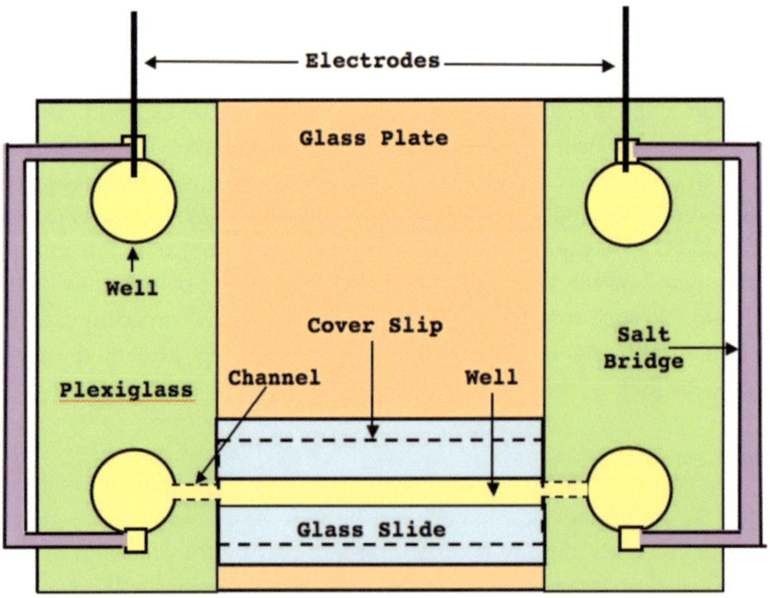


**Figure 5.** Schematic diagram of the system used to expose amoeba to an applied electric field.

We can estimate the tangential shear produced by a physiological strength electric field. For simplicity we model the amoeba glycocalyx as a uniform, negatively-charged, rectangular-solid gel that fills the space between the amoeba body and the substrate. We assume that the amoeba is already elongated in the direction of the field. The length, L, and width, W, of the rectangular solid are those of the amoeba which we take to be L = 0.2 mm and W = 0.05 mm. We take the thickness of the glycocalyx to be the length of the amoeba minipodia, T = 8 μm [55]. A typical estimate for the charge density of the glycocalyx is $\rho = 25$ mEq/l or $2.5 \times 10^6$ C/m$^3$ [57]. The electric force on this model glycocalyx is then $F = \rho LWTE$. The shear stress, $\sigma$, is that force divided by the cross-section to which it is applied or

$$\sigma = \rho LWTE/WT = \rho LE \qquad (5)$$

For an electric field of 200 V/m, $\sigma$ is 0.1 MPa. Our calculated shear stress produced by the field should then be sufficient to bend the glycocalyx. Bending the glycocalyx increases its separation from the substrate. Increasing the separation increases the cell's adhesion and, thus, its motility.

The response to a shear stress depends on the mechanical properties of the glycocalyx. The shear stress modulus and the shear loss modulus of the glycocalyx increase with the frequency of the applied stress [58]. Consequently the bending produced by a 160 Hz electric field should be less than that produced by a 1.6 Hz field. As a result, the increase in motility should be less for a 160 Hz field that for a 1.6 Hz field, as reported in Ref. [37].

### 5.3. Galvanotaxis experimentation to validate the model

#### *5.3.1. Experimental setup*

Amoeba were purchased from Carolina Biological Company (Burlington, NC, USA), and cultured in the container in which they were shipped for at least 3 days before they were used in experiments. The details of the design and construction of the apparatus (**Figure 5**) used for all the experiments are found in Hart and Palisano [37]. Briefly, the amoeba cells were isolated from the wells containing the Pt wire source of current and placed in the trough that we connected to the wells to which the electrical current was introduced by salt bridges.

Because of the complex geometry it was necessary to model numerically the field inside the trough for a given voltage applied to the wells using COMSOL Multiphysics (COSMOL Inc., Burlington, MA, USA). COSMOL Multiphysics was also used to model the temperature increase of electric fields during the 40-minute time interval of experiments and indicated that the temperature increase produced was than 1°C. The field and temperature calculations were confirmed by measurements with a voltmeter and thermocouple probe, respectively.

In all experiments, each amoeba served as it owns control. During the first 20 minutes of recording, the movement of the amoeba cells occurred in the absence of an electric field. In the last 40 minutes an electric field was applied. Cell movements were recorded for the entire 60-minute experiment and snapshots were taken from the movie at 1-minute intervals and converted to a

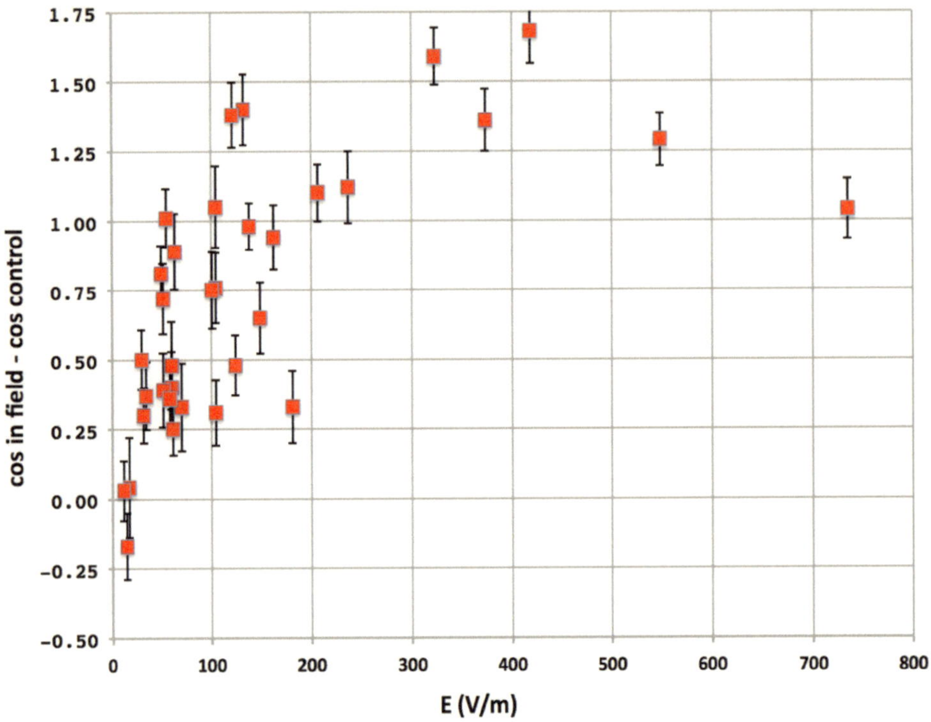


**Figure 6.** Increase in directionality in the DC field: Cosine during the last 10 minutes in the field minus the cosine during the last 10 minutes of control for each individual cell, averaged over all the cells. Error bars represent the standard error of the mean.

time-lapse movie. The positions of each amoeba at each 1-minute interval were digitized from the movie using LoggerPro (Venier Software & Technology, Beaverton, OR, USA) software.

*5.3.2. Galvanotaxis results*

We applied DC electric fields ranging from 15 V/m to 800 V/m. The number of cells tracked in an experiment typically ranged between 25 and 35. The position of each amoeba is tracked at 1-minute intervals during a 20-minute control period during which no field is applied and in the subsequent 40-minute interval during which a field is applied. During each 1-minute interval, an amoeba travels a distance, d. The sum of all these distances over a particular interval of time is the total distance traveled, D, during that interval. The average speed or motility of an individual amoeba in each interval is D/T where T is the duration of the interval, here 10 minutes. We consider the D/T ratio for the last 10 minutes of control as the control motility and that ratio for the final 10 minutes of the field as the final motility. The field is directed along the x-axis. During any 10-minute interval the distance traveled by an amoeba in the direction of the field is $\Delta x$. The orientation of the amoeba movement relative to the field is parameterized by angle $\Phi$ where $\cos \Phi = \Delta x/D$. We refer to $\cos \Phi$ as the "directionality" of the amoeba. A value of $\cos \Phi = +1$ indicates migration parallel to the field whereas −1

http://dx.doi.org/10.5772/intechopen.71683

represents migration directly opposite the field. We consider the cosine for the last 10 minutes of control as the control directionality and the cosine for the final 10 minutes of the field as the field directionality.

**Figure 6** compares the difference of the field and control directionalities of an individual cell, averaged over all the cells, as a function of the applied field. Hence, each cell serves as its own control. As evidenced by the increase in directionality, the cells begin to detect fields greater than about 30 V/m. The increase in directionality improves as the field increases and becomes relatively steady for fields above about 200 V/m. There is noticeable variation in the increases for low fields, but the overall trend is apparent. For weaker fields,

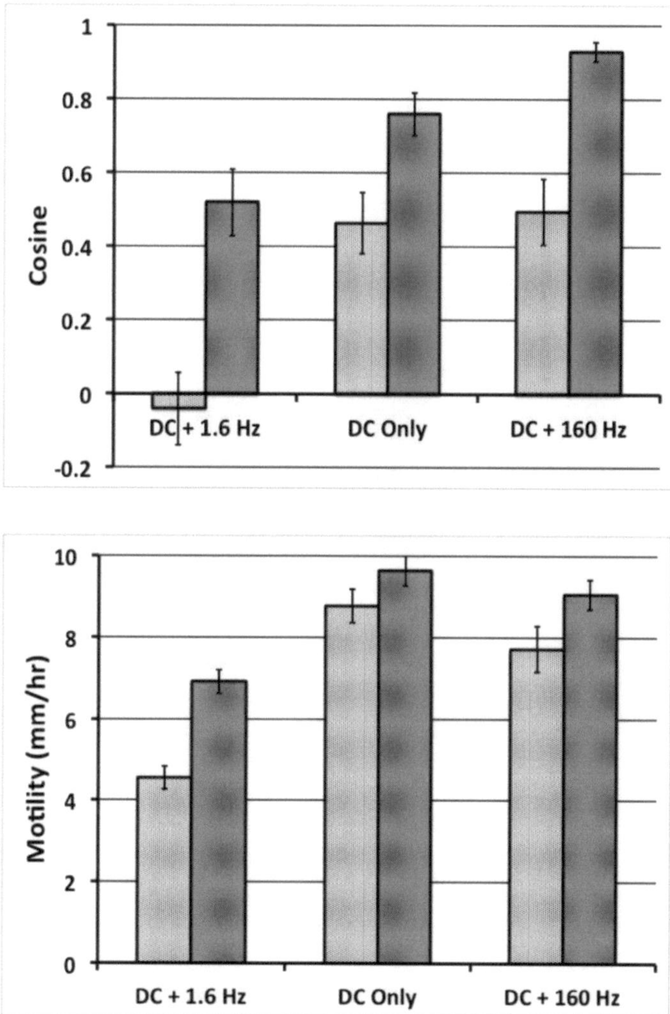


**Figure 7.** The effect on directionality and motility of superimposing a smaller AC field on a DC field. In each case the DC field was about 65 V/m and the AC field was about 32 V/m. The light shading indicates values obtained during the first 10 minutes the field was applied; the dark shading, the last 10 minutes. Error bars represent the standard error of the mean. (A) Illustrates the differences in directionality; (B) the differences in speed.

nearly 40 minutes are required to detect an increase in directionality. Thus, comparisons of directionality increase for different applied fields are made for the last 10 minutes of field application.

Measurement of cell motility indicated that cells with low control motility increased their speeds much more dramatically than did cells with a high initial motility [37]. A clear motility increase of about 50% began above about 60 V/m and improved out to about 200 V/m beyond which field there was considerable variability. The motility increase became smaller with greater control motility. There was no clear, corresponding variation of the directionality increase with control motility.

One of the goals of this research is to examine how widespread is the applicability of the electromechanical model. Hart et al. [36] demonstrated that it described well the loss of directionality of keratinocytes when a weaker, low-frequency AC field was superimposed on a stronger DC field. **Figure 7** indicates that a similar response was observed in *Amoeba proteus*, a free-living cell. The directionality and motility of amoeba were compared when exposed to a 65 V/m DC field alone, a 32 V/m, 1.6 Hz AC field superimposed on a 65 V/m DC field or a 32 V/m, 160 Hz AC field superimposed on a 65 V/m DC field. In each case the average electric field was 65 V/m DC. The three sets of cells all came from the same shipment and were thus exposed to similar conditions prior to the experiments.

**Figure 7A** shows that directionality is eliminated for the added 1.6 Hz field during the first 10 minutes, but is comparable and significant for the pure DC and added 160 Hz fields. The directionality is improved for all three cases during the final 10 minutes. Although greatly improved, the directionality for the added 1.6 Hz case is still less than the other two cases. These results are similar to those obtained for keratinocytes [36]. **Figure 7B** illustrates that the motility during the first 10 minutes for the added 1.6 Hz case is much lower than for the other applied fields. As with directionality, the motility for the added 1.6 Hz case has improved for the last 10 minutes, but is still less than the motilities of the other two cases. These results are consistent with the motility model described above.

## 6. Numerical simulations

The applicability of the methods described above requires knowledge of the detailed electric field distribution at the site of its application. Because the field distribution depends on the electrical properties and physical shapes of the various cells and tissues involved, numerical methods must be used for the calculation [59]. These methods have evolved considerably over the years. For example, spreadsheet cells can be used to represent small volume elements in a limb. The value of the cell is the electrical potential of the element, Vo. That value is related to the potential of the six surrounding cells in the three-dimensional model. Setting the current entering the cell equal to the current leaving relates Vo to the potentials of the surrounding cells via the inter-cell resistivities. The electrical anisotropies of the tissue can be introduced in this manner. The fundamental equations can be introduced into many similar cells simultaneously and solved by the method of successive over-relaxation in an Excel spreadsheet. This approach was used to determine how the electric field at the site of a tibia fracture evolved as the injury healed during an electrical fracture-healing treatment [60]. The method can even be extended to calculate the electric field distribution induced in a complex distribution of cells in a culture dish [61].

More recently, commercial software programs are used to determine more precisely and rapidly the field distribution in complicated situations. As noted above in Section 5.3.1, COMSOL Multiphysics (COSMOL Inc., Burlington, MA, USA) was used to determine both the electric field and temperature distributions in our experimental setup. Electrode placement for the administration of electroporation pulses requires the detailed numerical analysis of the resulting fields at and surrounding the tumor site [8]. Full-body animal and human virtual models are now available for very detailed electric field computations [62]. The models contain information regarding the electrical, thermal and flow properties of the various tissues. In this way a comprehensive analysis of the field distribution and its associated physical effects can be determined for various electrode placements.

## 7. Conclusions

Electric fields have a wide variety of applications in biology and medicine. Physiological strength fields are used to improve the healing of wounds, the stimulation of neurons and the positioning and activation of cells on scaffolds for tissue engineering purposes. The brief, strong pulses used in electroporation are used to improve the insertion of drugs into tumors and DNA into cell nuclei. Numerical simulations must be used to select the proper field strengths to be applied in a clinical setting. Some fundamental issues are still being addressed; in particular, how do cells actually detect the fields? The electromechanical model described here is consistent with the experimental evidence whereas other models are not.

## Author details

Francis X. Hart[1]* and John R. Palisano[2]

*Address all correspondence to: fhart@sewanee.edu

1 The Department of Physics, The University of the South, Sewanee, TN, USA

2 The Department of Biology, The University of the South, Sewanee, TN, USA

## References

[1] Pullar CE, editor. The Physiology of Bioelectricity in Development, Tissue Regeneration, and Cancer. Boca Raton, FL: CRC Press; 2011. 342 p

[2] McCaig CD, Rajnicek AM, Song B, Zhao M. Controlling cell behavior electrically: Current views and future potential. Physiological Reviews. 2005;**85**:943-978. DOI: 10.1152/physrev.00020.2004

[3] Robinson KR, Messerli MA. Left/right, up/down: The role of endogenous electric fields as directional signals in development, repair and invasion. BioEssays. 2003;**25**:759-766. DOI: 10.1002/bies.10307

[4] Levin M. Molecular bioelectricity: How endogenous voltage potentials control cell behavior and instruct pattern regulation in vivo. Molecular Biology of the Cell. 2014;**25**:3835-3850. DOI: 10.1091/mbc.E13-12-0708

[5] Meng S, Rouabhia M, Zhang Z. Electrical stimulation modulates osteoblast proliferation and bone protein production through heparin-bioactivated conductive scaffolds. Bioelectromagnetics. 2013;**34**:189-199. DOI: 10.1002/bem.21766

[6] Hart FX. Bioimpedance in the clinic. Zdravstveni Vestnik. 2009;**78**:782-790

[7] Pullar CE, Isseroff RR. Cyclic AMP mediates keratinocyte directional migration in an electric field. Journal of Cell Science. 2005;**118**:2023-2034. DOI: 10.1242/jcs.02330

[8] Yarmush ML, Golberg A, Sersa G, Kotnik T, Miklavic D. Electroporation-based technologies for medicine: Principles, applications and challenges. Annual Review of Biomedical Engineering. 2014;**16**:295-320. DOI: 10.1146/annurev-bioeng-071813-104622

[9] Rosazza C, Meglic SH, Zumbusch A, Rols M-P, Miklavcic D. Gene electrotransfer: A mechanistic perspective. Current Gene Therapy. 2016;**16**:98-129

[10] Pullar CE, Isseroff RR, Nuccitelli R. Cyclic AMP-dependent protein kinase a plays a role in directed migration of human keratinocytes in a dc electric field. Cell Motility and the Cytoskeleton. 2001;**50**:207-217

[11] Nuccitelli R. A role for endogenous electric fields in wound healing. Current Topics in Developmental Biology. 2003;**58**:1-26

[12] Kloth LC. Electrical stimulation technologies for wound healing. Advances in Wound Healing. 2014;**3**:81-90. DOI: 10.1089/wound.2013.0459

[13] Pullar CE. The biological basis for electric stimulation as a therapy to heal chronic wounds. Journal of Wound Technology. 2009;**N6**:20-24

[14] Hoare JJ, Rajnicek AM, McCaig CD, Barker RN, Wilson HM. Electric fields are novel determinants of human macrophage functions. Journal of Leukocyte Biology. 2016;**99**:1141-1151. DOI: 10.1189/jlb.3A0815-390R

[15] The Parkinson Foundation. Deep Brain Stimulation http://www.parkinson.org/understanding-parkinsons/treatment/surgery-treatment-options/Deep-Brain-Stimulation

[16] Ho CH et al. Functional electrical stimulation and spinal cord injury. Physical Medicine and Rehabilitation Clinics of North America. 2014;**25**:631-ix. DOI: 10.10.1016/j.pmr.2014.05.001

[17] Rejc A, Angeli C, Harkema S. Effects of lumbosacral spinal cord epidural stimulation for standing after chronic complete paralysis in humans. PLoS ONE; **10**(7): e0133998. https://doi.org/10.1371/journal.pone.0133998

[18] Solopova IA, Sukhotina IA, Zhvansky DS, Ikoeba GA, Vissarionov SV, Baindurashvili AG, Edgerton VR, Gerasimenko YP, Moshonkina TR. Effects of spinal cord stimulation on motor functions in children with cerebral palsy. Neuroscience Letters. 2017;**639**:192-198. DOI: 10.1016/j.neulet.2017.01.003

[19] DeSantana JM, Walsh DM, Vance C, Rakel BA, Sluka KA. Effectiveness of transcutaneous electrical nerve stimulation for treatment of hyperalgesia and pain. Current Rheumatology Reports. 2008;**10**:492-499

[20] Hart FX, Evely K, Finch C. The use of a spreadsheet program to calculate the electric field/current distributions induced in irregularly-shaped inhomogeneous biological structures by low-frequency magnetic fields. Bioelectromagnetics. 1993;**14**:161-172

[21] Griffin M, Bayat A. Electrical stimulation in bone healing: Critical analysis by evaluating levels of evidence. Eplasty. 2011;**11**:303-353

[22] Aleem IS, Aleem I, Evaniew N, Busse JW, Yaszemski M, Agarwal A, Einhorn T, Bhandari M. Efficacy of electrical stimulators for bone healing: A meta-analysis of randomized sham-controlled trials. Scientific Reports. 2016;**6**:31724. DOI: 10.1038/srep31724

[23] Ueno S, Sekino M. Biomagnetics: Principles and Applications of Biomagnetic Stimulation and Imaging. Boca Raton, FL: CRC Press; 2017. 342 pp

[24] Miklavcic D, Pavselj N, Hart FX. Electric properties of tissues. In: Akay M, editor. Wiley Encyclopedia of Biomedical Engineering. Vol. 6. New York: Wiley. 2006. p. 3578-3589

[25] Radisic M, Park H, Shing H, Consi T, Schoen FJ, Langer R, Freed LE, Vunjak-Novakovic G. Functional assembly of engineered myocardium by electrical stimulation of cardiac myocytes cultured on scaffolds. PNAS. 2004;**101**:18129-18134. www.pnas.org/cgi. DOI: 10.1073/pnas.0407817101

[26] Weng B, Liu X, Shepherd R, Wallace GG. Inkjet printed polypyrrole/collagen scaffold: A combination of spatial control and electrical stimulation of PC12 cells. Synthetic Metals. 2012;**162**:1375-1380. DOI: 10.1016/j.synnthmet.2012.05.022

[27] Bodhak S, Bose S, Kinsel WC, Bandyopadhyay A. Investigation of *in vitro* bone cell adhesion and proliferation on Ti using direct current stimulation. Materials Science and Engineering. 2012;**C32**:2163-2168. DOI: 10.1016/j.msec.2012.05.032

[28] Hoi Ting HA, Cheng I, Chowdhury MF, Radisic M. Interactive effects of surface topography and pulsatile electrical field stimulation on orientation and elongation of fibroblasts and cardiomyocytes. Biomaterials. 2007;**28**:4277-4293. DOI: 10.1016/j.biomaterials.2007.06.001

[29] Rajnicek AM, Foubister LE, McCaig CD. Alignment of corneal and lens epithelial cells by co-operative effects of substratum topography and DC electric fields. Biomaterials. 2008;**29**:2082-2095. DOI: 10.1016/j.biomaterials.2008.01.015

[30] Arinzeh TL, May J, Huang GP. Structural support for damaged tissue repair. American Scientist. 2017;**105**:298-305

[31] Hronik-Tupaj M, Kaplan DL. A review of the responses of two and three-dimensional engineered tissues to electric fields. Tissue Engineering: Part B. 2012;**18**:167-180. DOI: 10.1089/ten.teb.2011.0244

[32] Rems L, Miklavcic D. Tutorial: Electroporation of cells in complex materials and tissue. Journal of Applied Physics. 2016;**119**:201101-1-21. DOI: 10.1063/1.4949264

[33] Miklavcic D, Davalos RV. Electrochemotherapy (ECT) and irreversible electroporation (IRE)–Advanced techniques for treating deep-seated tumors based on electroporation. BioMedical Engineering Online. 2015;**14**(Suppl 3):I1

[34] Roustit M, Blaise S, Cracowski J-L. Trials and tribulations of skin iontophoresis in therapeutics. British Journal of Clinical Pharmacology. 2013;**77**:63-71. DOI: 10.1111/bcp.12128

[35] Napotnik TB, Rebersek M, Vernier PT, Mali B, Miklavcic D. Effects of high voltage nanosecond electric pulses on eukaryotic cells (*in vitro*): A systematic review. Bioelectrochemistry. 2016;**110**:1-12. DOI: 10.1016/j.bioelectrochem.2016.02.011

[36] Hart FX, Laird M, Riding A, Pullar CE. Keratinocyte galvanotaxis in combined DC and AC electric fields supports an electromechanical transduction sensing mechanism. Bioelectromagnetics. 2013;**34**:85-94. DOI: 10.1002/bem.21748

[37] Hart FX, Palisano JR. Glycocalyx bending by an electric field increases cell motility. Bioelectromagnetics. 2017;**38**:482-493. DOI: 10.1002/bem.22060

[38] Jaffe LF, Nuccitelli R. Electrical controls of development. Annual Review of Biophysics and Bioengineering. 1977;**6**:445-476

[39] McLaughlin S, Poo M-M. The role of electro-osmosis in the electric-field-induced movement of charged macromolecules on the surface of cells. Biophysical Journal. 1981; **34**:85-93

[40] Djamgoz MB, Mycielska M, Madeja Z, Fraser SP, Korohoda W. Directional movement of rat prostate cancer cells in direct-current electric field: Involvement of voltage-gated Na+ channel activity. Journal of Cell Science. 2001;**114**:2697-2705

[41] Hart FX. Cytoskeletal forces produced by extremely low-frequency electric fields acting on extracellular glycoproteins. Bioelectromagnetics. 2010;**31**:77-84. DOI: 10.1002/bem.20525

[42] Weinbaum S, Zhang X, Han Y, Vink H, Cowin SC. Mechanotransduction and flow across the endothelial glycocalyx. Proceedings of the National Academy of Sciences. 2003;**100**:7988-7995 www.pnas.org/cgi/doi/10.1073/pnas.1332808100

[43] Alberts B, Johnson A, Lewis J, Raff M, Roberts K, Walter P. Molecular Biology of the Cell. 6th ed. New York: Garland Press; 2015. p. 582

[44] Tarbell JM, Shi Z-D. Effect of the glycocalyx layer on transmission of interstitial flow shear stress to embedded cells. Biomechanics and Modeling in Mechanobiology. 2013; **12**:111-121. DOI: 10.1007/s10237-012-0385-8

[45] Barker AL, Konopatskaya O, Neal CR, Macpherson JV, Whatmore JL, Winlove CP, Unwin PR, Shore AC. Observation and characterization of the glycocalyx of viable human endothelial cells using confocal laser scanning microscopy. Physical Chemistry Chemical Physics. 2004;**6**:1006-1011. DOI: 10.1039/b312189e

[46] Hart FX. The mechanical transduction of physiological strength electric fields. Bioelectromagnetics. 2008;**29**:447-455. DOI: 10.1002/bem.20411

[47] Finkelstein EI, Chao PG, Hung CT, Bulinski JC. Electric field-induced polarization of charged cell surface proteins does not determine the direction of galvanotaxis. Cell Motility and the Cytoskeleton. 2007;**64**:833-846. DOI: 10.1002/cm.20227

[48] Curry FE, Adamson RH. Endothelial glycocalyx: Permeability barrier and mechanosensor. Annals of Biomedical Engineering. 2012;**40**:828-839. DOI: 10.1007/s10439-011-0429-8

[49] Barnhart EL, Lee K-C, Keren K, Mogliner A, Theriot JA. An adhesion-dependent switch between mechanisms that determines motile cell shape. PLoS Biology. 2011;**9**:e1001059. DOI: 10.1371/journal.pbio.1001059

[50] Kolodziejczyk J, Klopocka WK, Lopatowska A, Grebecka L, Grebecki A. Resumption of locomotion by *Amoeba proteus* readhering to different substrata. Protoplasma. 1995; **189**:180-186

[51] Sabri S, Soler M, Foa C, Pierres A, Benoliel A-M, Bongrand P. Glycocalyx modulation is a physiological means of regulating cell adhesion. Journal of Cell Science. 2000; **11**:1589-1600

[52] Ebong EE, Macaluso FP, Spray DC, Tarbell JM. Imaging the endothelial glycocalyx in vitro by rapid freezing/freeze substitution transmission electron microscopy. Arteriosclerosis, Thrombosis, and Vascular Biology. 2011;**31**:1908-1915. DOI: 10.1161/ATVBAHA.111.225268

[53] Brewer JE, Bell LGE. Long-range electrostatic interactions between amoebae and anion-exchange particles. Experimental Cell Research. 1970;**61**:397-402

[54] Topf P-M, Stockem W. Protein and lipid composition of the cell surface complex from *Amoeba proteus* (Rhizopoda: Amoebida). European Journal of Protistology. 1996; **32**:156-170

[55] Grebecki A, Grebecka L, Wasik A. Minipodia, the adhesive structures active in locomotion and endocytosis of amoebae. Acta Protozoologica. 2001;**40**:235-247

[56] Cruz-Chu ER, Malafeev A, Pajarskas T, Pivkin IV, Koumoutsakos P. Structure and response to flow of the glycocalyx layer. Biophysical Journal. 2014;**106**:232-243. DOI: 10.1016/j.bpj.2013.09.060

[57] Chen B, Fu BM. An electrodiffusion-filtration model for effects of endothelial surface glycocalyx on microvessel permeability to macromolecules. Journal of Biomechanical Engineering. 2004;**126**:614-624. DOI: 10.1115/1.1800571

[58] Nijenhuis N, Mizuno D, Spaan JAE, Schmidt CF. Viscoelastic response of a model endothelial glycocalyx. Physical Biology. 2009;**6**:1-8. DOI: 10.1088/1478-3975/6/2/025014

[59] Hart FX. Investigation systems to study the biological effects of weak physiological electric fields. In: Pullar C, editor. The Physiology of Bioelectricity in Development, Tissue Regeneration and Cancer. Boca Raton, FL: CRC Press; 2011. pp. 17-38

[60] Hart FX. Changes in the electric field distribution at an injury site during healing under electrical stimulation. Journal of Bioelectricity. 1991;**10**:33-51

[61] Hart FX. Cell culture dosimetry for low frequency magnetic fields. Bioelectromagnetics. 1996;**17**:48-57

[62] IT'IS Foundation. EM Research. Virtual Population. https://www.itis.ethz.ch/virtual-population/

# Electric Field Features and Its Application for Air Gap Breakdown Voltage Prediction

Zhibin Qiu and Jiangjun Ruan

Additional information is available at the end of the chapter

http://dx.doi.org/10.5772/intechopen.72230

**Abstract**

Air gap breakdown voltage prediction by mathematical calculations instead of experiments has been a long sought goal in the area of high-voltage engineering. In this chapter, a prediction method is proposed based on the electric field features and support vector machine (SVM). Two sets of electric field features are defined on the shortest interelectrode path of sphere-sphere and rod (sphere)-plane gap to characterize their spatial structures, which can be extracted from the electric field calculation results by finite element method (FEM). A breakdown voltage prediction model is established by SVM, while the input parameters are the electric field features, and the output parameters are −1 and 1, respectively, characterizing withstanding and breakdown of an air gap under the applied voltage. The proposed method is used to predict the power frequency breakdown voltages of IEC standard sphere-sphere air gaps and the switching impulse discharge voltages of large sphere-plane air gaps. The prediction results coincide well with the experimental data, the mean absolute percentage error of the 260 sphere-sphere gaps is within 2% and that of the 16 sphere-plane gaps is 3.2%. The results preliminarily validate the validity and accuracy of the proposed method for air gap breakdown voltage prediction.

**Keywords:** electric field features, air gap discharge, breakdown voltage prediction, support vector machine (SVM), sphere-sphere gap, sphere-plane gap

## 1. Introduction

With the rapid development of numerical computation methods and calculation capability of computers, the multi-physics coupling analysis has been widely applied in structure design and condition assessment of electrical equipment [1–4]. The distributions of electric, magnetic, stress, and thermal field can be calculated by commercial software, and multi-physics problems in complex structures can be simulated using powerful numerical techniques, so as to guide the optimal design of machines and devices. However, up to now, the insulation design of high-voltage equipment still depends on experimental verification.

The electric field distributions of complex geometries can be simulated very well, that is because electric field can be defined according to Gauss's law and Maxwell's relation, with clear constitutive relation and governing equations [5, 6]. Several numerical methods were developed for electric field calculation, such as finite element method (FEM), finite volume method (FVM), boundary element method (BEM), and those combining the advantages of different methods, like the control volume-based finite element method (CVFEM) [7–9]. However, even though the electric field distribution can be calculated accurately, the insulation strengths of dielectrics cannot be calculated or predicted, that is because there are no governing equations to describe the discharge process. Hence, the insulation calculation or prediction can be viewed as the short board for virtual design and manufacturing of high-voltage electrical equipment.

The dielectrics used in electric power system include air, sulfur hexafluoride, transformer oil, electroceramics, and silicone rubber, etc. while the most commonly used is the air. Even though air discharge phenomena have been extensively studied for more than 100 years, both experimentally and theoretically, the air discharge theory is still imperfect, and therefore the air gap breakdown voltage prediction is still a great challenge and one of the most important issues to be solved for external insulation design of power transmission and transformation projects. The Townsend discharge theory [10], streamer discharge theory [11, 12], and leader discharge theory [13–16] lay the foundation for the analysis and interpretation of air discharge phenomena. These classical theories were summarized from experimental measurements and observations, and they can be used to describe air discharge process. However, due to the complexity and randomness of air discharge process, it is difficult to realize air gap discharge voltage calculation with different gap configurations or under different voltage waveshapes and atmospheric environment.

Currently, discharge tests are still the main method to obtain air gap breakdown voltages. Various empirical [17, 18] and semiempirical [19, 20] formulas have been summarized for applications. However, the experimental studies cannot exhaustively reproduce all the gap configurations in actual engineering [21]. The discharge tests are usually conducted on air gaps with typical electrodes, like the rod-plane gaps, to obtain the basic characteristics of long air gap discharge, and the fitting formulas obtained by the experimental results are used to predict the breakdown voltages of engineering gaps, such as the conductor-tower window gaps, by taking the gap factor into consideration. However, the gap factors of different gap configurations should be determined by time-consuming and costly full-scale tests. When generalized to other gap structures, the fitting results are with certain deviations.

In order to find an alternative to replace the discharge tests, many scholars are devoted to the studies of air discharge mechanisms and try to achieve air gap breakdown voltage prediction by theoretical calculation. Based on air discharge theories and some simplifications and assumptions, numerous physical models have been proposed to simulate the entire duration of the discharge, including the first corona inception, streamer propagation, leader inception and propagation, and final jump [22–30]. However, long air gap discharge researches are faced with some challenges, mainly including the inherent complexity of the discharge process and the scientific modeling, the limitations of observation and measurement of various physical quantities involved in the discharge process, and the uncertainty of the solutions obtained by these semiempirical models [31]. Hence, there are obvious errors between the computed results of these models and the experimental measurement results. At present, the research hotspot is still concentrated on further revealing the physical mechanisms of the complex discharge process, so as to establish more scientific simulation models, but the realization of discharge prediction still needs long-term studies.

The idea of predicting air gap breakdown voltage by some artificial intelligence algorithms has attracted interests of researchers for many years, but there are only a very few related studies and applications. The artificial neural network (ANN) [32, 33], fuzzy logic system [34, 35], and support vector machine (SVM) [36, 37] have been applied to predict the discharge voltages of air insulation gaps. In [36, 37], a method based on electric field features and SVM was proposed for discharge voltage prediction of air gaps, and it has been successfully applied to predict the breakdown voltages of air gaps with typical and atypical electrodes [38–40] and the corona onset voltages of rod-plane gaps, conductors, and valve hall fittings [41, 42]. Some features were extracted from the calculation results of the electric field distribution of an air gap to characterize its spatial structure, and the SVM was applied to establish the multi-dimensional nonlinear relationships between these features and the air gap discharge voltage. This method offers a possible way to achieve breakdown voltage calculation of air gaps, so as to guide the external insulation design of high-voltage electrical equipment.

Under a given applied voltage waveform and specific atmospheric environment, air gap breakdown voltage is determined by its structure, which can be characterized by the electrostatic field distribution. In previous studies, the electric field features were defined in the hypothetic discharge channel between two electrodes, on the surface of typical high-voltage electrode and on the shortest interelectrode discharge path, which contains different regions classified from the perspectives of volume, area, and line. However, the discharge channel and the electrode surface are difficult to be defined for air gaps with complicated configurations. Only the shortest interelectrode path can be defined for two-electrode air gaps with arbitrary structures. It is known that there is a one-to-one correspondence between the gap structure and the electric field distribution. If the distribution characteristics of a three-dimensional spatial electric field can be mapped to those along a one-dimensional path, it is possible to define some effective features on this path to characterize the gap structure and establish their relationships with the air gap breakdown voltage by SVM, so as to be applied for breakdown voltage prediction of air gaps with arbitrary structures.

In this chapter, the above ideas are achieved preliminarily. Two sets of electric field features are defined on the shortest path of sphere-sphere gaps and rod (sphere)-plane gaps, respectively, with the U-shaped curve and monotonously decreased curve of the electric field distribution along the shortest path. A prediction model is established by SVM, and the electric field features on the shortest path are taken as the input parameters, while the output parameters are −1 and 1, respectively, characterizing whether a gap will withstand or breakdown under a given applied voltage. This model is applied to predict the breakdown voltages of sphere-sphere air gaps and large sphere-plane air gaps, and the predicted results are compared with the experimental data given in IEC 60052 and other references to verify its validity and accuracy.

## 2. Electric field features

### 2.1. Electric field distributions along the shortest path

The sphere-sphere air gaps given in IEC 60052 [43] (or IEEE Std 4 [44]) and the large sphere-plane air gaps given in [45, 46] are taken as the samples for breakdown voltage prediction studies in this chapter. The schematic diagrams of the sphere-sphere and the sphere-plane gaps are shown in **Figure 1**, where $d$ is the gap distance. For sphere-sphere gaps, the sphere diameter $D$ ranges from 5 to 200 cm, and the experimental breakdown voltages of these gaps are given in [43, 44], with different sphere diameters and gap distances. For the large sphere-plane air gap, it is composed of the sphere electrode with a mounting rod tube and the grounded plane electrode. The experimental 50% discharge voltages of these gaps, with the sphere diameter ranging from 25 to 95 cm and the gap distance ranging from 2 to 5 m, are given in [45], under the standard switching impulse voltage.

Since the sphere-sphere and the sphere-plane air gaps are with axisymmetric structures, the two-dimensional axisymmetric models are established by ANSYS, a finite element analysis software, to calculate their electric field distributions. The high-voltage sphere electrodes are applied unit voltage 1 V, and the grounded sphere electrode or the grounded plane electrode is applied zero potential. Taking a sphere-sphere gap with $D = 10$ cm and $d = 3$ cm and a sphere-plane gap with $D = 45$ cm and $d = 3$ m as examples, the cloud charts of their electric field distributions are shown in **Figure 2**. It can be seen that the maximum electric field strength appears at the bottom of the high-voltage sphere electrode, both for the short sphere-sphere gap and the long sphere-plane gap.

Taking the 3 cm short sphere-sphere gaps and the 3 m long sphere-plane gaps with different sphere diameters as examples, their electric field distributions along the shortest path are

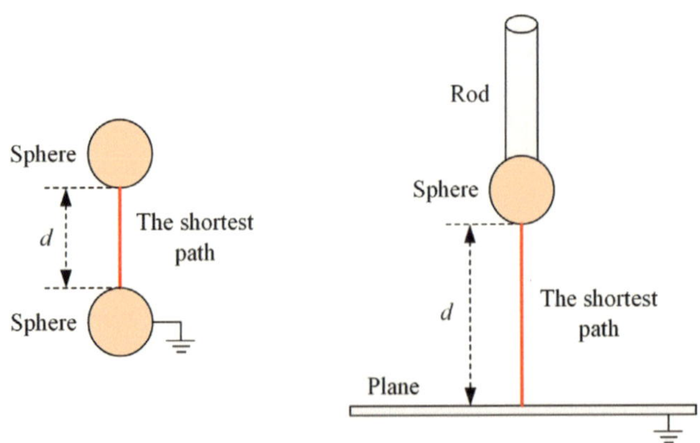


**Figure 1.** Schematic diagrams of sphere-sphere and large sphere-plane air gaps.

http://dx.doi.org/10.5772/intechopen.72230

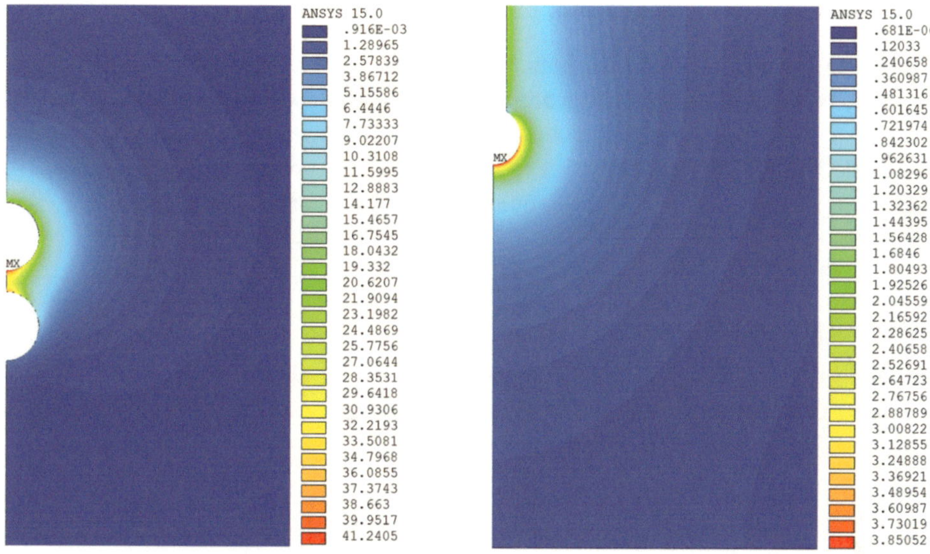


(a) Sphere-sphere gap ($D$=10 cm, $d$=3 cm)

(b) Sphere-plane gap ($D$=45 cm, $d$=3 m)

**Figure 2.** Electric field distribution cloud charts of sphere-sphere and sphere-plane gaps.

shown in **Figure 3**. For sphere-sphere gaps, the electric field distribution along the shortest path is a U-shaped curve, while the field strengths near two sphere electrodes are higher than those in the middle of the path. For sphere-plane gaps, the electric field distribution along the shortest path is a monotonously decreased curve, while the field strength gradually decreases along the path from the sphere electrode to the plane electrode. The field strength reduces quickly within 1 m from the sphere electrode, while on the path from position of 1 m to the plane electrode, the field strength changes a little. In addition, the maximum field strength decreases with the increase of the sphere diameter, both for sphere-sphere and sphere-plane gaps.

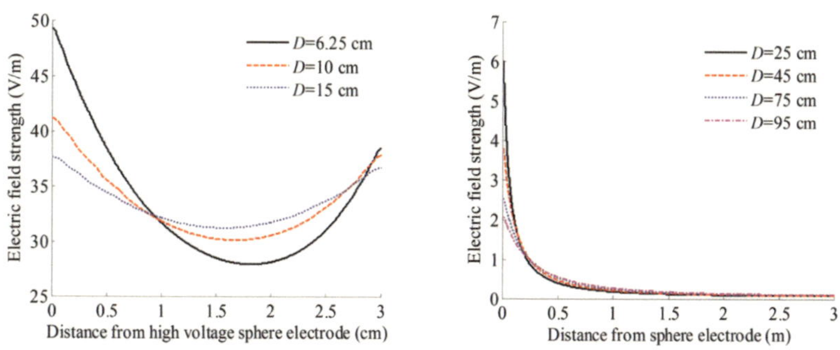


(a) Sphere-sphere gap ($d$=3 cm)

(b) Sphere-plane gap ($d$=3 m)

**Figure 3.** Electric field distributions along the shortest path of the short sphere-sphere gaps and long sphere-plane gaps.

### 2.2. Definitions of electric field features

According to the electric field distribution characteristics of sphere-sphere gap and sphere-plane gaps, two sets of electric field features are, respectively, defined for these two different gap types. These features mainly include the electric field strength, electric field gradient, square of electric field strength, electric field integral (electric potential), path length, and some scaling parameters related to the above quantities used to characterize the electric field inhomogeneity.

#### 2.2.1. *Sphere-sphere air gap*

Some equally spaced sampling points are selected along the shortest path, and the values of their electric field strength are extracted to calculate the features. The electric field features for sphere-sphere air gap are defined as follows:

(1) Electric field strength, including the maximum values ($E_{\text{hmax}}$ and $E_{\text{lmax}}$), respectively, on the surface of the high-voltage and grounded sphere electrode; the minimum value $E_{\text{min}}$ and the average value $E_{\text{a}}$; and the variance and standard deviation of the electric field strength along the shortest path ($E_{\text{std2}}$ and $E_{\text{std}}$):

$$E_{\text{hmax}} = \max E_i \quad (i = 1, 2, \cdots, m) \tag{1}$$

$$E_{\text{lmax}} = \max E_i \quad (i = m+1, m+2, \cdots, n) \tag{2}$$

$$E_{\text{min}} = \min E_i \quad (i = 1, 2, \cdots, n) \tag{3}$$

$$E_{\text{a}} = \sum_{i=1}^{n} E_i / n \tag{4}$$

$$E_{\text{std2}} = \frac{1}{n}\sum_{i=1}^{n} (E_i - E_{\text{a}})^2 = \frac{1}{n}\sum_{i=1}^{n} E_i^2 - E_{\text{a}}^2 \tag{5}$$

$$E_{\text{std}} = \sqrt{E_{\text{std2}}} = \sqrt{\frac{1}{n}\sum_{i=1}^{n} E_i^2 - E_{\text{a}}^2} \tag{6}$$

where $n$ is the number of the sampling points along the shortest path, $E_i$ is the electric field strength of the $i$th point, and $m$ is the point number whose electric field strength is the minimum value.

(2) Electric field gradient, including the maximum gradient $E_{\text{ghm}}$ on the path from the point of $E_{\text{hmax}}$ to that of $E_{\text{min}}$, the maximum gradient $E_{\text{glm}}$ on the path from the point of $E_{\text{min}}$ to that of $E_{\text{lmax}}$, and the average gradient $E_{\text{ga}}$ on the whole shortest path:

$$E_{\text{ghm}} = \max(|-\text{grad}E_i|) \quad (i = 1, 2, \cdots, m) \tag{7}$$

$$E_{\text{glm}} = \max(|-\text{grad}E_i|) \quad (i = m, m+1, \cdots, n) \tag{8}$$

$$E_{\mathrm{ga}} = \sum_{i=1}^{n} (|-\mathrm{grad}E_i|)/n \tag{9}$$

where | | is the absolute value sign.

(3) Square of electric field strength ($W$ and $W_{\mathrm{a}}$), respectively, means the integral of the field strength square on the shortest path and its expected value:

$$W = \int_0^d E^2 \mathrm{d}l \approx \sum_{i=1}^{n} E_i^2 d_i \tag{10}$$

$$W_{\mathrm{a}} = \frac{W}{d} \approx \frac{1}{n} \sum_{i=1}^{n} E_i^2 \tag{11}$$

where $d$ is the gap distance and $d_i$ is the length of each segment between two sampling points on the shortest path, $d_i = d/(n\text{-}1)$.

(4) Electric field integral, that is, electric potential, including the field integral $V_x$ on the path whose field strength exceeds $E_x = x\%E_{\mathrm{hmax}}$, the potential $V_{\mathrm{h}}$ between the high-voltage sphere electrode and the point of $E_{\mathrm{min}}$, and the potential $V_{\mathrm{l}}$ between the point of $E_{\mathrm{min}}$ and the grounded sphere electrode:

$$V_x = \int_{E_i \geq E_x} E_i \mathrm{d}l \approx \sum_{E_i \geq E_x} E_i d_i \tag{12}$$

$$V_{\mathrm{h}} = \int_{E_{\mathrm{min}} \leq E_i \leq E_{\mathrm{hmax}}} E_i \mathrm{d}l \approx \sum_{E_{\mathrm{min}} \leq E_i \leq E_{\mathrm{hmax}}} E_i d_i \tag{13}$$

$$V_{\mathrm{l}} = \int_{E_{\mathrm{min}} \leq E_i \leq E_{\mathrm{lmax}}} E_i \mathrm{d}l \approx \sum_{E_{\mathrm{min}} \leq E_i \leq E_{\mathrm{lmax}}} E_i d_i \tag{14}$$

(5) Path length $L_x$, including the length $L_{Ex}$ of the path whose electric field strength exceeds $E_x = x\%E_{\mathrm{hmax}}$, the length $L_{gx}$ of the path whose electric field gradient exceeds $E_{gx} = x\%E_{\mathrm{ghm}}$, the length $L_{Wx}$ of the path where the sum of the electric field strength square equals to $W_x = x\%W$, and the distance $L_{\mathrm{min}}$ from the high-voltage sphere electrode to the point of $E_{\mathrm{min}}$:

$$L_x = \sum_{i=1}^{p} d_i \tag{15}$$

where $p$ is the number of the points on the shortest path which meet the related conditions about the electric field strength, gradient, and square and $d_i$ is the length of each segment.

(6) Electric field inhomogeneity, characterized by some scaling parameters related to the above five kinds of features. For electric field strength, these parameters include the field distortion factor $E_{\mathrm{d}}$; the ratio of $E_{\mathrm{lmax}}$ to $E_{\mathrm{hmax}}$, namely, $E_{\mathrm{rlh}}$; and the ratio of $E_{\mathrm{min}}$ to $E_{\mathrm{hmax}}$, namely, $E_{\mathrm{rm}}$:

$$E_{\rm d} = E_{\rm hmax}/E_{\rm a} \tag{16}$$

$$E_{\rm rlh} = E_{\rm lmax}/E_{\rm hmax} \tag{17}$$

$$E_{\rm rm} = E_{\rm min}/E_{\rm hmax} \tag{18}$$

For electric field square, the scaling parameter is a ratio $W_{rx}$ of the sum of the electric field strength square on the path where $E_i \geq E_x = x\%E_{\rm hmax}$ to $W$:

$$W_{\rm rx} = \frac{\sum_{E_i \geq E_x} E_i^2 d_i}{W}, \quad E_x = x\%E_{\rm hmax} \tag{19}$$

For electric field integral, the scaling parameters are the ratios of $V_x$, $V_{\rm h}$, and $V_{\rm l}$ to the applied voltage $U$, namely, $V_{rx}$, $V_{\rm rh}$, and $V_{\rm rl}$:

$$V_{\rm rx} = V_x/U \tag{20}$$

$$V_{\rm rh} = V_{\rm h}/U \tag{21}$$

$$V_{\rm rl} = V_{\rm l}/U \tag{22}$$

For path length, the scaling parameters are the ratios of $L_{Ex}$, $L_{gx}$, $L_{Wx}$, and $L_{\rm min}$ to the gap distance $d$, namely, $L_{{\rm r}Ex}$, $L_{{\rm rg}x}$, $L_{{\rm r}Wx}$, and $L_{\rm rmin}$. They all have the following expression:

$$L_{\rm rx} = L_x/d \tag{23}$$

The $x\%$ is set as 90 and 75%; therefore, the electric field features are summarized in **Table 1**. There are altogether 38 features for sphere-sphere air gap.

| Category | Features | Number |
|---|---|---|
| Electric field strength | $E_{\rm hmax}$, $E_{\rm lmax}$, $E_{\rm min}$, $E_{\rm a}$, $E_{\rm std2}$, $E_{\rm std}$ | 6 |
| Electric field gradient | $E_{\rm ghm}$, $E_{\rm glm}$, $E_{\rm ga}$ | 3 |
| Square of electric field strength | $W$, $W_{\rm a}$ | 2 |
| Electric field integral | $V_{90}$, $V_{75}$, $V_{\rm h}$, $V_{\rm l}$ | 4 |
| Path length | $L_{E90}$, $L_{E75}$; $L_{g90}$, $L_{g75}$; $L_{W90}$, $L_{W75}$; $L_{\rm min}$ | 7 |
| Electric field inhomogeneity | $E_{\rm d}$, $E_{\rm rlh}$, $E_{\rm rm}$; $W_{\rm r90}$, $W_{\rm r75}$; $V_{\rm r90}$, $V_{\rm r75}$, $V_{\rm rh}$, $V_{\rm rl}$; $L_{{\rm r}E90}$, $L_{{\rm r}E75}$, $L_{\rm rg90}$, $L_{\rm rg75}$, $L_{{\rm r}W90}$, $L_{{\rm r}W75}$, $L_{\rm rmin}$ | 16 |

**Table 1.** Electric field features for sphere-sphere air gap.

### 2.2.2. *Sphere-plane air gap*

For sphere-plane air gap with a monotonously decreased curve of the electric field distribution along the shortest path, the electric field features are defined as follows:

1. The maximum, the minimum, and the average value of the electric field strength ($E_{max}$, $E_{min}$, and $E_a$) and the variance and standard deviation of the electric field distribution along the shortest path ($E_{std2}$ and $E_{std}$). Their calculation formulas are similar to or the same with Eqs. (1)–(6).

2. The maximum and the average value of the electric field gradient ($E_{gm}$ and $E_{ga}$), whose calculation formulas are similar to or the same with Eqs. (7)–(9).

3. The square of electric field strength ($W$ and $W_a$), which can be calculated, respectively, by Eqs. (10) and (11).

4. The electric field integral of the path on which the field strength exceeds $x\%E_{max}$ ($V_x$) and its ratio to the applied voltage $U$ ($V_{rx}$). $V_x$ and $V_{rx}$ can be calculated, respectively, by Eqs. (12) and (20).

5. The length of the path on which the electric field strength exceeds $x\%E_{max}$ ($L_x$) and the ratio of $L_x$ to the gap length $d$ ($L_{rx}$). $L_x$ and $L_{rx}$ can be calculated, respectively, by Eqs. (15) and (23).

For long sphere-plane air gaps, $x\%$ includes 90, 75, 50, and 25%. Hence, there are altogether 25 features extracted from the shortest path used to characterize the electric field distribution of the sphere-plane air gap.

## 3. Breakdown voltage prediction method

### 3.1. Basic ideas

The proposed method for air gap breakdown voltage prediction is based on electric field features and SVM. A support vector classifier (SVC) is used to establish the prediction model. The withstanding and breakdown of an air gap under a given voltage are, respectively, denoted as −1 and 1. The input data of the prediction model are the abovementioned electric field features extracted from the electric field calculation results under different applied voltages, and the outputs are −1 and 1, respectively, means whether the gap will withstand or breakdown under the applied voltage, so as to transform the breakdown voltage prediction from a regression problem to a binary classification problem.

To be specific, if the critical breakdown voltage is $U_b$, the interval $[(1-a)U_b, U_b]$ is defined as withstand voltage interval, and $[U_b, (1 + a)U_b]$ is defined as breakdown voltage interval. The value of $a$ is determined by experience, which is set as 0.1 in this chapter. Set the step size as $0.01U_b$, and then the applied voltage values are $0.9U_b$, $0.91U_b$, …, $U_b$, $1.01U_b$, …, $1.1U_b$. The SVM model should be trained by some training samples to make it have the learning ability for accurate classification. By this binary classification method, one training sample can be extended to 21 samples to train the SVM model. For test samples, if the model outputs −1 to 1 under the applied voltage from $U_{b0}$-d$U$ to $U_{b0}$, then $U_{b0}$ is the predicted breakdown voltage.

### 3.2. Brief introduction of SVM

SVM is a machine learning algorithm developed on the basis of VC dimension in statistical learning theory and the principle of structural risk minimization [47]. The fundamentals of SVM were detailed and introduced in [47, 48]. Here, a brief introduction is provided.

Set a known training sample as $T = \{(x_i, y_i)\}$, in which $x_i \in R^k$, $y_i \in \{-1, 1\}$, $i = 1, 2, \ldots, n$. The implementation of SVM is to solve an optimization problem based on the maximum margin principle, which finds an optimal separating hyperplane to divide the sample data into two diverse classes. By application of the kernel trick, SVM transforms the original sample data into a high-dimensional Hilbert space $H$ using nonlinear mapping. The training sample data are transformed to $T_\Phi = \{(\Phi(x_i), y_i)\}$, where $\Phi(x_i) \in H$, $y_i \in \{-1, 1\}$, $i = 1, 2, \ldots, n$. In this feature space, the sample data can be linearly separated, and the decision function can be expressed as.

$$f(x) = \operatorname{sgn}\left(w^{\mathrm{T}} \cdot \phi(x_i) + b\right) \tag{24}$$

where $w$ and $b$ are, respectively, the weight vector and bias term of the separating hyperplane ($w \in R^k$) and $b$ is a real number. The symbol "sgn" is the signum function. When $w^{\mathrm{T}} \cdot x_i + b > 0$, the output is 1, while $w^{\mathrm{T}} \cdot x_i + b < 0$, the output is $-1$. Meanwhile, $w^{\mathrm{T}} \cdot x_i + b = 0$ is the classification hyperplane.

The optimization problem can be expressed as

$$\begin{cases} \min\limits_{w,\, b,\, \xi} \ \dfrac{1}{2}\|w\|^2 + C\sum\limits_{i=1}^{n} \xi_i \\ \text{s.t.} \quad y_i\left(w \cdot \phi(x_i) + b\right) \geq 1 - \xi_i, \quad \xi_i \geq 0, i = 1, 2, \cdots, n \end{cases} \tag{25}$$

where $\xi_i$ is the slack variable and $C$ is the penalty factor, which determines the balance between the maximization of the margin and the minimization of the classification error [48].

In order to make the solution of the primal problem more simple and practicable, the Lagrange function is introduced:

$$L(w, b, \xi, \alpha, \beta) = \frac{1}{2}\|w\|^2 + C\sum_{i=1}^{n} \xi_i - \sum_{i=1}^{n} \alpha_i \{y_i[(w \cdot x_i) + b] - 1 + \xi_i\} - \sum_{i=1}^{n} \beta_i \xi_i \tag{26}$$

where $\alpha = (\alpha_1, \ldots, \alpha_n)^{\mathrm{T}}$ and $\beta = (\beta_1, \ldots, \beta_n)^{\mathrm{T}}$ are the Lagrange multiplier vectors. By solving the partial derivatives of Eq. (26) for $w$, $b$, and $\xi_i$, according to the extremum conditions, the primal optimization problem (25) can be transformed to the following dual problems:

$$\begin{cases} \max\limits_{\alpha,\, \beta} -\dfrac{1}{2}\sum\limits_{i=1}^{n}\sum\limits_{j=1}^{n} \alpha_i \alpha_j y_i y_j (x_i \cdot x_j) + \sum\limits_{j=1}^{n} \alpha_j \\ \text{s.t.} \quad \sum\limits_{i=1}^{n} \alpha_i y_i = 0 \\ \qquad C - \alpha_i - \beta_i = 0, \ i = 1, 2, \cdots, l \\ \qquad \alpha_i \geq 0, \ \beta_i \geq 0, \ i = 1, 2, \cdots, l \end{cases} \tag{27}$$

By solving Eq. (27), the decision function can be obtained.

It can be seen that the function of transformation $\Phi$ is realized by inner product $(\Phi(x_i)\cdot\Phi(x_j))$. The kernel function can be expressed as

$$K(x_i, x_j) = (\phi(x_i) \cdot \phi(x_j)) \tag{28}$$

Hence, if the function $K$ is selected, it is not necessary to choose the transformation $\Phi$. $K(x_i, x_j)$ is used in training and classification instead of $\Phi(x)$. The generalization performance of SVM is determined by properly selecting kernel functions. In this chapter, the radial basis function (RBF) kernel is selected as the kernel function of SVM for its good generalization performance and high computational efficiency:

$$K(x_i, x_j) = \exp\left(-\gamma\|x_i - x_j\|^2\right),\ \ \gamma > 0 \tag{29}$$

where $\gamma$ is the kernel parameter.

The penalty factor $C$ and the kernel parameter $\gamma$ determine the classification performance of SVM. They can be optimized by grid search (GS) method or genetic algorithm (GA) based on $K$-fold cross validation or leave-one-out (LOO) cross validation, so as to obtain the optimal predictive model [36–39].

### 3.3. Implementation procedures of the prediction method

The flow chart of the prediction method is shown in **Figure 4**. The implementation procedures are depicted as follows.

Firstly, the training samples with known gap structures and experimental data of breakdown voltage are used to train the SVM model. The electric field features of each training sample are extracted from the FEM calculation results of the electric field distribution. These features are normalized to [0, 1] by

$$\overline{x}_i = \frac{x_i - x_{\min}}{x_{\max} - x_{\min}} \tag{30}$$

where $x_i$ is a feature, $\overline{x}_i$ is its normalized value, and $x_{\min}$ and $x_{\max}$ are, respectively, the minimum and maximum values of $x_i$. After normalization, the electric field features are taken as the input data to train the SVM model, while the outputs are $-1$ and 1, respectively, correspond to the applied voltage in the withstand interval and the breakdown interval. Based on cross validation, the optimal penalty factor $C$ and kernel parameter $\gamma$ are searched by GS method or GA to obtain an optimal prediction model. Trained by the known experimental data, SVM establishes the multidimensional nonlinear relationships between the electric field features and the air gap breakdown voltage.

Then, the optimal SVM model is used to predict the breakdown voltages of test samples. For an air gap, an estimated breakdown voltage is set in the range $[U_{\min}, U_{\max}]$; the golden section search method is applied for the breakdown voltage prediction [49]. For each applied voltage,

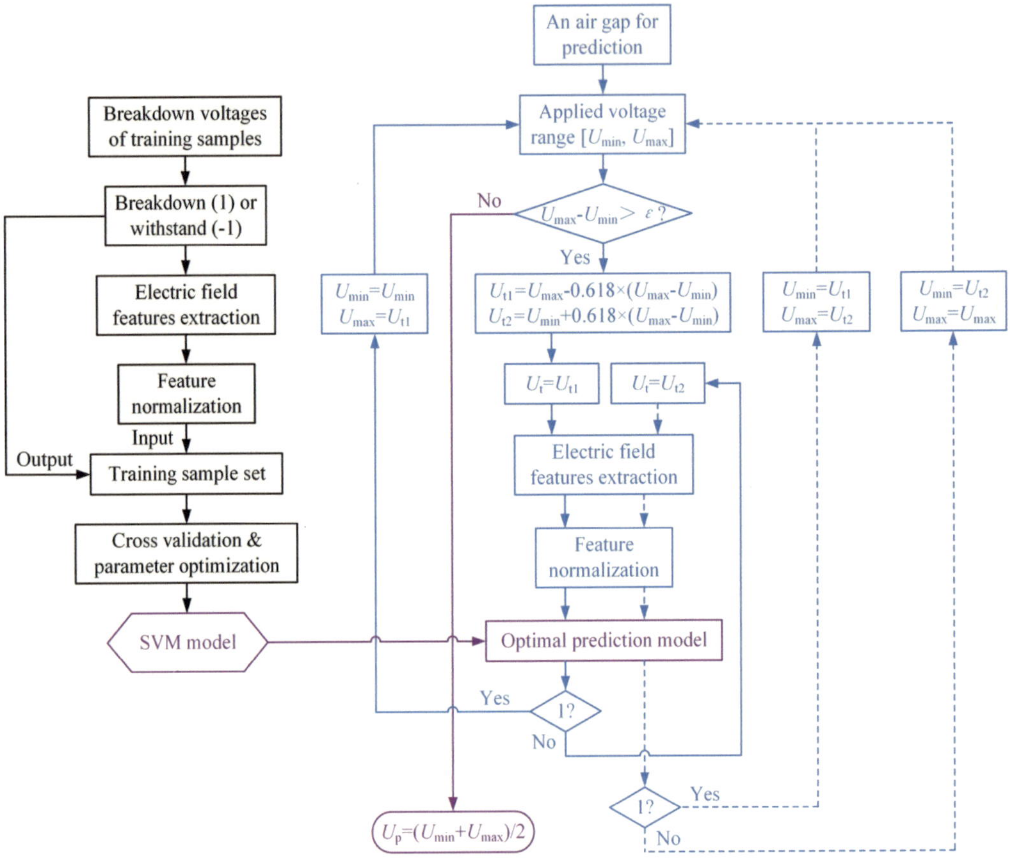


**Figure 4.** Flowchart of the prediction method.

the electric field features are extracted and input to the prediction model to judge whether the output is −1 or 1. For example, the first applied voltage is $U_{t1} = U_{max}-0.618 \times (U_{max}-U_{min})$, and then the electric field features of this air gap are calculated under $U_{t1}$ and input to the SVM model. According to the output of the model, namely, 1 or −1, the search interval will be narrowed to $[U_{min}, U_{t1}]$ to generate another applied voltage value, or otherwise, $U_{t2} = U_{min} + 0.618 \times (U_{max}-U_{min})$ is applied to calculate the electric field features which will be input to the SVM model to judge whether the output is −1 or 1. So repeatedly, the search interval is narrowed constantly, and the iterative predictions are conducted until the convergence condition $U_{max}-U_{min} < \varepsilon$ is satisfied, where $\varepsilon$ is the convergence precision. The breakdown voltage prediction result is the average value of the last two applied voltages.

## 4. Breakdown voltage prediction of air gaps

The proposed method is applied to predict the power frequency breakdown voltages of sphere-sphere air gaps and the switching impulse discharge voltages of sphere-plane air gaps. The predicted results are compared with the experimental data given in references.

### 4.1. Power frequency breakdown voltage prediction of sphere-sphere air gaps

#### 4.1.1. Training and test samples

The sample data of sphere-sphere air gaps are selected from IEC 60052 [43] and IEEE Std 4 [44]. The values of sphere diameter $D$ are 5, 6.25, 10, 12.5, 15, 25, 50, 75, 100, 150, and 200 cm. The shortest gap distance is 1 cm, for those with $D$ ranging from 5 to 25 cm, and the longest gap distance is 100 cm for that of $D$ = 200 cm. There are altogether 271 sample data with different sphere diameters and different gap distances.

The proper selection of training samples is of vital importance for the generalization performance of the SVM model. Here, the training samples are selected according to the electric field nonuniform coefficient $f$, that is, the ratio of the maximum field strength $E_{max}$ to the average field strength $E_a = U/d$, where $U$ is the applied voltage. The maximum field strength of each sphere-sphere gap can be calculated by FEM, and therefore the electric field nonuniform coefficient $f$ can be obtained.

According to the calculation results, all of these samples are slightly nonuniform electric field, and the values of $f$ range from 1.00 to 1.55. Taking 0.05 as the step size, the values of $f$ can be divided into 11 intervals. The samples belong to each $f$ interval are collected together, and the sample sizes corresponding to each $f$ interval are summarized in **Table 2**. According to the electric field nonuniform coefficient $f$, the 271 samples are divided into 11 groups. The training sample set is constituted by random selection of one sample from each group. Hence, there are 11 training samples, and the other 260 samples are taken as the test samples to verify the validity of the prediction method. Since the training sample selection is conducted by computer program which has a certain randomness, three different selection results are successively taken as the training sample set to train the SVM model, and the three times of prediction results are compared to validate the accuracy. The three groups of training samples are shown in **Table 3**.

| *f* interval | Sample size |
|---|---|
| 1.00–1.05 | 58 |
| 1.05–1.10 | 57 |
| 1.10–1.15 | 41 |
| 1.15–1.20 | 23 |
| 1.20–1.25 | 19 |
| 1.25–1.30 | 18 |
| 1.30–1.35 | 15 |
| 1.35–1.40 | 13 |
| 1.40–1.45 | 9 |
| 1.45–1.50 | 7 |
| 1.50–1.55 | 11 |

**Table 2.** Sample size of each $f$ interval.

| Training sample set 1 | | Training sample set 2 | | Training sample set 3 | |
|---|---|---|---|---|---|
| $D$ (cm) | $d$ (cm) | $D$ (cm) | $d$ (cm) | $D$ (cm) | $d$ (cm) |
| 5 | 1.4 | 5 | 2.2 | 5 | 1.6 |
| 5 | 2.2 | 6.25 | 3.0 | 10 | 2.0 |
| 10 | 4.5 | 12.5 | 1.4 | 15 | 7.0 |
| 15 | 6.0 | 12.5 | 2.8 | 25 | 10 |
| 25 | 5.5 | 12.5 | 5.0 | 25 | 12 |
| 25 | 8.0 | 25 | 1.5 | 50 | 13 |
| 50 | 8.0 | 50 | 17 | 100 | 24 |
| 50 | 17 | 75 | 19 | 100 | 36 |
| 75 | 5.5 | 150 | 45 | 150 | 7.5 |
| 75 | 10 | 200 | 38 | 150 | 20 |
| 200 | 100 | 200 | 90 | 150 | 65 |

**Table 3.** Three groups of training sample set.

#### *4.1.2. Prediction results and analysis*

Based on fivefold cross validation, the grid search method is applied to determine the optimal parameters of the SVM model. The search ranges of the penalty factor $C$ and the kernel parameter $\gamma$ are, respectively, set as $[2^3, 2^9]$ and $[2^{-8}, 2^{-2}]$, and the step sizes are both $2^{0.1}$. Taking the training sample set 1, for example, the parameter optimization results of $C$ and $\gamma$ by GS method are shown in **Figure 5**. It can be seen that the best $C = 90.5097$, $\gamma = 0.0167$, under which the SVM model, has the highest classification accuracy for the training samples, that is, 98.2684%. The GS-optimized SVM models under the three groups of training sample set shown in **Table 3** are used for breakdown voltage prediction of the 260 test samples.

For each test sample, the breakdown voltage is predicted by the golden section search method, and the prediction results are compared with the experimental data given in [43, 44]. In order to evaluate the prediction accuracy of the proposed method and the SVM model, three error indices, including the root-mean-square error (RMSE), the mean absolute percentage error (MAPE), and the mean square percentage error (MSPE), are used to examine the errors of the prediction results, which can be calculated by

$$\mathrm{RMSE} = \sqrt{\frac{1}{n}\sum_{i=1}^{n}\left[U_{\mathrm{t}}(i) - U_{\mathrm{p}}(i)\right]^2} \tag{31}$$

$$\mathrm{MAPE} = \frac{1}{n}\sum_{i=1}^{n}\left|\frac{U_{\mathrm{t}}(i) - U_{\mathrm{p}}(i)}{U_{\mathrm{t}}(i)}\right| \tag{32}$$

$$\mathrm{MSPE} = \frac{1}{n}\sqrt{\sum_{i=1}^{n}\left[\frac{U_{\mathrm{t}}(i) - U_{\mathrm{p}}(i)}{U_{\mathrm{t}}(i)}\right]^2} \tag{33}$$

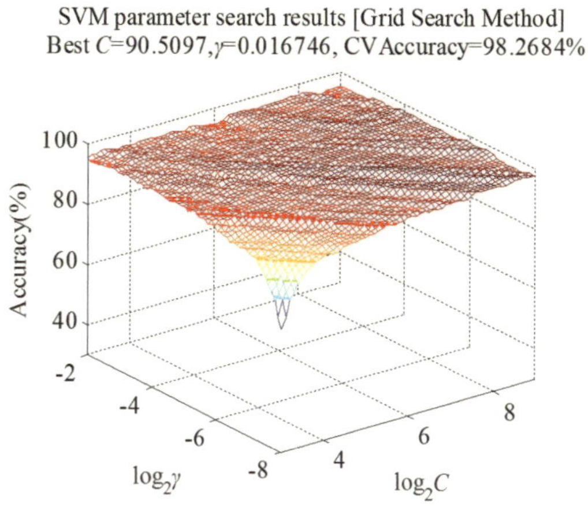


**Figure 5.** Parameter optimization results of $C$ and $\gamma$ by GS method.

where $n$ is the number of the test samples and $U_t(i)$ and $U_p(i)$ are, respectively, the experimental and predicted breakdown voltages of the $i$th test sample.

The optimal parameters and the error indices of the three times of prediction results are summarized in **Table 4**. It can be seen that the prediction results are with high accuracy, while the MAPEs of the three times of prediction are, respectively, 1.88, 2, and 1.4%. Taking the prediction results by training sample set 2, for example, the comparisons between the predicted and experimental breakdown voltages of sphere-sphere air gaps with different diameters are shown in **Figure 6**, where $U$ is the breakdown voltage, $D$ is the sphere diameter, $d$ is the gap distance, and T-value and P-value, respectively, mean the test value and the prediction value of the breakdown voltage. For better comparisons, the prediction results of training samples are also plotted in **Figure 6**.

It can be seen from **Figure 6** that the predicted results coincide well with the experimental data, the trends of the breakdown voltages with the gap distance are the same, and the errors are within an acceptable range. The results shown in **Table 4** and **Figure 6** validate the feasibility and accuracy of the proposed method for sphere-sphere air gap breakdown voltage prediction.

### 4.2. Switching impulse breakdown voltage prediction of sphere-plane air gaps

#### *4.2.1. Training and test samples*

The sample data of sphere-plane air gaps are selected from [45, 46]. The positive switching impulse discharge tests of sphere-plane air gaps with the sphere diameter of 25, 45, 75, and 95 cm were conducted in [45]. The applied voltage waveform is the standard 250/2500 μs switching impulse voltage. The experimental data were corrected to standard atmospheric condition. In order to make the SVM model generalize to sphere-plane gaps with different sphere diameters and gap lengths, seven test data shown in **Table 5** are selected as the training samples, where $D$ is the sphere diameter ranging from 25 to 95 cm, $d$ is the gap length ranging from 2 to 5 m, and $U_{50}$ is the 50% discharge voltage.

| Results | Training sample set 1 | Training sample set 2 | Training sample set 3 |
|---|---|---|---|
| $C$ | 90.5097 | 78.7932 | 13.9288 |
| $\gamma$ | 0.0167 | 0.0313 | 0.0625 |
| RMSE | 16.668 | 8.048 | 9.143 |
| MAPE | 0.0188 | 0.0200 | 0.0140 |
| MSPE | 0.0015 | 0.0140 | 0.0012 |

**Table 4.** Optimal parameters and error indices of the sphere-sphere air gap breakdown voltage prediction results.

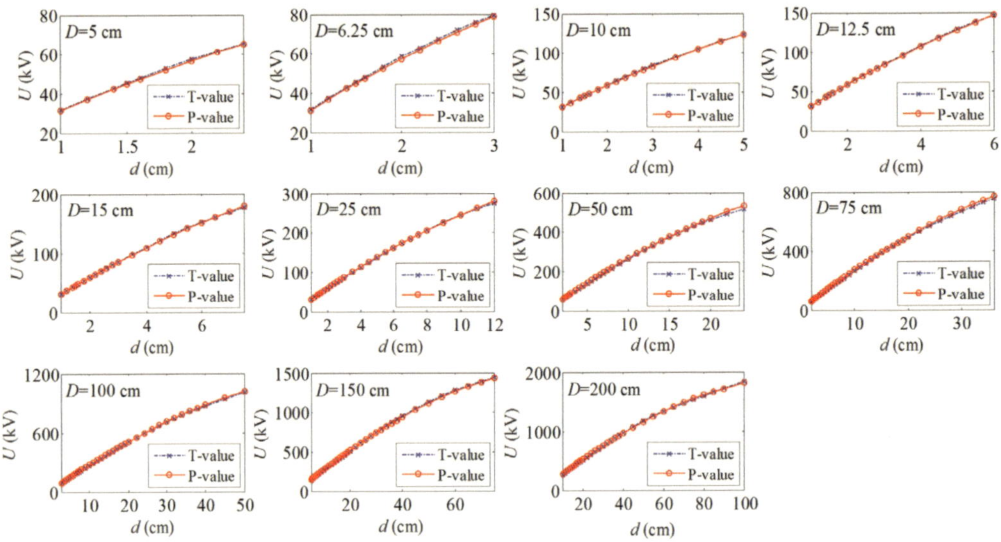


**Figure 6.** Comparisons between the predicted and experimental breakdown voltages of sphere-sphere air gaps (prediction results by training sample set 2).

The SVM model trained by the seven sample data is applied to predict the 50% discharge voltages of sphere-plane gaps with larger diameters, namely, 110, 150, and 200 cm. There are altogether 16 test samples. The predicted results will be compared with the experimental data cited from [46], as shown in **Table 5**. These experimental data also had been corrected to standard atmospheric condition.

### *4.2.2. Prediction results and analysis*

Based on LOO cross validation, the penalty factor $C$ and the kernel parameter $\gamma$ are optimized by the GA method. The population quantity is set as 20, the maximum generation is 200, and the crossover probability is 0.9. The search scopes of $C$ and $\gamma$ are, respectively, set as [10, 500] and [0.005, 0.25]. The fitness function is the classification accuracy of SVM for training samples. The parameter optimization results of $C$ and $\gamma$ by GA method are shown in **Figure 7**. It can be seen that the best $C$ = 85.3407, $\gamma$ = 0.0926, under which the SVM model has the highest classification accuracy for the training samples, that is, 97.2789%.

| Training sample set | | | | | |
|---|---|---|---|---|---|
| $D$ (cm) | $d$ (m) | $U_{50}$ (kV) | $D$ (cm) | $d$ (m) | $U_{50}$ (kV) |
| 45 | 2 | 864 | 25 | 3 | 1020 |
| | 3 | 1045 | 75 | | 1222 |
| | 4 | 1186 | 95 | | 1504 |
| | 5 | 1303 | — | — | — |
| **Experimental data of test samples** | | | | | |
| $D$ (cm) | $d$ (m) | $U_{50}$ (kV) | $D$ (cm) | $d$ (m) | $U_{50}$ (kV) |
| 110 | 3.9 | 1461 | 150 | 8.5 | 2408 |
| | 4.5 | 1527 | | 10 | 2497 |
| | 5.7 | 1675 | 200 | 3 | 2290 |
| | 6.5 | 1796 | | 4 | 2540 |
| | 7.5 | 1973 | | 5 | 2711 |
| 150 | 4 | 2101 | | 6 | 2745 |
| | 5.5 | 2242 | | 7 | 2836 |
| | 7 | 2316 | | 8 | 2927 |

**Table 5.** Training and test samples of the sphere-plane air gaps.

The GA-optimized SVM model is used to predict the $U_{50}$ of large sphere-plane air gaps by the golden section search method. The initial applied voltage interval [$U_{min}$, $U_{max}$] is set as 0–4000 kV, and the convergence precision $\varepsilon$ is set as 1 kV. The prediction results are summarized in **Table 6**, where $U_{50}$ is the experimental data extracted from [46], $U_p$ is the predicted discharge voltage, and $\delta$ is the relative error.

It can be seen from **Table 6** that the largest error of the prediction results is 9.7%, for the gap with $D$ = 110 cm and $d$ = 3.9 m. This is probably due to different experimental arrangements between [45, 46]. The $U_{50}$ of sphere-plane gaps with $D$ = 110 cm and $d$ = 4 m in [46] is even lower than that with $D$ = 95 cm and $d$ = 4 m in [45]. The SVM model is trained by the experimental data cited from [45], and therefore the prediction results of sphere-plane gaps with $D$ = 110 cm may be larger than the experimental data obtained in [46]. Overall, the prediction errors are acceptable in the view of engineering applications, while the MAPE of the 16 test samples is only 3.2%.

The $U_{50}$ prediction results and the experimental data are summarized in the same graph for a better comparison, as shown in **Figure 8**. It can be seen that the predicted values of the discharge voltage agree well with the experimental data, with similar trends and acceptable errors. The results verify the validity and accuracy of the proposed model for discharge voltage prediction of sphere-plane air gaps, with large sphere diameter and long gap length. Within the range of certain precision, the prediction method can be used to replace the experiments, so as to reduce the testing expenses.

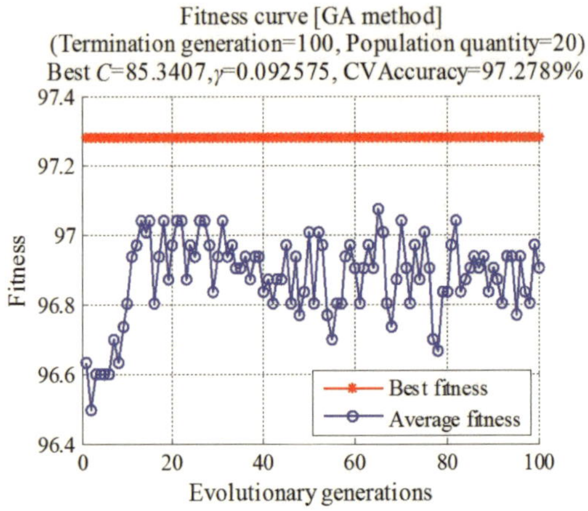


**Figure 7.** Parameter optimization results of C and $\gamma$ by GA method.

| $D$ (cm) | $d$ (m) | $U_{50}$ (kV) | $U_p$ (kV) | $\delta$ (%) |
|---|---|---|---|---|
| 110 | 3.9 | 1461 | 1602 | 9.7 |
| | 4.5 | 1527 | 1647 | 7.9 |
| | 5.7 | 1675 | 1745 | 4.2 |
| | 6.5 | 1796 | 1807 | 0.6 |
| | 7.5 | 1973 | 1910 | −3.2 |
| 150 | 4 | 2101 | 2036 | −3.1 |
| | 5.5 | 2242 | 2132 | −4.9 |
| | 7 | 2316 | 2266 | −2.2 |
| | 8.5 | 2408 | 2413 | 0.2 |
| | 10 | 2497 | 2560 | 2.5 |
| 200 | 3 | 2290 | 2348 | 2.5 |
| | 4 | 2540 | 2513 | −1.1 |
| | 5 | 2711 | 2595 | −4.3 |
| | 6 | 2745 | 2700 | −1.6 |
| | 7 | 2836 | 2778 | −2.0 |
| | 8 | 2927 | 2893 | −1.2 |

**Table 6.** 50% discharge voltage prediction results of the sphere-plane air gaps.

## 5. Conclusions

Two sets of electric field features defined on the shortest interelectrode path are, respectively, used to characterize the gap structure of the sphere-sphere air gap and the rod (sphere)-plane air gap. These features are taken as the input parameters of the SVM model, which is used to establish the breakdown voltage prediction model. The proposed method based on electric

http://dx.doi.org/10.5772/intechopen.72230

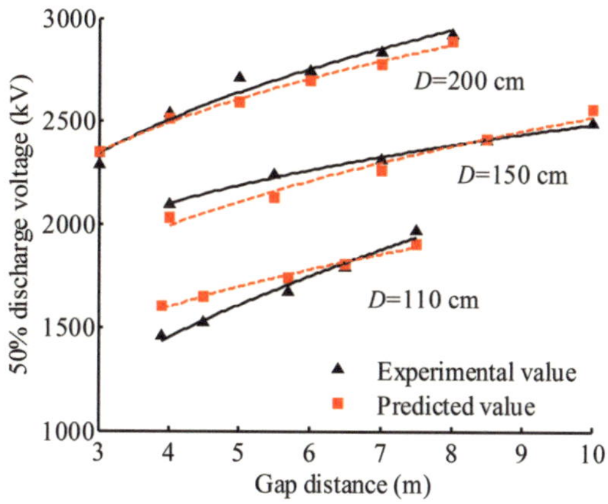


**Figure 8.** Comparison of the predicted and experimental 50% discharge voltages of large sphere-plane air gaps.

field features and SVM is applied to predict the breakdown voltages of sphere-sphere and sphere-plane air gaps. Some conclusions can be drawn as follows:

1. The proposed electric field features extracted from the shortest interelectrode path are effective to characterize the spatial structure of sphere-sphere and sphere-plane air gaps, and the multidimensional nonlinear relationships between these features and the air gap breakdown voltage can be established by SVM, so as to achieve breakdown voltage prediction of air gaps without considering the complex and random discharge process.

2. Trained by only 11 sample data selected randomly according to the electric field nonuniform coefficient *f*, the SVM model is able to accurately predict the power frequency breakdown voltages of IEC standard sphere-sphere air gaps. The mean absolute percentage errors of the 260 test samples, with three times of prediction by different training sample sets, are within 2%. The results validate the validity and accuracy of the proposed method for breakdown voltage prediction of sphere-sphere air gaps.

3. The proposed method is able to predict the switching impulse discharge voltages of sphere-plane air gaps, with large sphere diameters and long gap distances. The prediction results agree well with the experimental data, with similar trends and acceptable errors. The mean absolute percentage error of the 16 test samples is 3.2%, which is acceptable for engineering applications. The results verify the feasibility of the proposed model for discharge voltage prediction of large sphere-plane air gaps, which may be useful to replace the time-consuming and costly discharge tests.

The authors are still engaged in improving this model. The following work will be carried out in the future. Firstly, the electric field features will be simplified by some feature selection approaches to make it easier for applications. Secondly, the applications of this method will be extended to breakdown voltage prediction of more complex air gaps such as the practical engineering gaps. It should be noted that new problems will inevitably appear in different applications, and therefore this topic is worthy to be studied in-depth. We hope that it is possible to achieve breakdown voltage prediction of arbitrary engineering gap configurations in the future, so as to guide the insulation design of high-voltage electrical equipment by mathematical calculations rather than costly experiments.

## Acknowledgements

This work is supported by China Postdoctoral Science Foundation (2016 M602354). The authors would like to thank Dr. Shengwen Shu for his primary work on this topic during his study for PhD degree in Wuhan University. The experimental data of sphere-plane gaps are cited from [45, 46] and taken as the sample data in this paper. We would like to express thanks to those authors.

## Author details

Zhibin Qiu* and Jiangjun Ruan

*Address all correspondence to: whuerliuming@whu.edu.cn

School of Electrical Engineering, Wuhan University, Wuhan, China

## References

[1] Liao CB, Ruan JJ, Liu C, Wen W, ZY D. 3-D coupled electromagnetic-fluid-thermal analysis of oil-immersed triangular wound core transformer. IEEE Transactions on Magnetics. 2014;**50**(11):8401904. DOI: 10.1109/TMAG.2014.2330953

[2] Liu XM, Yang YM, Yang F, Jadoon A. Numerical research on the losses characteristic and hot-spot temperature of laminated core joints in transformer. Applied Thermal Engineering. 2017;**110**:49-61. DOI: 10.1016/j.applthermaleng.2016.08.158

[3] Zhang XC, Li WL, Kou BQ, Cao JC, Cao HC, Gerada C, Zhang H. Electrothermal combined optimization on notch in air-cooled high-speed permanent-magnet generator. IEEE Transactions on Magnetics. 2015;**51**(1):8200210. DOI: 10.1109/TMAG.2014.2332437

[4] Wang QY, Yang X, Tian HD, Liu P, Peng ZR. A novel dissipating heat structure of converter transformer RIP bushings based on 3-D electromagnetic-fluid-thermal analysis. IEEE Transactions on Dielectrics and Electrical Insulation. 2017;**24**(3):1938-1946. DOI: 10.1109/TDEI.2017.006027

[5] Sheikholeslami M, Soleimani S, Ganji DD. Effect of electric field on hydrothermal behavior of nanofluid in a complex geometry. Journal of Molecular Liquids. 2016;**213**:153-161. DOI: 10.1016/j.molliq.2015.11.015

[6] Sheikholeslami M, Ganji DD. Impact of electric field on nanofluid forced convection heat transfer with considering variable properties. Journal of Molecular Liquids. 2017;**229**:566-573. DOI: 10.1016/j.molliq.2016.12.107

[7] Sheikholeslami M, Chamkha AJ. Electrohydrodynamic free convection heat transfer of a nanofluid in a semi annulus enclosure with a sinusoidal wall. Numerical Heat Transfer, Part A: Applications. 2016;**69**(7):781-793. DOI: 10.1080/10407782.2015.1090819

[8] Sheikholeslami M, Hayat T, Alsaedi A, Abelman S. Numerical analysis of EHD nanofluid force convective heat transfer considering electric field dependent viscosity. International Journal of Heat and Mass Transfer. 2017;**108**(Part B):2558–2565. DOI: 10.1016/j.ijheatmasstransfer.2016.10.099

[9] Sheikholeslami M, Bhatti MM. Active method for nanofluid heat transfer enhancement by means of EHD. International Journal of Heat and Mass Transfer. 2017;**109**:115-122. DOI: 10.1016/j.ijheatmasstransfer.2017.01.115

[10] Townsend JS. The Theory of Ionization of Gases by Collision. New York: Van Nostrand Company; 1910. p. 88

[11] Reather H. Electron Avalanches and Breakdown in Gases. London: Butterworths; 1964. p. 191

[12] Loeb LB, Meek JM. The Mechanism of the Electric Spark. USA: Stanford University Press; 1941. p. 188

[13] Les Renardières Group. Research on long air gap discharges at Les Renardières. Electra. 1972;(23):53–157

[14] Les Renardières Group. Research on long air gap discharges at Les Renardières–1973 results. Electra. 1974;(35):49–156

[15] Les Renardières Group. Positive discharges in long air gap discharges at Les Renardières–1975 results and conclusions. Electra. 1977;(53):31–153

[16] Les Renardières Group. Negative discharges in long air gap discharges at Les Renardières–1978 results. Electra. 1981;(74):67–216

[17] Gallet G, Leroy G, Lacey R, Kromer I. General expression for positive switching impulse strength valid up to extra long air gaps. IEEE Transactions on Power Apparatus and Systems. 1975;**94**(6):1989-1993. DOI: 10.1109/T-PAS.1975.32045

[18] Kishizima I, Matsumoto K, Watanabe Y. New facilities for phase-to-phase switching impulse tests and some test results. IEEE Transactions on Power Apparatus and Systems. 1984;**103**(6):1211-1216. DOI: 10.1109/TPAS.1984.318451

[19] Carrara G, Thione L. Switching surge strength of large air gaps: A physical approach. IEEE Transactions on Power Apparatus and Systems. 1976;**95**(2):512-514. DOI: 10.1109/T-PAS.1976.32131

[20] Rizk FAM. Switching impulse strength of air insulation: Leader inception criterion. IEEE Transactions on Power Delivery. 1989;**4**(4):2187-2195. DOI: 10.1109/61.35646

[21] Chen WJ, Zeng R, He HX. Research progress of long air gap discharges. High Voltage Engineering. 2013;**39**(6):1281-1295. DOI: 10.3969/j.issn.1003-6520.2013.06.001

[22] Gallimberti I. A computer model for streamer propagation. Journal of Physics D: Applied Physics. 1972;**5**(12):2179-2189. DOI: 10.1088/0022-3727/5/12/307

[23] Gallimberti I. The mechanism of the long spark formation. Journal de Physique Colloques. 1979;**40**(C7):193-250. DOI: 10.1051/jphyscol:19797440

[24] Gallimberti I, Bacchiega G, Bondiou-Clergerie A, Lalande P. Fundamental processes in long air gap discharges. Comptes Rendus Physique. 2002;**3**(10):1335-1359. DOI: 10.1016/S1631-0705(02)01414-7

[25] Fofana I, Béroual A. A model for long air gap discharge using an equivalent electrical network. IEEE Transactions on Dielectrics and Electrical Insulation. 1996;**3**(2):273-282. DOI: 10.1109/94.486779

[26] Goelian N, Lalande P, Bondiou-Clergerie A, Bacchiega GL, Gazzani A, Gallimberti I. A simplified model for the simulation of positive-spark development in long air gaps. Journal of Physics D: Applied Physics. 1997;**30**(17):2441-2452. DOI: 10.1088/0022-3727/30/17/010

[27] Becerra M, Cooray V. A self-consistent upward leader propagation model. Journal of Physics D: Applied Physics. 2006;**39**(16):3708-3715. DOI: 10.1088/0022-3727/39/16/028

[28] Arevalo L, Cooray V, Montano R. Numerical simulation of long sparks generated by positive switching impulses. Journal of Electrostatics. 2009;**67**(2–3):228-234. DOI: 10.1016/j.elstat.2008.12.022

[29] Xie YH. A Physical Model for the Simulation of Positive Discharge Development in Long Air Gaps [Thesis]. Wuhan, China: Huazhong University of Science & Technology; 2013. p. 106

[30] Zhou X, Zeng R, Li ZZ, Zhuang CJ. A one-dimensional thermo-hydrodynamic model for upward leader inception considering gas dynamics and heat conduction. Electric Power Systems Research. 2016;**139**:16-21. DOI: 10.1016/j.epsr.2015.11.028

[31] Zeng R, Zhuang CJ, ZQ Y, Chen S, Li ZZ, Chen WJ. Challenges and achievement in long air gap discharge research. High Voltage Engineering. 2014;**40**(10):2945-2955. DOI: 10.13336/j.1003-6520.hve.2014.10.001

[32] Mokhnache L, Boubakeur A. Prediction of the breakdown voltage in a point-barrier-plane air gap using neural networks. In: 2001 Annual Report Conference on Electrical Insulation and Dielectric Phenomena; 14–17 October 2001; Kitchener, Ontario, Canada. New York: IEEE; 2001. pp. 369-372. DOI: 10.1109/CEIDP.2001.963559

[33] Mokhnache L, Boubakeur A, Feliachi A. Breakdown voltage prediction in a point-barrier-plane air gap arrangement using self-organization neural networks. In: IEEE Power Engi neering Society General Meeting; 6–10 June 2004; Denver, Colorado, USA. New York: IEEE; 2004. pp. 569–572. DOI: 10.1109/PES.2004.1372865

[34] Bourek Y, Mokhnache L, Said NN, Kattan R. Study of discharge in point-plane air interval using fuzzy logic. Journal of Electrical Engineering & Technology. 2009;**4**(3):410-417. DOI: 10.5370/jeet.2009.4.3.410

[35] Bourek Y, Mokhnache L, Said NN, Kattan R. Determination of ionization conditions characterizing the breakdown threshold of a point-plane air interval using fuzzy logic. Electric Power Systems Research. 2011;**81**(11):2038-2047. DOI: 10.1016/j.epsr.2011.06.012

[36] Shu SW. Study on Prediction for Corona Onset and Breakdown Voltages of Air Gap Based on Electric Field Features and Support Vector Machine [Thesis]. Wuhan, China: Wuhan University; 2014. p. 139

[37] Qiu ZB. Study on Energy Storage Features and Breakdown Voltage Prediction of Air Gap [Thesis]. Wuhan, China: Wuhan University; 2016. p. 141

[38] Qiu ZB, Ruan JJ, Huang DC, PU ZH, Shu SW. A prediction method for breakdown voltage of typical air gaps based on electric field features and support vector machine. IEEE Transactions on Dielectrics and Electrical Insulation. 2015;**22**(4):2125-2135. DOI: 10.1109/TDEI.2015.004887

[39] Qiu ZB, Ruan JJ, Huang DC, Wei MT, Tang LZ, Huang CP, WJ X, Shu SW. Hybrid prediction of the power frequency breakdown voltage of short air gaps based on orthogonal design and support vector machine. IEEE Transactions on Dielectrics and Electrical Insulation. 2016;**23**(2):795-805. DOI: 10.1109/TDEI.2015.005398

[40] Qiu ZB, Ruan JJ, Huang CP, Xu WJ, Tang LZ, Huang DC, Liao YF. A method for breakdown voltage prediction of short air gaps with atypical electrodes. IEEE Transactions on Dielectrics and Electrical Insulation. 2016;**23**(5):2685-2694. DOI: 10.1109/TDEI.2016.7736827

[41] Qiu ZB, Ruan JJ, Huang DC, Shu SW, Du ZY. Prediction study on positive DC corona onset voltage of rod-plane air gaps and its application to the design of valve hall fittings. IET Generation, Transmission & Distribution. 2016;**10**(7):1519-1526. DOI: 10.1049/iet-gtd.2015.0192

[42] Qiu ZB, Ruan JJ, Huang DC, Shu SW, Prediction DZY. On corona onset voltage of DC conductors and valve hall fittings. Transactions of China Electrotechnical Society. 2016; **31**(12):80-89. DOI: 10.3969/j.issn.1000-6753.2016.12.010

[43] IEC 60052. Voltage measurement by means of standard air gaps; 2012

[44] IEEE Std 4. IEEE standard for high-voltage testing techniques; 2013

[45] Wang X. The Comparison of Critical Radius of Rod-Plane Gap at Different Altitudes and Research on Altitude Correction [Thesis]. China Electric Power Research Institute: Beijing, China; 2010. p. 69

[46] Chen S, Zhuang CJ, Zeng R, Ding YJ, ZY S, Liao WM. Improved gap factor of large sphere-plane and its application in calculating air gap clearance in UHVDC converter station. High Voltage Engineering. 2013;**39**(6):1360-1366. DOI: 10.3969/j.issn.1003-6520.2013.06.011

[47] Vapnik VN. The Nature of Statistical Learning Theory. 2nd ed. New York: Springer-Verlag; 2000. p. 314

[48] Abe S. Support Vector Machines for Pattern Classification. 2nd ed. London: Springer-Verlag London Limited; 2010. p. 471

[49] Qiu ZB, Ruan JJ, WJ X, Huang CP. Energy storage features and a predictive model for switching impulse flashover voltages of long air gaps. IEEE Transactions on Dielectrics and Electrical Insulation. 2017;**24**(5):2703-2711

Chapter 10

# Standard Electric and Magnetic Field for Calibration

Walter Gustavo Fano

Additional information is available at the end of the chapter

http://dx.doi.org/10.5772/intechopen.72687

**Abstract**

The electric and magnetic field from an electromagnetic wave usually are a matter of interest for regulatory standards to accomplish electromagnetic compatibility. Also in research works of technology in radio frequency, it is very important to analyze the electromagnetic fields. The electric and magnetic probes are studied and discussed in this chapter. The most important electric field probe is the dipole antenna, usually an electrically short dipole antenna. The loop antenna is used as magnetic field probe to measure the magnetic field. The current distribution and the electric circuit model for the dipole and the loop antennas are obtained. The antenna factor is also defined and calculated for the same antenna. This antenna factor allows to determine the incident electric field to the dipole antenna and the incident magnetic field to the loop antenna. The special cases of the electrically short dipole and the electrically short loop antennas are discussed. The Helmholtz coils are usually used to establish a known and uniform magnetic field zone for various applications. In the area where a uniform magnetic field is generated, sensor and magnetic field probe calibrations can be made in a low-frequency range.

**Keywords:** Maxwell equations, dipole antennas, loop antennas, electric field probe, magnetic field probe, Helmholtz coils, effective area, effective length, antenna factor

## 1. Introduction

This section explains the concepts that correspond to the maximum effective area and the effective length of a receiving antenna, when an electromagnetic wave is incident on the antenna. The concepts of electromagnetic waves, plane wave propagation in space, electric and magnetic fields, Poynting vector and power density incident on an antenna, dipole and loop antennas are required.

### 1.1. Effective area and effective length

In **Figure 1a**, an incident electromagnetic wave captures by receiving antenna, that is connected to a load impedance $Z_L$, can be observed. The electromagnetic wave is polarized in the z

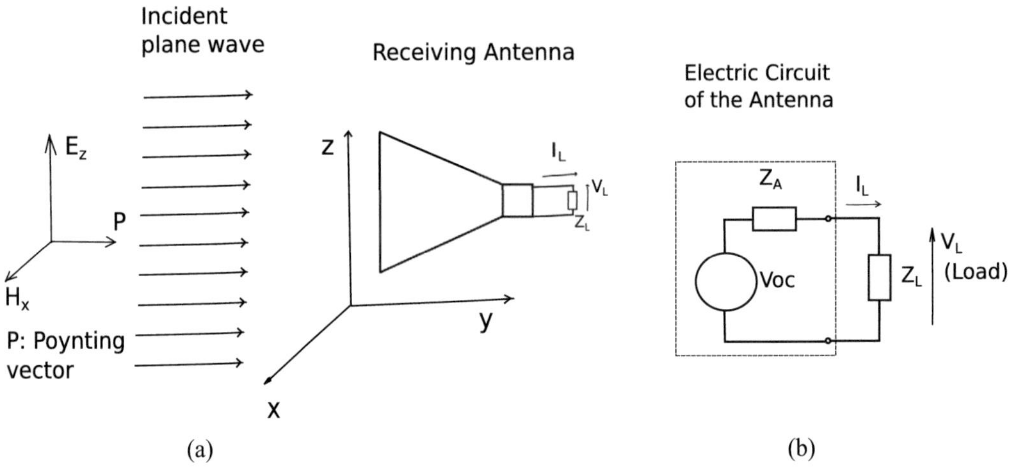


**Figure 1.** (a) Receiving antenna, with incident electromagnetic plane wave, parallel to the antenna. (b) the Thevenin equivalent electric circuit of the receiving antenna.

direction, which leads to a, and the antenna is oriented to the maximum gain direction. The incident power density of the electromagnetic wave [1] can be expressed in terms of the electric field as:

$$P_i = \frac{|E_i|^2}{2Z_{00}} \tag{1}$$

where $Z_{00}$ is the intrinsic vacuum impedance, $P_i$ is the time average of incident power density, and $|E_i|$ is the intensity of the incident electric field.

The maximum effective area of an antenna can be defined as the ratio between the power received at the antenna terminals and the power density incident to the antenna, for linear polarization:

$$A_{eff} = \frac{W_R}{P_i} \tag{2}$$

In this case, the antenna is oriented to receive maximum power, and the load impedance $Z_L$ is matched to the impedance of the receiving antenna $Z_A$:

$$W_R = P_i A_{eff} \tag{3}$$

From Eqs. (1) and (3):

$$W_R = \frac{|E_i|^2}{2Z_{00}} A_{eff} \tag{4}$$

The equivalent electric circuit of a receiving antenna in **Figure 1b**, where the incident electromagnetic wave is polarized in the z direction, and coincides with the polarization of the receiving antenna. The effective length of a receiving antenna can be defined as the ratio between the voltage induced at the terminals of the open circuit antenna and the incident electric field:

$$L_{eff} = \frac{|V_{oc}|}{|E_i|} \tag{5}$$

where $|V_{oc}|$ is the voltage received at the terminals of the open circuit antenna and $|E_i|$ is the incident electric field.

The total apparent power in the load can be expressed thus:

$$V_L \cdot I_L = |V_L||I_L|cos\phi + |V_L||I_L|sin\phi \tag{6}$$

where $V_L$ is the voltage at the load, $I_L$ is the current flowing through the load, and $\phi$ is the phase difference between $V_L$ and $I_L$.

The active received power at the antenna terminals is:

$$W_R = \frac{|V_L|^2 cos\phi}{Z_L} \tag{7}$$

The relationship between the load voltage and the open circuit voltage is:

$$|V_L| = |V_{oc}|\frac{|Z_L|}{|Z_L + Z_A|} \tag{8}$$

From Eqs. (7), (8) and (5):

$$W_R = \frac{|E_i L_{eff}|^2 |Z_L| cos\phi}{|Z_L + Z_A|^2} \tag{9}$$

From Eqs. (4) and (9):

$$L_{eff} = \sqrt{\frac{|Z_L + Z_A|^2 A_{eff}}{|Z_L| 2 Z_{00} cos\phi}} \tag{10}$$

This effective length is a general expression for any antenna rather than what is usually defined in the bibliography for dipole antennas [1].

## 2. Electric field probes

Consider a linear dipole antenna with an incident electromagnetic plane wave, polarized parallel to the antenna, which is captured by the antenna. The electric circuit of the antenna without load is shown in **Figure 2**. The voltage induced in the antenna terminals is denoted by $V_{oc}$.

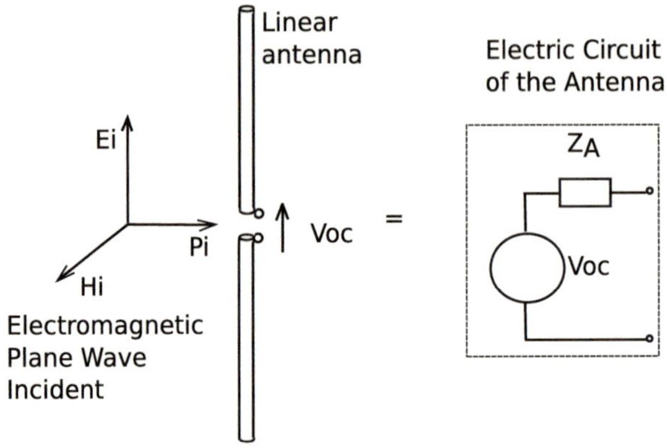


**Figure 2.** Receiving antenna, with incident electromagnetic plane wave, parallel to the antenna.

The effective length of a linear dipole antenna in receiving mode can be defined as a relation of electric field strength incident to the antenna and the voltage on the antenna, as shown in **Figure 3**, [2]. This can be expressed as:

$$|V_{oc}| = |E_i| L_{eff} \tag{11}$$

where $|V_{oc}|$ is the amplitude of the open circuit voltage on the antenna, $|E_i|$ is the amplitude of the incident electric field parallel to the antenna, and $L_{eff}$ is the effective length of the linear antenna.

The received voltage is usually measured by the spectrum analyzer, which has the input impedance $Z_L = 50\Omega$, as shown in **Figure 4**. This receiving voltage and the receiving power can be expressed as:

$$|V_r| = \sqrt{W_r |Z_L|} \tag{12}$$

where $W_r$ is the power received by the spectrum analyzer and $Z_L$ is the impedance of the spectrum analyzer.

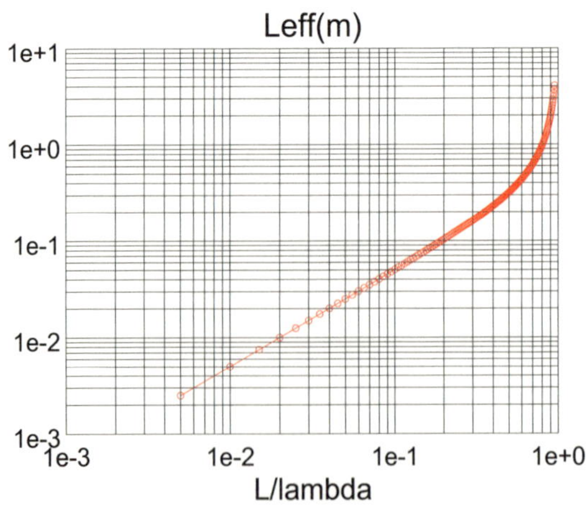


**Figure 3.** Effective length of the probe (short dipole): $L = 0.48\lambda$ (30 mm).

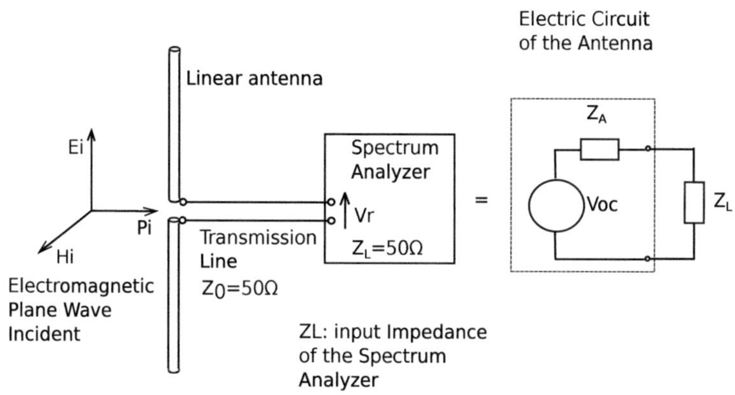


**Figure 4.** The receiving antenna with spectrum analyzer connected, and the incident electromagnetic plane wave, has the electric field parallel to the antenna.

The open circuit voltage can be expressed as:

$$|V_{oc}| = |V_r| \left( \frac{|Z_A + Z_L|}{|Z_L|} \right) \tag{13}$$

where $Z_L = 50\Omega$

From Eqs. (12) and (13):

$$|V_{oc}| = \sqrt{W_r |Z_L|} \left( \frac{|Z_A + Z_L|}{|Z_L|} \right) \tag{14}$$

Using the Eq. (11), the electric field incident to the probe is:

$$|E_i| = \frac{|V_{oc}|}{L_{eff}} \tag{15}$$

From Eqs. (10), (14), and (15), the electric field incident to the dipole can, therefore, be obtained as:

$$|E_i| = \frac{|V_{oc}|}{L_{eff}} = \sqrt{\frac{W_r |Z_L| Z_{00} cos\phi}{A_{eff}}} \tag{16}$$

where $Z_L = 50\Omega$ is the impedance of the spectrum analyzer; $Z_A$ is the input impedance of the antenna; $Z_{00} = 120\pi\Omega$ is the intrinsic impedance of the vacuum; $\phi$ is the phase of the impedance $Z_L$; $A_{eff}$ is the effective area of the receiving antenna; and $W_r(W)$ is the power measured by the spectrum analyzer.

The natural parameter of a receiving antenna is the capture area, or effective area, and the natural parameter of a transmitting antenna is the gain [3]. Other parameters such as the transmitting antenna area, or the receiving antenna gain, are obtained by means of the natural parameters and the effective area of the isotropic radiator.

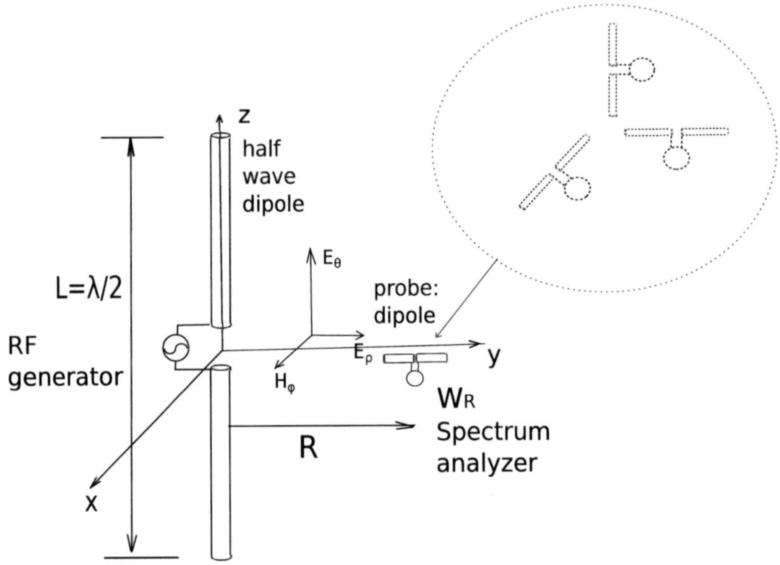


**Figure 5.** Measurement setup for E field components $E_x$, $E_y$, and $E_z$.

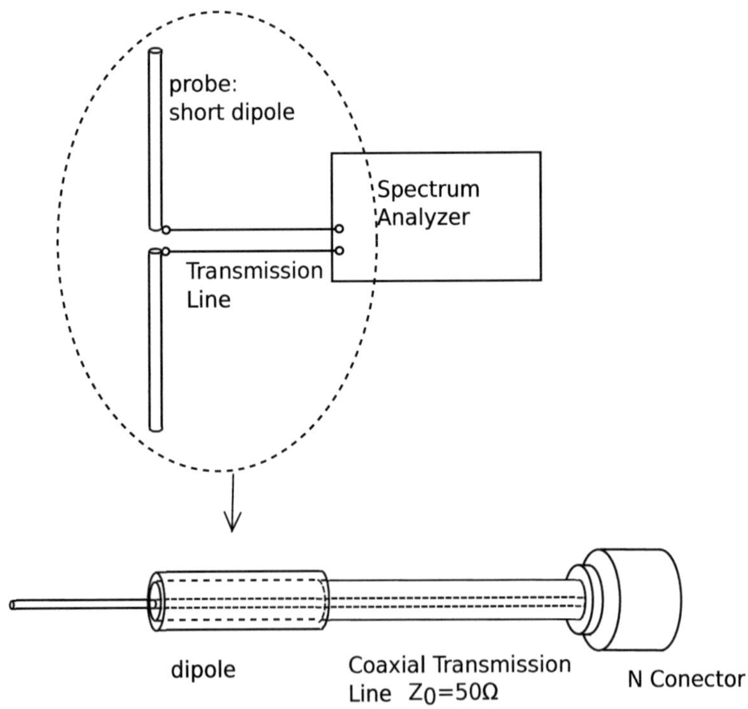


**Figure 6.** Simple electric field probe circuit.

### 2.1. Example of a radiating field from a dipole

An example of the electric field measurement using a dipole probe is shown in **Figure 5**. A half wave dipole is radiating energy, and the probe (dipole) measures the induced voltage for the three axis: $E_\theta$, $E_\rho$, $E_\phi$. **Figure 6** shows the electric field probe, composed of a dipole antenna, a coaxial transmission line, and the connections. This probe is connected to the spectrum analyzer in order to obtain the received voltage.

**Figure 7a** shows the received power measured with a dipole probe and **Figure 7b** shows the electric field calculated by means of Eq. (16). To obtain the electric field, it is necessary to know the effective length of the electric field probe ($L_{eff}$), the impedance of the electric field probe ($Z_A$), and the impedance of the spectrum analyzer ($Z_L$).

**Figure 8a** and **8b** shows the curve of the electric field component $E_\rho$ and $E_\phi$ as a function of distance.

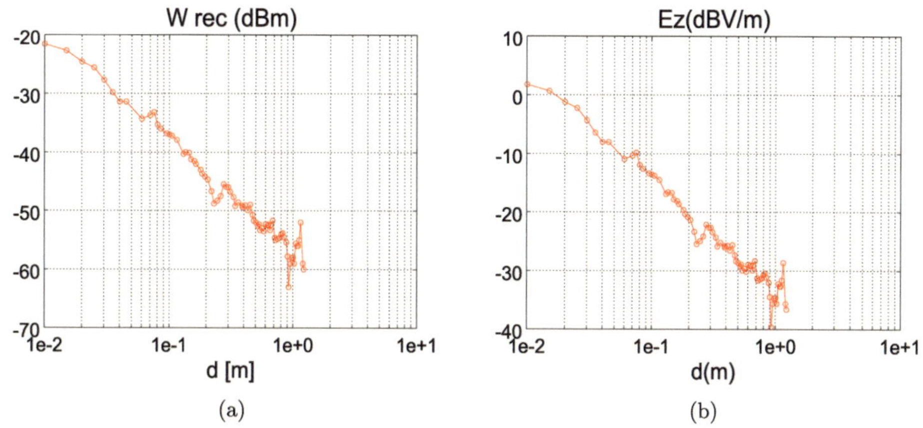


**Figure 7.** (a) Power measurement at the probe antenna terminals and (b) measured electric field component $E_z$.

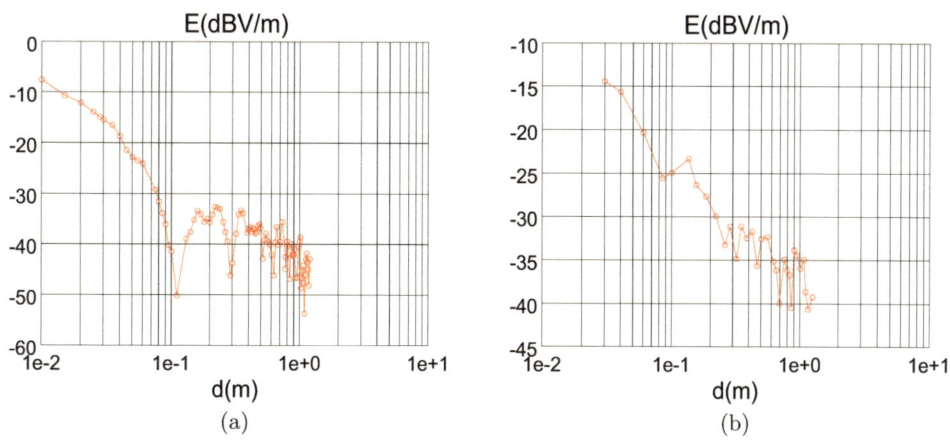


**Figure 8.** (a) Measured electric field component $E_\rho$ and (b) measured electric field component $E_\phi$.

## 3. Magnetic field probes

A very useful method for measuring the magnetic field of electromagnetic waves is by means of a loop antenna, which can be used with one turn or several turns, where a radiating field is received. A simple loop antenna is shown in **Figure 9a** [1] and a shielded loop antenna is shown in **Figure 9b** [4].

In order to obtain the induced voltage at the terminals of the loop antenna, consider a short loop antenna, where the wavelength of the electromagnetic wave received is much greater than the perimeter of the one-turn antenna. The Maxwell-Faraday Law expression in differential form is [1]:

$$\nabla\times \vec{E}= -\frac{\partial \vec{B}}{\partial t} \qquad (17)$$

where $B[T]$ is the magnetic flow density, $E[V/m]$ is the electric field, and $t[s]$ is the time.

If the electromagnetic wave has a harmonic variation as a function of the time, the electric and magnetic field can be expressed as [5]:

$$E = E_0 e^{j\omega t} \qquad (18)$$

$$H = H_0 e^{j\omega t} \qquad (19)$$

From Eq. (17), making the integration at both sides:

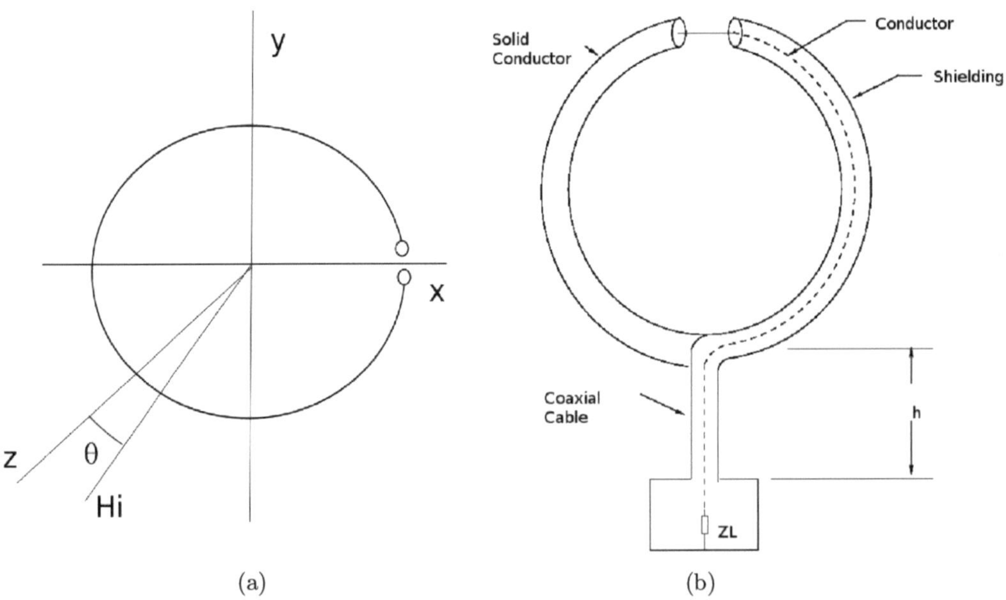


**Figure 9.** (a) Ideal Loop. (b) Shielding loop antenna.

$$\int_{s}\left(\nabla\times \vec{E}\right)\cdot \vec{ds} = -\int_{s}\frac{\partial \vec{B}}{\partial t}\cdot \vec{ds} \tag{20}$$

From Eq. (20), applying the Stokes theorem, the electromotive force (FEM) is [1]:

$$fem_{induced} = -\frac{\partial}{\partial t}\int_{s}\vec{B}\cdot \vec{ds} \tag{21}$$

The loop antennas can be classified (into two classes) electrically small and electrically large antennas. The first is useful for the measurement of the magnetic field induced by the radiated electromagnetic field [4]. The electrically large loop antennas are those whose circumference is about a free-space wavelength ($C \cong \lambda$). In this chapter, the electrically small loop antennas are discussed. The electrically small loop antennas are those whose overall length (number of turns by the length of the circumference) is usually less than one-tenth of a wavelength. This can be written as [2]:

$$n\cdot C < \lambda/10 \tag{22}$$

where n is the number of turns, $C[m]$ is the perimeter of the circumference of the loop, and $\lambda[m]$ is the wavelength.

In the small loop antennas, the current in the conductor of the antenna can be considered constant, i.e., $I(x, y, z) = I_0$ [2, 4].

### 3.1. Example

A small loop antenna of radius = 10 cm, with 50 turns, which bandwidth of interest is 200 kHz, then the number of turns by the perimeter of the circumference is

$$n\cdot C = 31.4\,\mathrm{m} \tag{23}$$

The wavelength for the highest frequency is

$$\lambda = \frac{c}{f} = \frac{3\cdot 10^{8}\,\mathrm{m/s}}{200\cdot 10^{3}} = 1500\,\mathrm{m} \tag{24}$$

Therefore,

$$\frac{\lambda}{10} = 150\,\mathrm{m} \tag{25}$$

The condition of Eq. (22) is

$$31.4 < 150\,\mathrm{m} \tag{26}$$

Then, the condition of Eq. (22) has been satisfied for electrically small loops of a bandwidth of 200 kHz [2, 4].

The magnetic field intensity inside the loop antenna shown in **Figure 15** can be considered constant, then, Eq. (21) can be expressed as:

$$fem_{induced} = -j\omega B cos\theta A \tag{27}$$

where $\omega = 2\pi f[\text{rad/s}]$, $f[\text{Hz}]$ is the frequency, $B\left[\text{Wb/m}^2\right]$ is the magnetic flux density, $A\left[\text{m}^2\right]$ is the area of the loop, and $\theta[\text{rad}]$ is the angle between z y H.

As the constitutive relation between the incident magnetic field to the antenna $H$ and the magnetic flux density $B$ [6]:

$$B = \mu_0 H \tag{28}$$

where $\mu_0$ is the magnetic permeability of the vacuum. Then,

$$fem_{induced} = -j\omega\mu_0 H cos\theta A \tag{29}$$

For the loop antenna of n turns, the induced FEM can be expressed as:

$$fem_{induced} = -jn\omega\mu_0 H cos\theta A \tag{30}$$

The amplitude of the induced FEM can be named open circuit voltage of the loop antenna ($V_{oc}$):

$$V_{oc} = n\omega\mu_0 H cos\theta A \tag{31}$$

If the loop antenna has a ferrite core, the magnetic permeability of the ferrite is $\mu = \mu_0\mu_r$, where $\mu_r$ is the relative magnetic permeability of antenna loop with a ferrite core, and the magnetic flux density is $B = \mu_r\mu_0 H$.

### 3.2. Electric circuit of the loop antenna

The electric circuit of a loop antenna of n turns in receiving mode can be expressed by impedance $Z_A = R + jX$ in series with the ideal voltage generator. The spectrum analyzer can be represented by means of an impedance $Z_L = 50\,\Omega$. This equivalent Thevenin electric circuit is shown in **Figure 18**, where $Voc$ is the open circuit voltage, $Z_L[\Omega]$ is the impedance of the spectrum analyzer, and $Z_A[\Omega]$ is the impedance of the loop antenna.

The measured voltage $V_m$ by the spectrum analyzer can be written as:

$$V_{oc}\frac{Z_L}{Z_L + Z_A} = V_m \tag{32}$$

From Eqs. (31) and (32), the magnetic field can be expressed like a function of the measured voltage $V_m$:

$$H = V_m \frac{Z_L + Z_A}{Z_L} \frac{1}{n\omega A\mu_0 cos\theta} \tag{33}$$

Therefore, the magnetic flux density results in (**Figure 10**):

$$B = \mu_0 H = V_m \frac{Z_L + Z_A}{Z_L} \frac{1}{n\omega A cos\theta} \tag{34}$$

### 3.3. Antenna factor

The definition of the antenna factor $K_i$, relates the magnetic field $H$, of the electromagnetic wave incident to the loop antenna with the current flowing in the loop $I_m$ [7]:

$$K_i = \frac{H}{I_m} \tag{35}$$

where

$$I_m = \frac{V_m}{Z_L} \tag{36}$$

From Eqs. (35) and (36), the antenna factor is

$$K_i = (Z_L + Z_A) \frac{1}{n\omega A\mu_0 cos\theta} \tag{37}$$

The antenna factor is important because it relates the current measured in the loop antenna and the incident unknown magnetic field.

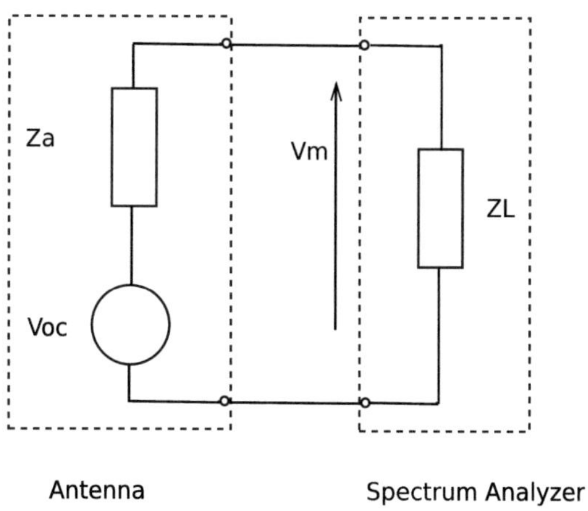


**Figure 10.** Electric circuit model of the loop antenna used for the measurement.

Another way to define the antenna factor is:

$$K_v = \frac{H}{V_{oc}} \tag{38}$$

where the subindex shows if it is defined by the voltage (v) or current (i).

Using Eq. (31) $K_v$ results:

$$K_v = \frac{1}{n\omega A\mu_0 cos\theta} \tag{39}$$

Note that the relation between $K_i$ and $K_v$ is

$$\frac{K_i}{K_v} = Z_L + Z_A \tag{40}$$

*3.3.1. Example*

The antenna factor *Kv* of three short loop antennas has been computed by means of Eq. (39), as shown in **Figures 11, 12,** and **13**. The antenna factor $K_v$ has a slope of −20 dB/dec and it is decreasing with $n$, $\omega$, $\mu$, and $A$. This behavior can be observed in **Table 1**.

Typically, the small loop antennas are used to measure the magnetic fields in a range of frequencies of 20 Hz–30 MHz [8]. The loop antenna can also be used to measure the magnetic field strength emission from a device [9].

From the equation of the open circuit voltage of Eq. (31), $V_{oc}$ increases with the area and the number of turns n. Both parameters can be used to select the voltage of reception and the sensibility of the measurement. The section of the wire selected has been chosen to reduce the losses as Joule effect.

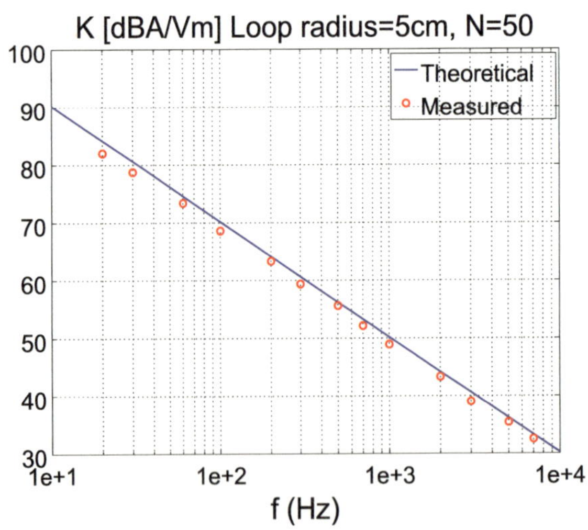


**Figure 11.** Antenna factor $K_v$ measured and calculated. Loop: radius = 0.05 m and N = 50.

http://dx.doi.org/10.5772/intechopen.72687

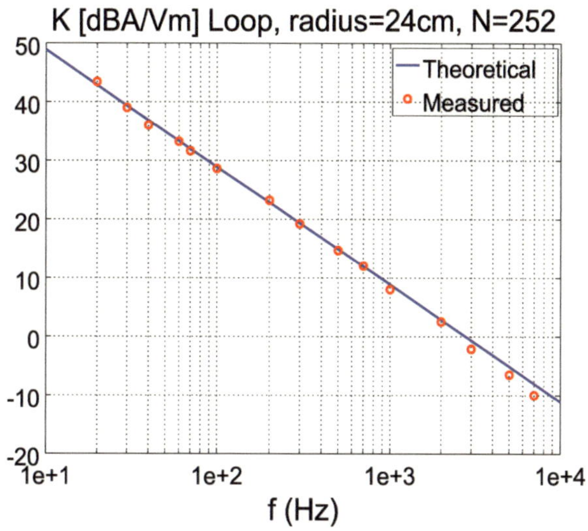


**Figure 12.** Antenna factor $K_v$ measured and calculated. Loop: radius = 0.24 m and N = 252.

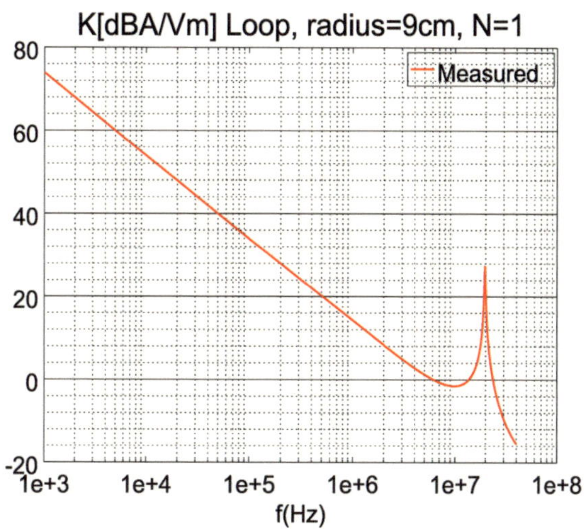


**Figure 13.** Antenna factor $K_v$ measured. Loop: radius = 0.09 m and N = 1.

| *a* | n | $K_v$ **(dBA/V$_m$)f = 1 kHz** |
|---|---|---|
| 0.05 | 50 | 50 |
| 0.24 | 252 | 8 |
| 0.09 | 1 | 75 |

**Table 1.** Antenna factors for three different loop antennas.

Currently, the measurement of magnetic field emitted by electronic devices is done to verify the electromagnetic compatibility [9, 10] or the emissions in the nature [11]. In the magnetic field measurement, a calibrated measurement system is required, where the probes are excited by the uniform magnetic field [12] to obtain the antenna factor [13]. The magnetic field to be measured should be in a near-field zone [14] or in a far-field zone [5].

## 4. Example

A measurement of the power supply of a personal computer is shown in **Figure 14a** and **14b**. The hotspots are indicated with dark grey where the emission is highest. The magnetic field probe made with a ferrite core, which has been used in this measurement, is shown in **Figure 15**.

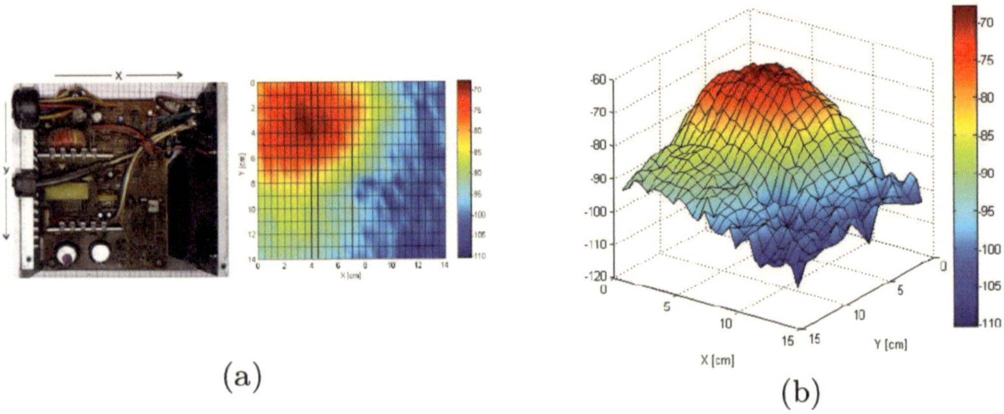


**Figure 14.** (a) Power supply in a 2D picture of emission. (b) Emission of the power supply versus x,y.

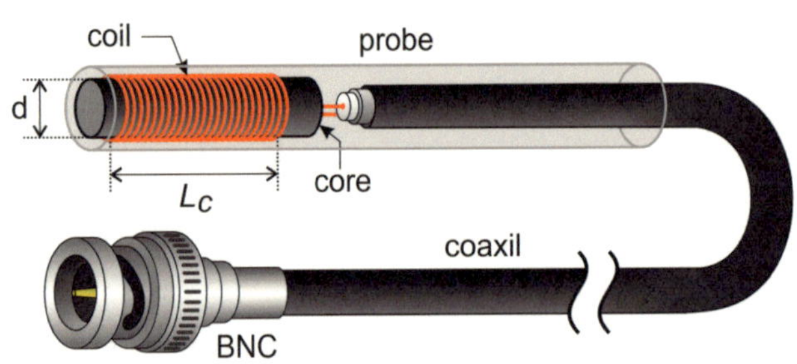


**Figure 15.** The magnetic field probe with a coil built with ferrite core.

## 5. Helmholtz coils

The Helmholtz coils are the most simple configuration to produce a magnetic field relatively constant. Helmholtz coils are two circular coils coaxial with the same radius, which is equal to the distance between the planes of the coils [15]. When the current flows in opposite directions, the configuration is called anti-Helmholtz coils. The cancellation of the magnetic field is important in various applications such as measurements, bioengineering investigations, and

http://dx.doi.org/10.5772/intechopen.72687

calibrations of probes and sensors. In Helmholtz coils expand the area of uniformity of the magnetic field, and reduce the variation of the field in the axes it is interesting. Based on the study of a two-coil system, it has been determined that three coils through which current flows produce magnetic field with better characteristics in intensity and uniformity than the standard Helmholtz coils. The polygonal coils are easier to be produced in the industry. [16]. When the number of sides of a polygon forming the coil is increased, the distribution on the intensity of magnetic field is more uniform [16, 17]. If greater uniformity of the magnetic field is required, a five-rectangular coils system for measuring biomagnetics can be used [16, 18]. Eight Helmholtz coils are used to calibrate a magnetometer in a uniform magnetic field system [19] (**Figures 16** and **17**).

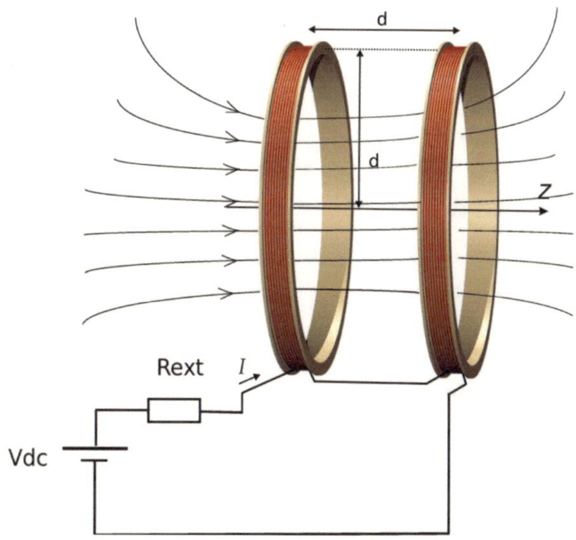


**Figure 16.** Picture of the Helmholtz coils.

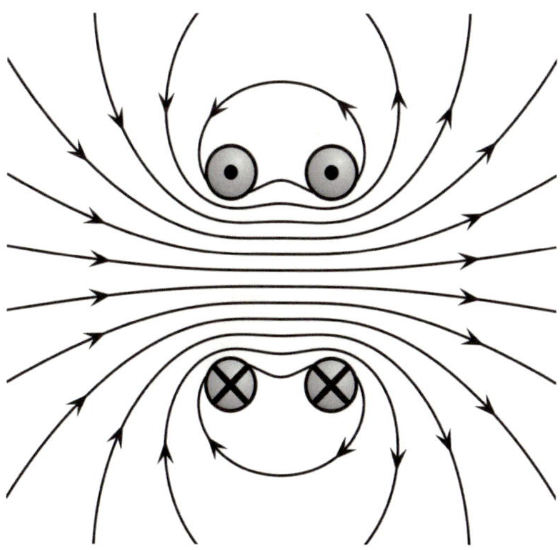

**Figure 17.** Magnetic field lines of the Helmholtz coils.

### 5.1. Helmholtz coils with DC current applied

In the magnetic flux density vector $\vec{B}$, produced by two coils, as shown in **Figure 16**, where a DC current I is flowing by the coils, two identical terms are obtained. The coils are placed at z=−d/2 and z=+d/2.

The magnetic flux density $\vec{B}$ of the Helmholtz coils with the radius equal to the distance between coils has first and second derivatives of zero at $z = 0$ $\frac{\partial B}{\partial z} = \frac{\partial^2 B}{\partial z^2} = 0$ [20]. Then, the magnetic flux density is uniform at $z = 0$:

$$\vec{B(r)} \mid_{z=0} = \frac{\mu_0 I}{\left(\frac{5}{4}\right)^{3/2} a} \hat{z} \tag{41}$$

The magnetic flux density can be expanded by Taylor series, and then the deviation from $\vec{B(z)} \mid_{z=0}$ is $B(z) = B(0)\left(1 \pm 1.510^{-4}\right)$ at $|z| < \frac{a}{10}$.

### 5.2. Helmholtz coils with AC current applied computed in any point in the space

Consider two coils connected to an AC generator, where a sinusoidal current $I(t)$ flows as shown in **Figure 18**. The Helmholtz coils are used in low frequencies: LF, VLF, and ULF. For these frequencies, the wavelength is much larger than the dimensions of the loop. A number of turns is used in each winding that make up the Helmholtz coils, where the number of turns of the coil (n), multiplied by the perimeter of the circumference (C), is much less than a wavelength $NC < \lambda$, which can be enunciated as [2]:

$$NC < \frac{\lambda}{10} \tag{42}$$

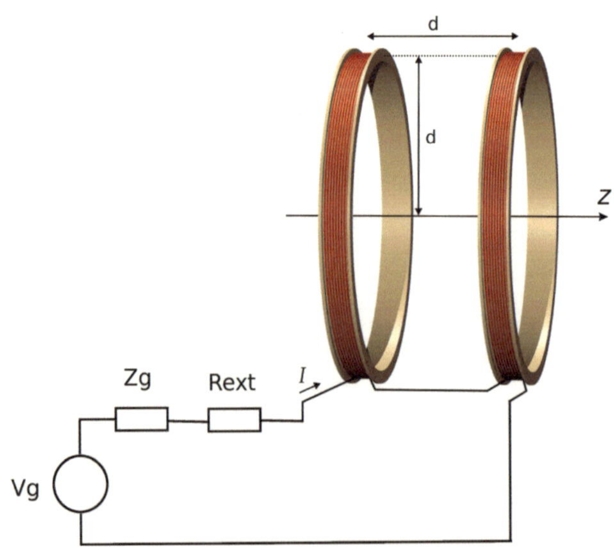


**Figure 18.** AC current connected to the Helmholtz coils.

To the frequencies that fulfill the relation (42), the current is practically constant in each point of the Helmholtz coils, and it will only be variable with the time. The vectorial $\vec{A}$ can be written as [2]:

$$\vec{A}=\frac{\mu_0}{4\pi}\int_v \frac{\overrightarrow{J(r)}\, dv'}{R} \tag{43}$$

For the particular case of the coil can be considered a linear wire where the current flows, the expression (43) is

$$\vec{A}=\frac{\mu_0}{4\pi}\oint_c I\frac{\overrightarrow{dl'}}{R} \tag{44}$$

where $\overrightarrow{dl'}=(-asen\phi', acos\phi', 0)d\phi'$

$\vec{r}=(rsen\theta, 0, rcos\theta)$

$r' = (acos\overrightarrow{\phi'}, asen\phi', 0)$

$R=|\vec{r}-\vec{r'}|=\sqrt{r^2+a^2-2rasen\theta cos\phi'}$

The current intensity I is constant at the perimeter of the loop [2], then, the vector $\widetilde{A}$ results in:

$$\vec{A}=\frac{\mu_0}{4\pi}\oint_0 \pi\frac{\widehat{\phi} Icos\phi' d\phi'}{\sqrt{r^2+a^2-2rasen\theta cos\phi'}} \tag{45}$$

that is, the Fredholm integral of the first kind [21], this results in:

$$\vec{A}=\frac{\mu_0 I}{\pi k}\sqrt{\frac{a}{\rho}}\cdot\left[\left(1-\frac{1}{2}k^2\right)\cdot K(k^2)-E(k^2)\right] \tag{46}$$

where $\rho=\sqrt{r^2-z^2}$, $k^2=\frac{4a\rho}{(a+\rho)^2+z^2}$, $K(k)=\int_0^{\pi/2}\frac{d\phi}{\sqrt{1-k^2sen^2\phi}}$ eliptic integral of the first kind. $E(k)=\int_0^{\pi/2}\sqrt{1-k^2sen^2\phi d\phi}$ eliptic integral of the second kind.

Then, $\vec{B}=\nabla\times\vec{A}$ can be computed [21]:

$$\vec{B_\rho}=\frac{\mu_0 I}{2\pi}\frac{z}{\rho\sqrt{(\rho+a)^2+z^2}}\cdot\left[\frac{\rho^2+a^2+z^2}{(a-\rho)^2+z^2}\cdot E(k^2)-K(k^2)\right] \tag{47}$$

$$\vec{B_z}=\frac{\mu_0 I}{2\pi}\frac{1}{\sqrt{(\rho+a)^2+z^2}}\cdot\left[\frac{a^2-\rho^2-z^2}{(a-\rho)^2+z^2}\cdot E(k^2)-K(k^2)\right] \tag{48}$$

The total magnetic flux density of the Helmholtz coils, where the coils are placed at $z=-d/2$ and $z=+d/2$ is

$$B_\rho = B_\rho(\rho, z + d) + B_\rho(\rho, z - d) \tag{49}$$

$$B_z = B_z(\rho, z + d) + B_z(\rho, z - d) \tag{50}$$

### 5.3. Example

The Helmholtz coils have been built with two coils of radius $a = 0.39\,\text{m}$, number of turns $N = 8$, space between coils $s = a = 0.39\,\text{m}$, with the AC current applied to the coils of $|I| \cong 15\,\text{mA}$ at $f = 1\,\text{kHz}$. The magnetic flux density can be calculated with Eq. (41), thus:

$$\vec{B}\Big|_{z=0} = \frac{4\pi 10^{-7}\,\text{H/m}15 \cdot 10^{-3}A}{\left(\frac{5}{4}\right)^{3/2} 0.39\text{ m}}\hat{z} \tag{51}$$

Results:

$$\vec{B}\Big|_{z=0} = 27.6\mu T\hat{z} \tag{52}$$

In **Figure 19a,** the magnetic flux density as a function of z for $\rho = 0$ has been plotted, and the relative error of the magnetic flux density as a function of $z$ and $\rho$ in a 3D plot is shown in **Figure 19b**.

In the zone where the magnetic flux density is uniform, at $z = 0$, the probes to be calibrated are placed, at a low frequency signals from DC up to 200 kHz, the Helmholtz coils are used. The errors of the magnetic flux density B can be calculated using Eqs. (47), (48), (49) and (50).

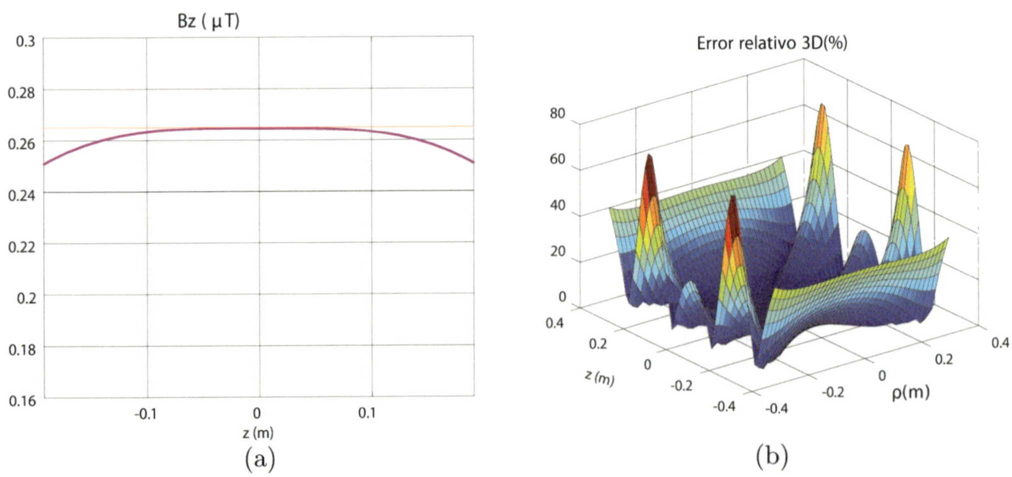


**Figure 19.** (a) Magnetic flux density $Bz(z)$ at $\rho = 0$ and (b) relative error of the magnetic flux density $B$ with respect to $B(z = 0)$.

## Author details

Walter Gustavo Fano

Address all correspondence to: gustavo.gf2005@gmail.com

Faculty of Engineering, Universidad de Buenos Aires, Buenos Aires, Argentina

## References

[1] Trainotti V, Fano WG. Ingenieria Electromagnetica. Vol. 1, 1st ed. Buenos Aires: Nueva Libreria; 2004. p. 512

[2] Balanis CA. Antenna Theory: Analysis and Design. Hoboken: Wiley-Interscience; 2005

[3] Trainotti V. Electromagnetic compatibility (emc) antenna gain and factor. IEEE Transactions on Electromagnetic Compatibility. 2017;**59**:1006-1015

[4] Johnson R, Jasik H. Antenna Engineering Handbook, Electronics: Electrical engineering. New York: McGraw-Hill; 1993

[5] Trainotti V, Fano WG, Dorado L. Ingenieria Electromagnetica. Vol. 2, 1st ed. Buenos Aires: Nueva Libreria; 2005. p. 1007

[6] Van Bladel J. Electromagnetic Fields. Hoboken: Wiley; 2007

[7] Osburn JDM. EMC antenna parameters and their relationships. Interference Technology; 2012. https://interferencetechnology.com/emc-antenna-parameters-and-their-relationships/#

[8] Montrose MA, Nakauchi EM. Test for EMC Compliance, Approach and Techniques. Hoboken: Wiley; 2004

[9] Fano WG. RF Emissions of compact fluorescent lights. Interference Technology. Vol. Europe EMC Guide; 2013. https://interferencetechnology.com/rf-emissions-of-compact-fluorescent-lights/

[10] Paul C. Introduction to Electromagnetic Compatibility, Wiley Series in Microwave and Optical Engineering. Hoboken: Wiley; 2006

[11] Maffia E, Trainotti V, Fano WG, Trench N. Medición de la resonancia de schumann. Latinmag Letter. 2011;**1**(A02):1-8

[12] Pahl RA, Rovey JL, Pommerenke DJ. Calibration of magnetic field probes at relevant magnitudes. In: Proceedings of the 2013 19th IEEE Pulsed Power Conference (PPC). 16-21 June 2013. San Francisco, CA, USA: IEEE; 2013. pp. 1-6

[13] Fano WG, Alonso R, Carducci LM. Near field magnetic probe applied to switching power supply. In: Proceedings of the 2016 IEEE Global Electromagnetic Compatibility Conference (GEMCCON). 7-9 Nov 2016. Mar del Plata, Argentina: IEEE; 2016. pp. 1-4

[14] Trzaska H. Electromagnetic Field Measurements in the Near Field. Atlanta: Noble Publishing; 2001

[15] Petrovik DM, Radic MD. Generalization of Helmholtz coil problem. Serbian Journal of Electrical Engineering. 2015;**12**:375-384

[16] Merritt R, Purcell C, Stroink G. Uniform magnetic field produced by three, four, and five square coils. Review of Scientific Instruments. 1983;**54**(7):879-882

[17] Lu X, Tian J. Simulation for the polygonal coil magnetic field. Simulation for the Polygonal Coil Magnetic Field. 2013;**6**:43-49

[18] Wang J, She S, Zhang S. An improved Helmholtz coil and analysis of its magnetic field homogeneity. Review of Scientific Instruments. 2002;**73**(5):2175-2179

[19] Baranov P, Baranova V, Uchaikin S, Pisarenko Y. Creating a uniform magnetic field using axial coils system for calibration of magnetometers. In: Proceedings of the 2016 Dynamics of systems, Mechanisms and Machines (Dynamics). 15-17 Nov. 2016. Omsk, Russia. pp. 1-5

[20] Fernandez JC. Ingenieria Electromagnetica. I Modelos estáticos y circuitales. Vol. 1, 1st ed. Buenos Aires: Eudeba; 2013. p. 690

[21] Dolezel I, Pavel K, Pavel S. Integral Methods in Low-Frequency Electromagnetics. Hoboken: Wiley; 2009

# Chapter 11

# Lifetime Assessment of Electrical Insulation

Eyad A. Feilat

Additional information is available at the end of the chapter

http://dx.doi.org/10.5772/intechopen.72423

**Abstract**

In this chapter, a review of the Weibull probability distribution, probability ranking, and the Weibull graphical estimation technique is presented. A review of single-stress and multiple-stress life models of electrical insulation is also introduced. The chapter also describes the graphical, linear and multiple linear regression techniques used in estimating the parameters of the aging models. The application of maximum likelihood estimation technique for estimating the parameters of combined life models of electrical insulation is illustrated.

**Keywords:** life models, aging, insulation, Weibull probability, maximum likelihood estimation, least square estimation

## 1. Introduction

A lifetime analysis of electrical insulation failure is an approach that relies on statistical analysis of data that are attributed to the breakdown of the electrical insulation due to the presence of degrading stresses, such as electrical, thermal and other environmental factors. The lifetime analysis can provide statistical information about the electrical insulation such as lifetime characteristics, probability of failures, lifetime percentiles or any time percentile under normal operating conditions. In this approach, the insulation life is determined by measuring the time-to-breakdown of identical specimens of the solid insulation subjected to life tests [1–5]. Obtaining life data under normal operating conditions is a very time consuming and costly process, rendering it impractical. Besides, it is important to observe modes of failure of the electrical insulation to better understand the prevailing mechanisms of breakdown. Consequently, electrical insulation design engineers and material scientists devised methods to force the insulation to fail in shorter periods of time. These methods seek to accelerate the failures of insulation samples by applying stresses at levels that exceed the levels that the insulation will encounter under normal operating service conditions. Acceleration is accomplished by testing the insulation (specimen or device) using single or combined high stress levels which could involve electrical, thermal or environmental stresses for either short periods (few seconds or minutes) or long periods (few hours or days) [6]. The accelerated test data are then considered as a base for extrapolation to obtain an estimate of the lifetime of the insulation when the device or material is operated at normal operating conditions for relatively long time periods (decades) of years [7–27].

Times-to-breakdown obtained by accelerated life (aging) tests are analyzed using an underlying lifetime probability distribution. The probability distribution can, correspondingly, be used to make predictions and estimates of lifetime measures of interest at the particular stress level. This is accomplished by projecting or mapping lifetime measures from high stress level to a service level. It can be assumed that there is some model (or function), which can be described mathematically, that maps the lifetime estimate from the high stress level to the service level, and can be as simple as possible (i.e. linear, exponential, etc.) [28–37]. The parameters of the lifetime models can be estimated by combining the proposed life models with the Weibull probability distribution function. Maximum likelihood estimation can be used to estimate the parameters of the combined Weibull-electrical-thermal models using experimental data obtained by measuring time-to-breakdown of the insulation.

## 2. Weibull probability distribution

The Weibull distribution has been widely recognized as the most common distribution in breakdown testing of solid dielectric insulation and in reliability studies [28–31, 37–40]. Its popularity is attributed to the many shapes it attains for various values of the shape parameter ($\beta$). It can model a large variety of data and life characteristics. The distribution also has the important properties of flexibility and a closed form solution for the integral of the Weibull probability density function (*pdf*). This latter property is important for easy determination of the Weibull cumulative probability function (*cdf*) and its corresponding parameters. There are three forms of the Weibull cumulative probability distribution functions, namely, the three-parameter Weibull distribution, the two-parameter Weibull distribution and the mixed Weibull distribution. In life data analysis, the two-dimensional Weibull distribution is often used to describe the time-to-breakdown of solid dielectric insulation with voltage, as it is a convenient way for deriving the V-t characteristic or life models. The two-parameter Weibull *pdf* and *cdf* distributions are defined as:

$$\begin{aligned} f(t;\alpha,\beta) &= \frac{\beta}{\alpha}\left(\frac{t}{\alpha}\right)^{\beta-1} \exp\left\{-\left(\frac{t}{\alpha}\right)^{\beta}\right\} \\ F(t;\alpha,\beta) &= 1-\exp\left\{-\left(\frac{t}{\alpha}\right)^{\beta}\right\} \qquad t\geq 0 \\ F(t;\alpha,\beta) &= 0 \qquad t<0 \end{aligned} \tag{1}$$

where $t$ is the time-to-breakdown; $\alpha$ is the scale parameter, $\alpha > 0$; $\beta$: is the shape parameter, $\beta > 0$.

$F(t)$ represents the proportion of samples initially tested that will fail by time $t$. The scale parameter ($\alpha$) represents the time required for 63.2% of the tested samples to fail. The shape parameter ($\beta$), or slope of the Weibull distribution, is a measure of the dispersion of the time-to-breakdown. The unit of $\alpha$ is time, while $\beta$ is dimensionless.

### 2.1. Specific characteristics of the Weibull probability distribution

The Weibull *pdf* and/or *cdf* distributions are characterized by the parameters $\alpha$ and $\beta$. The scale parameter $\alpha$ is usually a function of the applied voltage when the time is a random variable. A change in the scale parameter $\alpha$ has the same effect on the distribution as a change of the abscissa scale. If $\alpha$ is increased, while $\beta$ is kept the same, the distribution gets stretched out to the right and its height decreases, while maintaining its shape and location as shown in **Figure 1** [30]. On the other hand, if $\alpha$ is decreased, while $\beta$ is kept the same, the distribution gets pushed in toward the left (i.e. toward its beginning, or 0) and its height increases. For $\beta < 1$ the distribution has the reversed *J* shape. If $\beta = 1$, the Weibull distribution becomes a two-parameter exponential distribution as shown in **Figure 2**. For $\beta = 2$, it becomes the Rayleigh distribution. For $\beta$ <2.6, the Weibull *pdf* is positively skewed (has a right tail). For $2.6 < \beta < 3.7$, its coefficient of skewness approaches zero (no tail); consequently, it may approximate the normal *pdf*, and for $\beta$ >3.7, it is negatively skewed (left tail) [30].

An interpretation of the shape parameter indicates that if $\beta$ >1, the dielectric fails as a result of wear and the failure rate increases with time. In such a case, the failure rate increases with time. If $\beta$ <1, the failure rate decreases with time. When such behavior is encountered, this indicates that the sample has some units with technical defects.

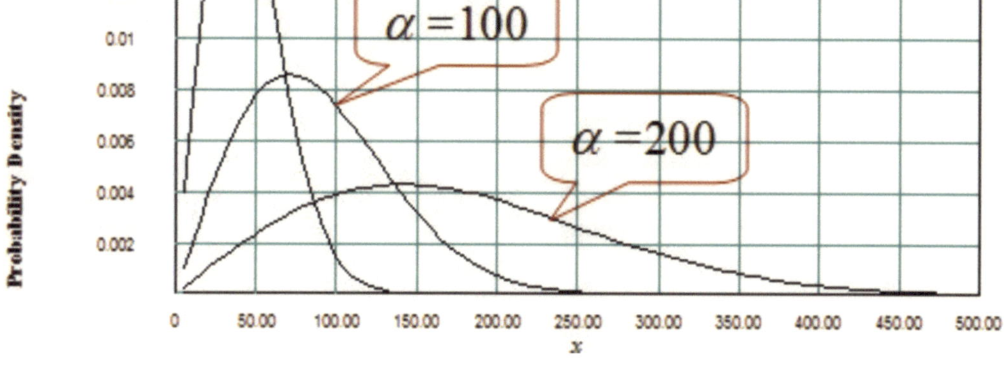


**Figure 1.** Effect of the scale parameter on Weibull probability distribution.

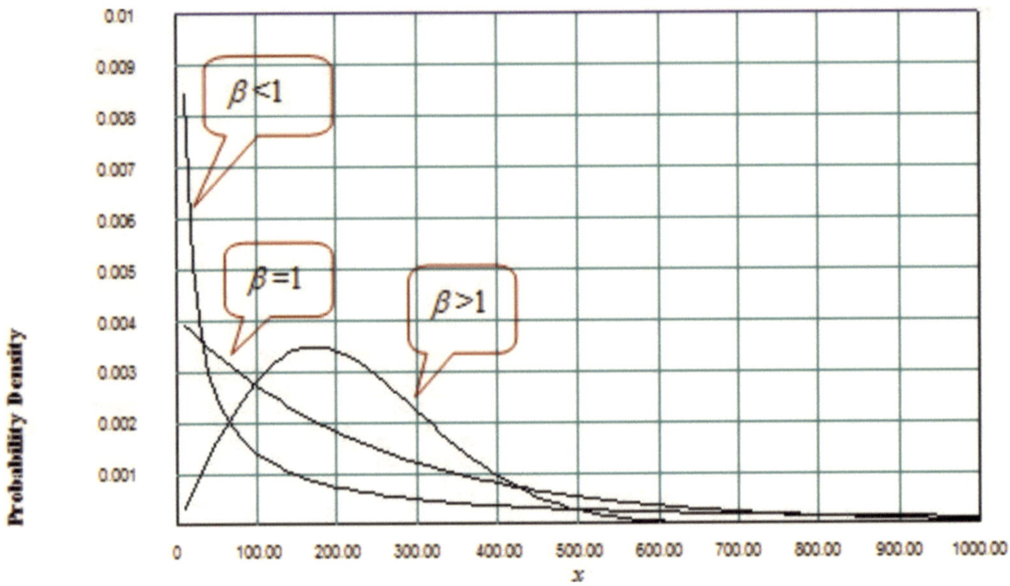


**Figure 2.** Effect of the shape parameter on the Weibull probability distribution.

### 2.2. Estimating the cumulative probability of failure data

The cumulative probability of failure $F(t_i)$ is estimated by a value $P_i$, based on the order statistic rank $i$ and the sample size $n$. A variety of equations for approximating $F(t_i)$ has been suggested for the plotting position $P_i$ [28, 29]:

1. Mean rank approximation

$$P_i = E\{F(t_i)\} = \frac{i}{n+1} \tag{2}$$

2. Median rank approximation

$$P_i = \text{median of } F(t_i) = \frac{i-0.3}{n+0.4} \tag{3}$$

3. Mode rank approximation

$$P_i = \text{mode}\{F(t_i)\} = \frac{i-1}{n-1} \tag{4}$$

4. Sample cumulative distribution function (cdf) approximation

$$P_i = \text{sample } cdf = \frac{i-0.5}{n} \tag{5}$$

The choice of the rank approximation is generally immaterial except when the sample size is small. However, both the mean and median rank approximations are most widely used. In this work, the median rank approximation is used in estimating the probability of failure. Note that the median rank approximation is also known as Benard's approximation.

### 2.3. Parameter estimation of Weibull distribution

The Weibull distribution parameters $\alpha$ and $\beta$ are the variables that govern the characteristics of the Weibull *pdf*. Once the Weibull distribution has been selected to represent the failure data, the associated parameters can be determined from the experimental data. Weibull parameters can be estimated graphically on a probability plotting paper [28–31] or analytically using either Least Squares (LS) or Maximum Likelihood (ML) estimation techniques. An overview of these techniques [1–4] will be presented next.

### 2.4. Graphical technique

A graphical technique is the simplest method for estimating the Weibull parameters using probability plotting [29, 38]. The breakdown results at each individual stress level are plotted on a specially constructed probability plotting paper (Weibull Probability Paper) as shown in **Figure 3**. A probability plotting looks at the *cdf* of the Weibull distribution and attempts to linearize it by employing a Weibull probability paper. The ordinate axis has a nonlinear scale corresponding to the cumulative probability of failure. The abscissa has a log scale of the time-to-breakdown. To plot the data, the data are ordered from smallest to largest and then a cumulative probability of breakdown to each point is assigned. If the plotted data fit a straight line, the slope of the line can be obtained and thus the shape parameter $\beta$ can be obtained [14]. The scale parameter $\alpha$ can be determined by finding the time-to-breakdown corresponding to

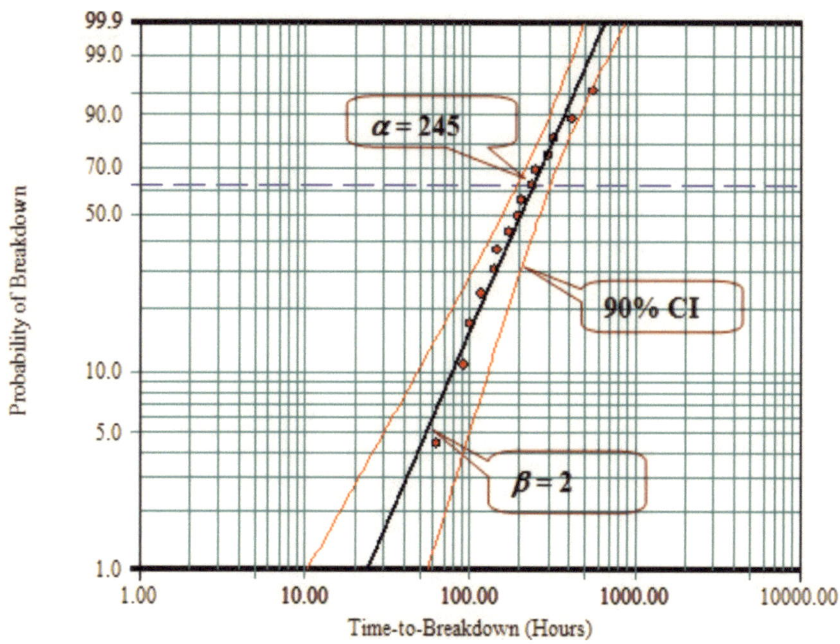


**Figure 3.** Cumulative probability distribution with confidence interval plotted on Weibull paper.

a cumulative probability of 63.2%. Using this simple, but time consuming approach, the parameters of the Weibull distribution can be determined. This procedure is repeated for each stress level of the accelerated aging tests.

Estimating the parameters of the Weibull distribution by a graphical technique using a probability plotting method has some shortfalls. A manual probability plotting is not always consistent in the results. Plotting a straight line through a set of points is a subjective procedure; it differs from person to person. In addition, the probability plot must be constructed for each stress level. This, as a result, takes tremendous time and effort to plot the data. Furthermore, sufficient failures must be observed at each stress level, which is not always possible.

### 2.5. Least squares technique (LS)

The least squares technique is a linear regression estimation technique that fits a straight line to a set of data points, in an attempt to estimate the parameters associated with the straight lines. The parameters are estimated such that the sum of the squares of the vertical deviations from the points to the line is minimized according to [1–4]

$$J = \sum_{i=1}^{N} \left(\tilde{a} + \tilde{b}x_i - y_i\right)^2 = \min\left(\tilde{a}, \tilde{b}\right) \sum_{i=1}^{N} \left(\tilde{a} + \tilde{b}x_i - y_i\right)^2 \tag{6}$$

where $\tilde{a}$and $\tilde{b}$are the LS estimates of $a$ and $b$, and $N$ is the number of data points. To obtain $\tilde{a}$ and $\tilde{b}$, the performance index $J$ is differentiated with respect to $a$ and $b$ as shown below.

$$\frac{\partial J}{\partial \tilde{a}} = 2\sum_{i=1}^{N} \left(\tilde{a} + \tilde{b}x_i - y_i\right) \tag{7}$$

$$\frac{\partial J}{\partial \tilde{b}} = 2\sum_{i=1}^{N} \left(\tilde{a} + \tilde{b}x_i - y_i\right)x_i \tag{8}$$

Setting Eqs. (7) and (8) equal to zero yields

$$\sum_{i=1}^{N} \left(\tilde{a} + \tilde{b}x_i - y_i\right) = \sum_{i=1}^{N} (\tilde{y}_i - y_i) = -\sum_{i=1}^{N} (y_i - \tilde{y}_i) = 0 \tag{9}$$

$$\sum_{i=1}^{N} (a + bx_i - y_i)x_i = \sum_{i=1}^{N} (\tilde{y}_i - y_i)x_i = -\sum_{i=1}^{N} (y_i - \tilde{y}_i)x_i = 0 \tag{10}$$

Solving Eqs. (9) and (10) simultaneously yields

$$\tilde{a} = \frac{\sum_{i=1}^{N} y_i}{N} - \tilde{b}\frac{\sum_{i=1}^{N} x_i}{N} \tag{11}$$

http://dx.doi.org/10.5772/intechopen.72423

and

$$\widetilde{b} = \frac{\sum_{i=1}^{N} x_i y_i - \frac{\sum_{i=1}^{N} x_i \sum_{i=1}^{N} y_i}{N}}{\sum_{i=1}^{N} x_i^2 - \frac{\left(\sum_{i=1}^{N} x_i\right)^2}{N}} \tag{12}$$

The least squares estimation technique is good for functions that can be linearized. Its calculations are easy and straightforward, and the correlation coefficient provides a good measure of the goodness-of-fit of the chosen distribution. However, for some complex distributions, it is difficult and sometimes impossible to implement.

### 2.6. Maximum likelihood estimation (MLE)

The maximum likelihood parameter estimation seeks to determine the parameters that maximize the probability (likelihood) of the failure data. Statistically, the MLE is considered to be more robust and yields estimators with good statistical properties. The MLE has the following statistical properties:

1. The ML estimators are consistent and asymptotically efficient.
2. The probability distribution of the estimators is asymptotically normal.
3. For small sample sizes, the ML estimators are considered to be more precise than those obtained by LS method. Moreover, the ML estimators can converge into a solution even with only one failure.
4. The MLE technique applies to most models and to different types of data.
5. The ML estimates are unique, and as the size of the sample increases, the estimates statistically approach the true values of the population.

The theory of the MLE method is described as follows: Let $t$ be a continuous random variable representing the time-to-breakdown and characterized by the two-parameter Weibull distribution with *pdf*:

$$f(t;\alpha,\beta) = \left(\frac{\beta}{\alpha}\right)\left(\frac{t}{\alpha}\right)^{\beta-1} \exp\left[-\left(\frac{t}{\alpha}\right)^{\beta}\right] \tag{13}$$

where $\alpha$ and $\beta$ are unknown constant parameters which need to be estimated. For an experiment with $N$ independent observations, $t_1$, $t_2$, ..., $t_N$ in a given sample, then the likelihood function associated with this sample is the joint density of the $N$ random variables, and thus is a function of the unknown Weibull parameters ($\alpha$, $\beta$). The likelihood function is defined by [1–4]:

$$L\left(t_1, t_2, \cdots, t_N | \widetilde{\alpha}, \widetilde{\beta}\right) = L = \prod_{i=1}^{N} f\left(t_i; \widetilde{\alpha}, \widetilde{\beta}\right) = \prod_{i=1}^{N} \frac{\widetilde{\beta}}{\widetilde{\alpha}} \left(\frac{t_i}{\widetilde{\alpha}}\right)^{\widetilde{\beta}-1} \exp\left[-\left(\frac{t_i}{\widetilde{\alpha}}\right)^{\widetilde{\beta}}\right] \tag{14}$$

The logarithmic likelihood function is given by

$$\Lambda = \ln L = \sum_{i=1}^{N} \ln f\left(t_i; \widetilde{\alpha}, \widetilde{\beta}\right) = \sum_{i=1}^{N} \ln \left[\frac{\widetilde{\beta}}{\widetilde{\alpha}} \left(\frac{t_i}{\widetilde{\alpha}}\right)^{\widetilde{\beta}-1} \exp\left(-\left(\frac{t_i}{\widetilde{\alpha}}\right)^{\widetilde{\beta}}\right)\right] \tag{15}$$

The parameter estimates $\left(\widetilde{\alpha}, \widetilde{\beta}\right)$ are obtained by maximizing $L$ or $\Lambda$ which is much easier to work with than $L$. The ML estimators are determined by taking the partial derivatives of $\Lambda$ with respect to $\widetilde{\alpha}$ and $\widetilde{\beta}$ set them to zero, that is, $\frac{\partial \Lambda}{\partial \widetilde{\alpha}} = 0$, $\frac{\partial \Lambda}{\partial \widetilde{\beta}} = 0$, where,

$$\frac{\partial \Lambda}{\partial \widetilde{\beta}} = \sum_{i=1}^{N} \frac{1}{\widetilde{\beta}} + \sum_{i=1}^{N} \ln\left(\frac{t_i}{\widetilde{\alpha}}\right) - \sum_{i=1}^{N} \left(\frac{t_i}{\widetilde{\alpha}}\right)^{\beta} \ln\left(\left(\frac{t_i}{\widetilde{\alpha}}\right)\right) \tag{16}$$

$$\frac{\partial \Lambda}{\partial \widetilde{\alpha}} = -\sum_{i=1}^{N} \frac{\widetilde{\beta}}{\widetilde{\alpha}} + \frac{\widetilde{\beta}}{\widetilde{\alpha}} \sum_{i=1}^{N} \ln\left(\frac{t_i}{\widetilde{\alpha}}\right)^{\widetilde{\beta}} \tag{17}$$

The resulting equations give the best estimates $\widetilde{\beta}$ and $\widetilde{\alpha}$.

$$\frac{1}{\widetilde{\beta}} = -\frac{1}{N} \sum_{i=1}^{N} \ln t_i + \sum_{i=1}^{N} \frac{(t_i)^{\widetilde{\beta}} \ln t_i}{\sum_{i=1}^{N} (t_i)^{\widetilde{\beta}}} \tag{18}$$

$$\widetilde{\alpha} = \left[\frac{1}{N}\left(\sum_{i=1}^{N} (t_i)^{\widetilde{\beta}}\right)\right]^{(1/\widetilde{\beta})} \tag{19}$$

Eq. (18) is written in terms of $\widetilde{\beta}$ only, and can only be solved by an iterative technique, such as the Newton-Raphson iterative technique. Once $\widetilde{\beta}$, is obtained, $\widetilde{\alpha}$ can be determined using Eq. (19).

Although the methodology for the MLE is simple, the implementation is mathematically intense. The present high-speed computers, however, have made the obstacles of the mathematical complexity of the MLE an easy process. A specialized statistical commercial package Weibull++ is used throughout this work to find the ML estimates of the Weibull distribution parameters [39].

### 2.7. Failure time percentiles

Once the Weibull distribution parameters are obtained, the failure time percentiles, $t_p$, can be derived from Eq. (2) as follows (by substituting $F(t; \alpha, \beta) = p$)

http://dx.doi.org/10.5772/intechopen.72423

$$t_p = \widetilde{\alpha}[-\ln(1-p)]^{1/\widetilde{\beta}} \tag{20}$$

where $t_p$ is the time-to-breakdown for which a sample will fail with a probability of failure $p$, and $\alpha = L(V, T, f)$ is function of the applied stresses, (e.g. voltage, temperature, frequency, etc.). If $p = 0.632$, then $t_p = \alpha$, the scale parameter, or the life by which 63.2% of the samples will fail. Likewise, if $p = 0.50$, then $t_p = \breve{t}$, the median life, or the life by which half of the samples will fail.

## 3. Life models

There are two approaches for studying the electrical breakdown and estimating the insulation lifetimes (under normal operating conditions) of polymeric insulating materials. One approach is based on phenomenological studies which require a complete understanding of the breakdown mechanism. This approach requires physical and/or chemical tests to be performed on the insulating material that may yield to the development of mathematical models functional with the lifetime. An example of this is relating the lifetime of the insulation to the length of trees formed in the insulation bulk as a result of treeing mechanism [40]. The other widely known approach relies on a statistical analysis of failures that are attributed to the breakdown of the electrical insulation due to the presence of degrading stresses, such as electrical, thermal, and other environmental factors [6–36]. In this approach, the insulation life is determined by measuring the time-to-breakdown of identical specimens of the solid insulation subject to life tests. Life tests, however, show that the times-to-breakdown are widely variable. This variation is best modeled by the Weibull probability distribution.

Conducting life tests at realistic working stresses is not possible due to the time constraint, given that most electrical insulation is expected to serve for several decades. Instead, breakdown data are obtained, without paying much attention to the details of the breakdown mechanism, by conducting accelerated life (aging) tests in laboratory experiments so that the insulation life is severely reduced [41, 42]. The main goal of life tests is to establish mathematical models for the aging process and the stresses causing it [32–36]. The constants of these models need to be estimated from life tests where the lifetimes at a variety of stress levels are measured. Once the constants are estimated, the life at any particular stress including normal operating conditions can, in principle, be estimated.

### 3.1. Single-stress life models

Life models of single stress include the inverse power law and exponential law models for an electrical stress, and the Arrhenius model for a thermal stress [32–34].

#### 3.1.1. *Life models for electrical stresses*

An electrical stress is considered as one of the main factors causing deterioration of electrical insulation. There are two empirical models that relate the test of an electrical stress to the time-to-breakdown. One is the inverse power law model, and the second is the exponential law model. The parameters of both models are obtained from experimental data taken at several different high voltage levels with other conditions unchanged. The electrical life models mathematically describe the aging in a solid dielectric insulation that experiences an electric stress [43]. The life models do not characterize the exact type of aging mechanism that takes place. The life models are totally empirical and have no physical meaning other than defining the degradation rate as power or exponential. However, the models have proven to fit reasonably well with experimental data.

#### 3.1.1.1. *Inverse power law*

The inverse power law (IPL) model is one of the most frequently used in the aging studies under an electrical stress. The inverse power law model is given by:

$$L(V) = k\ V^{-n} \tag{21}$$

where $L$, the time-to-breakdown, is usually a Weibull scale parameter $\alpha$ at 63.2% probability, or any other percentile, $V$ is the applied voltage, and $k$, $n$, are constants to be determined for the specific tested material or device. The inverse power law is considered valid if the data being plotted on a log–log graph fits a straight line [44].

#### 3.1.1.2. *Exponential law*

The exponential law is also commonly used for lifetime calculations. The exponential model is given by:

$$L(V) = c\exp{(-bV)} \tag{22}$$

where $L$ is the time-to-breakdown, $V$ is the applied voltage, and $c$ and $b$ are constants to be determined from the experimental data. The exponential model is verified by plotting the data points on a semi-log graph. The model is considered valid if a straight line is obtained [44].

### 3.1.2. *Life model for a thermal stress*

The life of electrical insulation is seriously affected by a thermal stress. This effect can only be recognized by a thermal life test. The life of electrical insulation under a thermal stress is empirically expressed by the well-known Arrhenius equation. This equation describes the thermal aging of materials and shows the dependency of the chemical reaction rate as a function of the temperature. The Arrhenius equation is given by [43]:

$$L(T) = A\exp\left(\frac{B}{T}\right) \tag{23}$$

where $L$ is the time-to-breakdown, $T$ is the absolute temperature, and $A$, $B$ are constants to be determined experimentally.

http://dx.doi.org/10.5772/intechopen.72423

### 3.2. Multi-stress life models

Multi-stress Life models were developed to predict the life of the insulation under combined electrical and thermal stresses. In general these models are limited to the common electrical and thermal aging stresses acting simultaneously [43]. Mainly, some models have empirical nature such Simoni's, Ramu's, and Fallou's [11, 12, 32–36]. These models account for the interactions of electrical and thermal stresses by using a multiplicative law, in which the life under a combined stress is related to the product of the single-stress lives. One possible formula for this interaction can be manifested as the multiplication of the IPL model and Arrhenius relationship, which is given by:

$$L(V,T) = KV^{-n}\exp\left(\frac{B}{T}\right) \tag{24}$$

which is considered to be the basis for both Simoni's and Ramu's electrical-thermal life models. Alternatively, the electrical exponential model is associated with the Arrhenius relationship. This can be expressed as:

$$L(V,T) = C\exp\left(AV + \frac{B}{T}\right) \tag{25}$$

which constitutes the Fallou's electrical-thermal life model.

Another probabilistic life model based on IPL was presented by Montanari et al. [34]. A brief overview of the above mentioned electrical-thermal life models of Simoni, Ramu, Fallou, and the probabilistic model by Montanari will be presented. The above models were developed for a relatively simple dielectric system involving polymer films or slabs. A variety of multi-stress life tests have been also developed for more complex insulation systems, for example, cables and rotating machines stator windings [16, 17]. However, lifetime models as a function of two or more stresses are rarely derived due to excessive cost and time needed for collecting the failure data.

Regarding the frequency as an aging factor little research has been published in developing combined electrical-thermal-frequency life models. In some works, high frequency sinusoidal voltage was applied [41, 42]. The results of these works show that the frequency as an aging factor that causes insulation deterioration. The effect of the frequency is modeled by relating the variation of the parameters of the life model to the frequency [41].

#### *3.2.1. Simoni's model*

According to the Simoni's model, the insulation life, in relative terms with respect to a reference life determined by the absence of an electrical stress and at low temperature, is given by:

$$L(V,T) = t_o\left[\frac{V}{V_o}\right]^{-n}\exp\left(-B\Delta\left(\frac{1}{T}\right)\right) \tag{26}$$

where $t_o$ is the time-to-breakdown at room temperature and $V = V_o$, $\Delta\left(\frac{1}{T}\right) = \frac{1}{T} - \frac{1}{T_o}$, and $B$ and $n$ are constants which are determined experimentally.

#### 3.2.2. Ramu's model

The Ramu's model is obtained from a multiplication of classical single-stress laws, and is given by:

$$L(V,T) = K(T)[V]^{-n(T)} \exp\left(-B\Delta\left(\frac{1}{T}\right)\right) \tag{27}$$

where $K(T) = \exp\left(K_1 - K_2\Delta\left(\frac{1}{T}\right)\right)$, $n(T) = \exp\left(n_1 - n_2\Delta\left(\frac{1}{T}\right)\right)$, $K_1$, $K_2$, $n_1$, and $n_2$ are constants. $\Delta\left(\frac{1}{T}\right)$ is the same as that defined for the Simoni's model.

#### 3.2.3. Fallou's model

Fallou proposed a semi-empirical life model based on the exponential model for electrical aging:

$$L(V,T) = C\exp\left(AV + \left(\frac{B}{T}\right)\right) \tag{28}$$

where C, $A$, and $B$ are electrical stress constants and must be determined experimentally from time-to-breakdown curves at constant temperatures.

#### 3.2.4. Montanari's probabilistic model

The probabilistic life model of combined electrical and thermal stresses by Montanari *et al.* relates the failure probability $p$ to insulation life $L_{\mathrm{p}}$. It is based on substituting the scale parameter in the Weibull distribution with the life using the inverse power law. For a given time-to-breakdown probability $p$, the probabilistic model is given by:

$$L_p(V,T) = L_s(V/V_s)^n[-\ln(1-p)]^{(1/\beta(T))} \tag{29}$$

where $L_{\mathrm{p}}$ is a lifetime at probability $p$, $L_{\mathrm{s}}$ is a time-to-breakdown at reference voltage $V_s$, and $\beta$ is the shape parameter.

### 3.3. Estimating life model constants

Eqs. (21) to (23) describe several mathematical models which relate insulation life to a single aging stress, either voltage or temperature. Likewise, Eqs. (24) to (29) describe the electrical-thermal life models. In each model of the above life models, there are several parameters that are needed to be estimated from life testing data. The parameters are estimated from life tests where the lifetimes at a variety of stress levels are measured. Once the parameters are estimated, the life at any particular stress can, in principle, be estimated. This enables a method of estimating the life at normal stress based on failure data collected from accelerated life tests. Traditionally, the parameters of life models are calculated either graphically or analytically using graphical or regression analysis type methods.

#### 3.3.1. Graphical method

The graphical method for estimating the parameters of a life model involves generating two types of plots. First, the time-to-breakdown at each individual stress level is plotted on a probability paper appropriate to the assumed life distribution (i.e. Weibull, Lognormal). The parameters at each stress level are then estimated from the plot. Once the parameters of the life distribution have been estimated at each stress level using probability plotting methods, a second plot is created in which a characteristic lifetime is plotted versus stress on a paper that linearizes the assumed lifetime-stress relationship. For example, a log-log paper linearizes the inverse power law, a semi-log paper linearizes the exponential model, and a log-reciprocal linearizes the Arrhenius relationship. The lifetime characteristic can be any percentile, such as 10% lifetime, the scale parameter, the mean lifetime, etc. The parameters of the lifetime-stress relationship are then estimated from the second plot by solving for the slope and the intercept of the line [45].

In spite of the fact that the graphical method is simple and straightforward, the method suffers from some shortfalls such as:

- It is quite time consuming.
- The graphical method may fail in linearizing the lifetime-stress relationship when the data are plotted on the special paper.
- In accelerated life tests with small data, the separation and individual plotting of the data to obtain the parameters increase the underlying error.
- The estimated parameters, that are assumed constant, are likely to vary when the test is repeated. Confidence intervals on the estimated parameters cannot be established using the graphical methods.

#### 3.3.2. Regression analysis

Calculating the parameters of a life model using regression analysis is relatively straightforward. For most single-stress models, simple linear regression (SLR) is used to estimate the parameters. Similarly, multiple linear regression (MLR) is used for multi-stress models and complicated single-stress models. On the other hand, nonlinear regression methods are employed in cases where life models contain thresholds below certain values where aging does not occur. These nonlinear models are much more difficult to analyze [40]. Therefore, non-statisticians do not usually use nonlinear regression. For this reason, most life models assume that the threshold is close to zero, permitting the use of conventional regression analysis. The MLR method first requires the life model to be linearized into a form such as:

$$y = a_o + a_1 x_1 + a_2 x_2 + \cdots + a_k x_k \tag{30}$$

where $y$ is a dependent variable (life, or a mathematical transformation of life), $x_k$ are independent stresses (or transformations of stresses or combinations of stresses), and $a_o$, $a_1$, $a_2$, … $a_k$ are the constants to be determined. The method of least squares can be used to estimate the regression constants in a MLR model. The least squares technique involves finding the values of the constants that minimize the difference between the predicted lifetimes under a set of stress conditions and the actual lifetime measures under the same conditions. By relating a "characteristic" life of the underlying probability distribution to the aging stress, a set of mathematical equations can be used to calculate the parameters [45]. Each observation ($x_{i1}$, $x_{i2}$, …, $x_{ik}$, $y_i$), satisfies the model in (30), or

$$\begin{aligned} y_i &= a_o + a_1 x_{i1} + a_2 x_{i2} + \cdots + a_k x_{ik} \\ &= a_o + \sum_{j=1}^{k} a_j x_{ij} \qquad i = 1, 2, \cdots, n \end{aligned} \tag{31}$$

The least squares function is:

$$J = \sum_{i=1}^{n} \left( y_i - a_o - \sum_{j=1}^{k} \tilde{a}_j x_{ij} \right)^2 \tag{32}$$

The function $J$ is to be minimized with respect to $a_o$, $a_1$, $a_2$, …, $a_k$. The least squares estimates of $a_o$, $a_1$, $a_2$, …, $a_k$ must satisfy:

$$\frac{\partial J}{\partial a_o}\bigg|_{a_o, a_{o1}, a_2, \cdots, a_k} = -2 \sum_{i=1}^{n} \left( y_i - \tilde{a}_o - \sum_{j=1}^{k} \tilde{a}_j x_{ij} \right) = 0 \tag{33}$$

$$\frac{\partial J}{\partial a_j}\bigg|_{\tilde{a}_o, \tilde{a}_{o1}, \tilde{a}_2, \cdots, \tilde{a}_k} = -2 \sum_{i=1}^{n} \left( y_i - \tilde{a}_o - \sum_{j=1}^{k} \tilde{a}_j x_{ij} \right) x_{ij} = 0 \qquad j = 1, 2, \cdots, k \tag{34}$$

Simplifying Eqs.(33) and (34), the following least squares normal equations are obtained:

$$\begin{array}{l} n\tilde{a}_o + \tilde{a}_1 \sum_{i=1}^{n} x_{i1} + \tilde{a}_2 \sum_{i=1}^{n} x_{i2} + \cdots + \tilde{a}_k \sum_{i=1}^{n} x_{ik} = \sum_{i=1}^{n} y_i \\ \tilde{a}_o \sum_{i=1}^{n} x_{i1} + \tilde{a}_1 \sum_{i=1}^{n} x_{i2}^2 + \tilde{a}_2 \sum_{i=1}^{n} x_{i1} x_{i2} + \cdots + \tilde{a}_k \sum_{i=1}^{n} x_{i1} x_{ik} = \sum_{i=1}^{n} x_{i1} y_i \\ \vdots \qquad \vdots \qquad \vdots \qquad \vdots \qquad \vdots \qquad \vdots \\ \tilde{a}_o \sum_{i=1}^{n} x_{ik} + \tilde{a}_1 \sum_{i=1}^{n} x_{ik} x_{i1} + \tilde{a}_2 \sum_{i=1}^{n} x_{ik} x_{i2} + \cdots + \tilde{a}_k \sum_{i=1}^{n} x_{ik}^2 = \sum_{i=1}^{n} x_{ik} y_i \end{array} \tag{35}$$

The solution to the $k + 1$ normal equations will be the least square estimators of the regression parameters $a_o$, $a_1$, $a_2$, …, $a_k$ of the multi-stress life model. The normal equations can be solved by any method appropriate for solving a system of linear equations. Yet, it is much more convenient to solve the multiple linear regression problem using a matrix approach. For $k$ parameters and $n$ observations, the model relating the parameters to the dependent variable $y_n$ can be expressed in matrix notation as [45]:

$$\mathbf{y} = \mathbf{XA} \tag{36}$$

where,

$$\mathbf{y} = \begin{bmatrix} y_1 \\ y_2 \\ \vdots \\ y_n \end{bmatrix} \quad \mathbf{A} = \begin{bmatrix} a_o \\ a_1 \\ \vdots \\ a_k \end{bmatrix} \quad \mathbf{X} = \begin{bmatrix} 1 & x_{11} & x_{12} & \cdots & x_{1k} \\ 1 & x_{21} & x_{22} & \cdots & x_{2k} \\ \vdots & \vdots & \vdots & \vdots & \vdots \\ 1 & x_{n1} & x_{n2} & \cdots & x_{nk} \end{bmatrix}$$

In general, $\mathbf{y}$ is an $(n\times1)$ vector of the observations, $\mathbf{X}$ is an $(n\times k)$ matrix of the levels of stresses, and $\mathbf{A}$ is a $(k\times1)$ vector of the regression constants. The least squares estimators, $\widetilde{\mathbf{A}}$, is the solution for $\mathbf{A}$ that minimizes:

$$J = (\mathbf{y} - \mathbf{XA})'(\mathbf{y} - \mathbf{XA}) \tag{37}$$

by setting $\frac{\partial J}{\partial \mathbf{A}} = 0$. Accordingly,

$$\mathbf{X}'\mathbf{X}\widetilde{\mathbf{A}} = \mathbf{X}'\mathbf{y} \tag{38}$$

Eq. (38) is the least squares normal equations in matrix form, which is identical to the scalar form of the normal equations given in Eq. (35). The least squares estimators $\widetilde{\mathbf{A}}$ is obtained by multiplying both sides of Eq. (38) by the inverse of $\mathbf{X}'\mathbf{X}$ [45].

$$\widetilde{\mathbf{A}} = (\mathbf{X}'\mathbf{X})^{-1}\mathbf{X}'\mathbf{y} \tag{39}$$

In practice, MLR calculations are always performed using a computer. Many statistical analysis computer packages can quickly and accurately perform the necessary calculations. The more sophisticated commercial packages will also provide a plot of the stress versus lifetime [46].

If the times-to-breakdown are presented with a Weibull distribution, then the conventional linear regression is theoretically not applicable [47]. Only recently commercial computer programs have become available which enable linear regression with Weibull-life data [46, 48]. The calculation method is complicated, but depends on the use of information matrix using a large sample size normal approximation [1]. The regression analysis approach has only been standardized to estimate the parameters of the thermal life model. No other life models have been standardized to data.

Alternatively, another analytical approach will be used in this work to estimate the parameters of the proposed multi-stress life models using the life distribution-life-stress combined model. The MLE will be used for estimating the model parameters. An Accelerated Life Testing Analysis (*ALTA*) statistical computer package is used in this work to estimate the parameters of the proposed life model of the insulation [48]. This approach will be discussed in the next section. However, MLR will be used whenever the MLE is not possible.

## 4. Combined Weibull-life model

Three methods for estimating the parameters of accelerated life test models were presented in chapters two and three. First, the graphical method was illustrated using a probability plotting method for obtaining the parameters of the life distribution. The parameters of the life model were then estimated graphically by linearizing the model on a separate lifetime versus stress plot. However, not all life models can be linearized. Hence, instead of estimating the parameters of the life distribution and the life model graphically, an analytical technique based on least squares was presented. However, the accuracy of the graphical method and LS estimation is affected by the probability rank approximation. Furthermore, estimating the parameters of each individual distribution leads to accumulation of uncertainties, depending on the number of failures at each stress level. In addition, the slope (shape) parameters of each individual distribution are rarely equal (common). Using the graphical method or the LS technique, one must estimate a common shape parameter (usually the average) and repeat the analysis. By doing so, further uncertainties are introduced on the estimates, and these are uncertainties that cannot be qualified [48].

On the other hand, combining the life distribution and the life model relationships in one statistical model that describes both can be accomplished by including the life model into the *pdf* of the life distribution of failure data. Thus, the parameters of that combined model can be estimated using the complete likelihood function ($L$). Accordingly, a common shape parameter ($\beta$) is estimated from the combined model, thus eliminating the uncertainties of averaging the individual shape parameters. All uncertainties are accounted for in the form of confidence intervals that are quantifiable because they are obtained based on the overall model. Besides, the MLE technique is independent of any kind of probability ranks or a plotting method. Therefore, the MLE offers a very powerful method in estimating the parameters of life models [48].

The goal of the ML parameter estimation is to determine the parameters that maximize the probability (likelihood) of the life data. Statistically, the method of the ML is considered to be more robust and yields estimators with good statistical properties (unbiasedness, sufficiency, consistency, and efficiency). Due to its nature, the ML is a powerful tool in estimating the parameters of life models. In addition, the ML provides an efficient method for quantifying uncertainty through confidence intervals.

### 4.1. Maximum likelihood estimation of combined Weibull-life model

For life data analysis, the two-parameter Weibull distribution *pdf* is commonly used to represent the scatter of the failure data,

$$f(t;\alpha,\beta) = \frac{\beta}{\alpha}\left(\frac{t}{\alpha}\right)^{\beta-1} \exp\left[-\left(\frac{t}{\alpha}\right)^{\beta}\right] \tag{40}$$

where $t$ is the time-to-breakdown, $\alpha$ is the scale parameter (lifetime at 63.2%), and $\beta$ is the shape parameter or the slope of the Weibull cumulative distribution. The parameters of the life model can be statistically calculated by combining the life model to the Weibull distribution. For example, the combined Weibull-life model can be derived by setting the scale parameter $\alpha$=$L(V)$ for the electrical life model, or $\alpha$=$L(V,T)$ for the electrical-thermal multi-stress life model [13]. The maximum likelihood parameter estimation of combined Weibull-life model will be presented in this chapter for both single-stress and multi-stress life models. The inverse power law (IPL) of the electrical life model and exponential-Arrhenius of the electrical-thermal life model will be used as examples to show the procedure for deriving the parameters of the combined Weibull-life model.

### 4.2. Weibull-inverse power law electrical life model

The combined Weibull-IPL model can be derived by setting $\alpha = L(V) = kV^{-n}$, yielding the following Weibull *pdf*,

$$f(t, V) = \beta KV^n (KV^n t)^{\beta-1} \exp\left(-(KV^n t)^\beta\right) \tag{41}$$

where $K$ = $1/k$. This is a three-parameter model ($K$, $\beta$, $n$) where the parameters can be determined experimentally using life test data.

#### 4.2.1. Parameter estimation of Weibull-IPL model using MLE method

Substituting the IPL electrical life model into the Weibull-Log-Likelihood function yields ($\Lambda$) yields:

$$\Lambda = \sum_{j=1}^{M}\sum_{i=1}^{N} \ln\left[\tilde{\beta}\tilde{K}V_j^{\tilde{n}}\left(\tilde{K}V_j^{\tilde{n}}t_i\right)^{\tilde{\beta}-1} \exp\left(-\left(\tilde{K}V_j^{\tilde{n}}t_i\right)^{\tilde{\beta}}\right)\right] \tag{42}$$

where $M$ is the number of electrical life test groups; $N$ is the number of times-to-breakdown in $j^{th}$ life test; $V_j$ is the $j^{th}$ life voltage; $t_i$ is the $i^{th}$ time-to-breakdown in the $j^{th}$ group; $\tilde{\beta}$ is an estimate of the Weibull shape parameter; $\tilde{K}$ = $1/k$, $k$ is the IPL Parameter; $\tilde{n}$ is the second parameter of IPL.

The ML estimates of the parameters can be found by solving for $\tilde{\beta}, \tilde{K}, \tilde{n}$ such that [45].

$$\frac{\partial\Lambda}{\partial\tilde{\beta}} = 0, \quad \frac{\partial\Lambda}{\partial\tilde{K}} = 0, \quad \frac{\partial\Lambda}{\partial\tilde{n}} = 0.$$

where,

$$\frac{\partial\Lambda}{\partial\tilde{\beta}} = \sum_{j=1}^{M}\sum_{i=1}^{N}\frac{1}{\tilde{\beta}} + \sum_{j=1}^{M}\sum_{i=1}^{N}\ln\left(\tilde{K}V_j^{\tilde{n}}t_i\right) - \sum_{j=1}^{M}\sum_{i=1}^{N}\left(\tilde{K}V_j^{\tilde{n}}t_i\right)^{\tilde{\beta}}\ln\left(\tilde{K}V_j^{\tilde{n}}t_i\right) \tag{43}$$

$$\frac{\partial\Lambda}{\partial\tilde{K}} = \sum_{j=1}^{M}\sum_{i=1}^{N}\frac{\tilde{\beta}}{\tilde{K}} - \frac{\tilde{\beta}}{\tilde{K}}\sum_{i=1}^{M}\sum_{i=1}^{N}\left(\tilde{K}V_j^{\tilde{n}}t_i\right)^{\tilde{\beta}} \tag{44}$$

$$\frac{\partial \Lambda}{\partial \widetilde{n}} = \widetilde{\beta} \sum_{j=1}^{M} \sum_{i=1}^{N} \ln(V_j) - \widetilde{\beta} \sum_{j=1}^{M} \sum_{i=1}^{N} \ln(V_j) \left( \widetilde{K} V_j^{\widetilde{n}} t_i \right)^{\widetilde{\beta}} \tag{45}$$

### 4.3. Electrical-thermal life model

In this dissertation, a new electrical-thermal relationship has been proposed for predicting the lifetime of magnet wire insulation at service conditions when the voltage and temperature are the accelerated stresses in a test. This new combined model is given by:

$$L(V, T) = C \exp\left(\frac{A}{V} + \frac{B}{T}\right) \tag{46}$$

where $L$ is the lifetime at 63.2% probability of breakdown; $V$ is the voltage; $T$ is the temperature; $C$, $A$, and $B$: are constants to be estimated by analyzing the joint voltage-temperature life data.

The proposed lifetime relationship can be linearized by finding the natural logarithm of both sides of Eq. (46). A family of linear curves can be obtained by plotting the lifetime versus either of the stresses, $V$ or $T$, and keeping the other one constant. In this case, the constant $B$ represents the slope of the linearized Arrhenius equation when the voltage is constant, and $A$ represents the slope of the exponential electrical function model when the temperature is constant [45].

Considering that the lifetime is a random variable, the above model can be converted to a probabilistic model by setting the scale parameter $\alpha$ of the Weibull distribution equals to $L(V,T)$ of Eq. (46). Therefore, assuming the time-to-breakdown of the electrical insulation, under combined electrical and thermal stresses, is statistically distributed according to a Weibull distribution, then the Weibull *pdf* can be written as [45]

$$f(t, V, T) = \frac{\beta}{C} \exp - \left(\frac{A}{V} + \frac{B}{T}\right) \left(\frac{t}{C} \exp - \left(\frac{A}{V} + \frac{B}{T}\right)\right)^{\beta - 1} \exp - \left(\frac{t}{C} \exp - \left(\frac{A}{V} + \frac{B}{T}\right)\right)^{\beta} \tag{47}$$

This Weibull *pdf* will be used to estimate the parameters of the electrical-thermal life model.

#### 4.3.1. *Parameter estimation of Weibull-electrical-thermal life model using MLE*

The combined Weibull-electrical-thermal model has four parameters to be estimated using the joint voltage-temperature life data. Using the MLE method, the log-likelihood function of the combined Weibull-electrical-thermal *pdf* is given by [45]:

$$\Lambda = \ln(L) = \sum_{j=1}^{M} \sum_{i=1}^{P} \sum_{l=1}^{N} \ln \left[ \frac{\widetilde{\beta}}{\widetilde{C}} \exp - \left( \frac{\widetilde{A}}{V_i} + \frac{\widetilde{B}}{T_j} \right) \left( \frac{t_l}{\widetilde{C}} \exp - \left( \frac{\widetilde{A}}{V_i} + \frac{\widetilde{B}}{T_j} \right) \right)^{\widetilde{\beta} - 1} \exp - \left( \frac{t_l}{\widetilde{C}} \exp - \left( \frac{\widetilde{A}}{V_i} + \frac{\widetilde{B}}{T_j} \right) \right)^{\widetilde{\beta}} \right] \tag{48}$$

where $M$ is the number of thermal life test groups at voltage $V_i$; $P$ is the number of electrical life test groups at temperature $T_j$; $N$ is the number of times-to-breakdown in the $ji^{th}$ electrical-thermal life test; $V_i$ is the $i^{th}$ life voltage at the $j^{th}$ life temperature; $T_j$: is the $j^{th}$ life temperature at the $i^{th}$ life voltage; $t_l$ is the $l^{th}$ time-to-breakdown in the $ji^{th}$ group; $\widetilde{\beta}$ is the estimate of the

Weibull shape parameter; $\widetilde{C}$ is an estimate of a parameter of the combined E-T life model; $\widetilde{A}$ is an estimate of a parameter of the combined E-T life model; $\widetilde{B}$ is an estimate of a parameter of the combined E-T life model.

The parameter estimates $\left(\widetilde{\beta}, \widetilde{C}, \widetilde{A}, \widetilde{B}\right)$ can be found by solving:

$$\frac{\partial \Lambda}{\partial \widetilde{\beta}} = 0, \frac{\partial \Lambda}{\partial \widetilde{C}} = 0, \frac{\partial \Lambda}{\partial \widetilde{A}} = 0, \frac{\partial \Lambda}{\partial \widetilde{B}} = 0.$$

where,

$$\frac{\partial \Lambda}{\partial \widetilde{\beta}} = \sum_{j=1}^{M}\sum_{i=1}^{P}\sum_{l=1}^{N}\frac{1}{\widetilde{\beta}} + \sum_{j=1}^{M}\sum_{i=1}^{P}\sum_{l=1}^{N}\ln\left(\frac{t_l}{\widetilde{C}}\exp - \left(\frac{\widetilde{A}}{V_i} + \frac{\widetilde{B}}{T_j}\right)\right) - \sum_{j=1}^{M}\sum_{i=1}^{P}\sum_{l=1}^{N}\left(\frac{t_l}{\widetilde{C}}\exp - \left(\frac{\widetilde{A}}{V_i} + \frac{\widetilde{B}}{T_j}\right)\right)^{\widetilde{\beta}}\ln\left(\frac{t_l}{\widetilde{C}}\exp - \left(\frac{\widetilde{A}}{V_i} + \frac{\widetilde{B}}{T_j}\right)\right) \tag{49}$$

$$\frac{\partial \Lambda}{\partial \widetilde{C}} = \sum_{j=1}^{M}\sum_{i=1}^{P}\sum_{l=1}^{N} - \frac{\widetilde{\beta}}{\widetilde{C}} + \frac{\widetilde{\beta}}{\widetilde{C}}\sum_{j=1}^{M}\sum_{i=1}^{P}\sum_{l=1}^{N}\left(\frac{t_l}{\widetilde{C}}\exp - \left(\frac{\widetilde{A}}{V_i} + \frac{\widetilde{B}}{T_j}\right)\right)^{\tilde{\beta}} \tag{50}$$

$$\frac{\partial \Lambda}{\partial \widetilde{A}} = \sum_{j=1}^{M}\sum_{i=1}^{P}\sum_{l=1}^{N}\widetilde{\beta}\left(\frac{-1}{V_i}\right) + \sum_{j=1}^{M}\sum_{i=1}^{P}\sum_{l=1}^{N}\left(\frac{\beta}{V_i}\right)\cdot\left(\frac{t_l}{\widetilde{C}}\exp - \left(\frac{A}{V_i} + \frac{\widetilde{B}}{T_j}\right)\right)^{\tilde{\beta}} \tag{51}$$

$$\frac{\partial \Lambda}{\partial \widetilde{B}} = \sum_{j=1}^{M}\sum_{i=1}^{P}\sum_{l=1}^{N}\widetilde{\beta}\left(\frac{-1}{T_j}\right) + \sum_{j=1}^{M}\sum_{i=1}^{P}\sum_{l=1}^{N}\left(\frac{\beta}{T_j}\right)\cdot\left(\frac{t_l}{\widetilde{C}}\exp - \left(\frac{A}{V_i} + \frac{\widetilde{B}}{T_j}\right)\right)^{\tilde{\beta}} \tag{52}$$

### 4.4. Failure lifetime percentiles

Once the combined Weibull-Lifetime model parameters are estimated, the failure time percentiles, or the time-to-breakdown, $t_p$, as well as the life lines of the MW insulation at different breakdown probabilities, $p = F_p(t;\alpha,\beta)$ can be derived from Eq. (2) by substituting $\alpha=L(V,T)$. For the Weibull-IPL model, by substituting Eq. (21) into Eq. (20), the lifetime percentile is given by,

$$t_p = kV^{-\widetilde{n}}[-\ln(1-p)]^{1/\widetilde{\beta}} \tag{53}$$

where $\widetilde{k}$, $\widetilde{n}$, and $\widetilde{\beta}$ are the ML estimates of $k$, $n$, and $\beta$ of the combined Weibull-IPL model. Likewise, for the Weibull-Electrical-Thermal model, the lifetime percentile can be obtained by substituting Eq. (46) into Eq. (20). Thus yields,

$$t_p = \widetilde{C}\exp\left(\frac{\widetilde{A}}{V} + \frac{\widetilde{B}}{T}\right)[-\ln(1-p)]^{1/\tilde{\beta}} \tag{54}$$

where $\widetilde{C}$, $\widetilde{A}$, $\widetilde{B}$, and $\widetilde{\beta}$ are the ML estimates of $C$, $A$, $B$, and $\beta$ of the combined Weibull-Electrical-Thermal model.

## Nomenclature

| | |
|---|---|
| *ALTA* | accelerated life testing analysis |
| $\alpha$ | Weibull scale parameter |
| $\beta$ | Weibull shape parameter |
| *cdf* | cumulative distribution function |
| IPL | inverse power law |
| LS | least squares |
| $L$ | lifetime |
| ML | maximum likelihood |
| MLE | maximum likelihood estimation |
| MLR | multiple linear regression |
| *pdf* | probability density function |
| SLR | simple linear regression |
| $t$ | time-to-breakdown |
| $T$ | absolute temperature |
| $V$ | applied voltage |
| $V_s$ | reference voltage |

## Author details

Eyad A. Feilat

Address all correspondence to: e.feilat@ju.edu.jo

The University of Jordan, Amman, Jordan

## References

[1] Lawless JF. Statistical Models and Methods for Lifetime Data. 2nd ed. New York: John Wiley & Sons; 2003. DOI: 10.1002/9781118033005

[2] Nelson WB. Accelerated Testing: Statistical Models, Test Plans, and data Analyses. New York: John Wiley & Sons; 2008. DOI: 10.1002/9780470316795

[3] Mann RN, Schafer RE, Singpurwalla ND. Methods for Statistical Analysis of Reliability and Life Data. New York: John Wiley & Sons; 1974

[4] Meeker WQ, Escobar LA. Statistical Methods for Reliability Data. New York: John Wiley & Sons; 1998

[5] Occhini E. A statistical approach to the discussion of the dielectric strength in electric cables. IEEE Transactions on Power Apparatus and Systems. 1971;**90**:2671-2678. DOI: 10.1109/TPAS.1971.292920

[6] Feilat EA, Grzybowski S, Knight P, Doriott L. Breakdown and aging behavior of composite insulation system under DC and AC high voltages. WSEAS Transactions on Circuits and Systems. 2005;**4**:780-787

[7] Grzybowski S, Dobroszewski R, Grzegorski E: Accelerated endurance tests of polyethylene insulated cables. In: Proceedings of the 3rd International Conference on Dielectric Materials, Measurements and Applications; 10-13 September 1979; Birmingham, UK. p. 120–123

[8] Cacciari M, Montanari GC. Optimum design of life tests for insulating materials, systems and components. IEEE Transactions on Electrical Insulation. 1991;**26**:1112-1123. DOI: 10.1109/14.108148

[9] Mazznti G, Montanari GC, Simoni L. Insulation characterization in multistress conditions by accelerated life tests: an application to XLPE and EPR for high voltage cables. IEEE Electrical Insulation Magazine. 1997;**13**:24-34. DOI: 10.1109/57.637151

[10] Björklund A, Siberg H, Paloniemi P. Accelerated ageing of a partial discharge resistant enameled round wire. In: Proceedings of the Electrical Electronics Insulation Conference; 18-21 September 1995; Rosemont, Illinois. pp. 417–419

[11] Fallou B, Burguiere C, Morel JF. First approach on multiple stress accelerated life testing of electrical insulation. In: 1979 Annual Report of CEIDP; Pennsylvania, USA. pp. 621–628

[12] Ramu TS. On the estimation of life of power apparatus insulation under combined electrical and thermal stress. IEEE Transactions on Electrical Insulation. 1985;**20**:70-78. DOI: 10.1109/TEI.1985.348759

[13] Feilat EA, Grzybowski S, Knight P. Multiple stress aging of magnet wire by high frequency voltage pulses and high temperatures. In: Conference Records of the 2000 IEEE International Symposium on Electrical Insulation (ISEI2000); 2–05-04-2000; CA, USA. pp. 157–160

[14] Grzybowski S, Feilat EA, Knight P, Doriott L. Breakdown voltage behavior of PET thermoplastics at DC and AC Voltages. In: Proceedings of the IEEE SoutheastCon99; 25-28 March 1999; Lexington, Kentucky, USA. pp. 284–287

[15] Grzybowski S, Feilat EA, Knight P, Doriott L. Accelerated aging tests on magnet wires under high frequency pulsating voltage and high temperature. In: Annual Report of the 1999 Conference on Electrical Insulation and Dielectric Phenomena, (CEIDP99); 17-21 October 1999; Austin, Texas, USA. pp. 555–558

[16] Kim J, Kim W, Park H-S, Kang J-W. Lifetime assessment for oil-paper insulation using thermal and electrical multiple degradation. Journal of Electrical Engineering and Technology. 2017;**12**:840-845. DOI: 10.5370/JEET.2017.12.2.840

[17] Li Y, Tian M, Lei Z, Song J, Zeng J, Lin L, Li Y. Breakdown performance and electrical aging life model of EPR used in coal mining cables. In: Proceedings of the IEEE

International Conference on High Voltage Engineering and Application (ICHVE); 19-22 September 2016; Chengdu, China

[18] Maussion P, Picot A, Chabert M, Malec D. Lifespan and aging modeling methods for insulation systems in electrical machines: a Survey. In: Proceedings of the 2nd IEEE Workshop on Electrical Machines Design, Control and Diagnostics (WEMDCD'15); 26-27 March 2015; Turin, Italy

[19] Akolkar SM, Kushare BE. Remaining life assessment of power transformer. Journal of Automation and Control. 2014;**2**:45-48. DOI: 10.12691/automation-2-2-2

[20] Kerimli GM. On the mechanism of electrical aging of polymer insulation materials. Surface Engineering and Applied Electrochemistry. 2014;**50**:485-490. DOI: 10.3103/S1068375514060052

[21] Lahoud N, Faucher J, Malec D, Maussion P. Electrical aging of the insulation of low voltage machines: model definition and test with the design of experiments. IEEE Transactions on Industrial Electronics. 2013;**60**:4147-4155. DOI: 10.1109/TIE.2013.2245615

[22] Nasrat LS, Kassem N, Shukry N. Aging effect on characteristics of oil impregnated insulation paper for power transformers. Engineering. 2013;**5**:1-7. DOI: 10.4236/eng.2013.51001

[23] Koltunowicz TL, Cavallini A, Djairam D, Montanari GC, Smit JJ. The influence of square voltage waveforms on transformer insulation break down voltage. In: 2011 Annual Report Conference on Electrical Insulation and Dielectric Phenomena; 16–19-10-2011; Mexico

[24] Mazzanti G, Montanari GC, Dissado LA. Electrical aging and life models: the role of space charge. IEEE Transactions on Dielectrics and Electrical Insulation. 2005;**12**:876-890. DOI: 10.1109/TDEI.2005.1522183

[25] Liao RJ, Liang SW, Sun CX, Yang LJ, Sun HG. A comparative study of thermal aging of transformer insulation paper impregnated in natural ester and in mineral oil. European Transactions on Electrical Power. 2010;**20**:518-523. DOI: 10.1002/etep.336

[26] Montanari GC, Mazzanti G, Simoni L. Progress in electrothermal life modeling of electrical insulation during the last decades. IEEE Transactions on Dielectrics and Electrical Insulation. 2002;**9**:730-745. DOI: 10.1109/TDEI.2002.1038660

[27] Kaufhold M, Aninger H, Berth M, Speck J, Eberhardt M. Electrical stress and failure mechanism of the winding insulation in PWM-inverter fed low-voltage induction motors. IEEE Transactions on Industrial Electronics. 2000;**47**:396-402. DOI: 10.1109/41.836355

[28] Stone GC. The application of Weibull statistics to insulation aging tests. IEEE Transactions on Electrical Insulation. 1979;**14**:233-239. DOI: 10.1109/TEI.1979.298226

[29] Fothergill JC. Estimating the cumulative probability of failure data points to be plotted on Weibull and other probability papers. IEEE Transactions on Electrical Insulation. 1990; **25**:489-4992. DOI: 10.1109/14.55721

[30] Reliability Engineering Resources[Internet]. Available from: http://www.weibull.com

[31] Jacquelin J. Inference of sampling on Weibull parameter estimation. IEEE Transactionns on Dielectrics and Electrical Insulation. 1996;**3**:809-816. DOI: 10.1109/94.556564

[32] Cygan P, Laghari JR. Models for insulation aging under electrical and thermal multistress. IEEE Transactions on Electrical Insulation. 1990;**25**:923-934. DOI: 10.1109/14.59867

[33] Laghari JR. Complex electrical thermal and radiation aging of dielectric films. IEEE Transactions on Electrical Insulation. 1993;**28**:777-788. DOI: 10.1109/14.237741

[34] Montanari GC, Simoni L. Aging phenomenology and modeling. IEEE Transactions on Electrical Insulation. 1993;**28**:755-776. DOI: 10.1109/14.237740

[35] Gjaerde AC. Multifactor ageing models-origin and similarities. IEEE Electrical Insulation Magazine. 1997;**13**:6-13. DOI: 10.1109/CEIDP.1995.483610

[36] Montanari GC, Cacciari M. A probabilistic insulation life model for combined thermal-electrical stresses. IEEE Transactions on Electrical Insulation. 1985;**20**:519-522. DOI: 10.1109/TEI.1985.348776

[37] Dissado LA, Fothergill JC, Wolfe SV, Hill RM. Weibull statistics in dielectric breakdown; theoretical basis, applications and implications. IEEE Transactions on Electrical Insulation. 1984;**19**:227-233. DOI: 10.1109/TEI.1984.298753

[38] Gibbons DI, Vance LC. A simulation study of estimators for the 2-parameter Weibull distribution. IEEE Transactions on Reliability. 1981;**30**:61-66. DOI: 10.1109/TR.1981.5220965

[39] Weibull++ Life Data Analysis Reference, Version 11. Arizona, USA: ReliaSoft's; 2017

[40] Eichhorn RM. Treeing in solid extruded electrical insulation. IEEE Transactions on Electrical Insulation. 1977;**12**:2-8. DOI: 10.1109/TEI.1977.298001

[41] Kachen W, Laghari JR. Determination of aging-model constants under high frequency and high electric fields. IEEE Transactions on Dielectrics and Electrical Insulation. 1994; **6**:1034-1038. DOI: 10.1109/94.368660

[42] Cygan P, Krishnakumar B, Laghari JR. High field electrical and thermal aging of polypropylene films. In: Conference Record of the 1988 IEEE International Symposium on Electrical Insulation; 5-8 June 1988; Boston, MA, USA. pp. 188–191. DOI: 10.1109/ELINSL.1988.13901

[43] Grzybowski S, Bandaru S: Effect of multistress on the lifetime characteristics of magnet wires used in flyback transformer. In: Conference Record of the IEEE International Symposium on Electrical Insulation; 19-22 September 2004; Indianapolis, IN, USA

[44] Feilat E A, Grzybowski S, Knight P. Electrical aging models for fine gauge magnet wire enamel of flyback transformer. In: Proceedings of the IEEE Southeast Con 2000; 7-9 April 2000; Nashville, TN, USA. pp. 146–149

[45] Feilat EA. Lifetime characteristics of magnet wires under high frequency pulsating voltage and high temperature [Thesis]. Mississippi State: Mississippi State University; 2000

[46] SAS/STAT Guide for Personal Computers, Version 6, Chapter 21. The LIFEREG Procedure. SAS Institute, 1987

[47] Draper NR, Smith H. Applied Regression Analysis. John Wiley & Sons; 2014. DOI: 10.1002/9781118625590

[48] Accelerated Life Testing Analysis, Version 11. Tucson, Arizona, USA. ReliaSoft's: 2017

Chapter 12

# The Primary Role of the Electric Near-Field in Brain Function

Salvatore Domenic Morgera

Additional information is available at the end of the chapter

http://dx.doi.org/10.5772/intechopen.71945

**Abstract**

The origin and spatial-temporal structure of the endogenous (internal) electric near-fields associated with the neurological network activity of the brain are described. Recent discoveries have elevated the importance of the endogenous fields to a leading role of primary phenomena, as opposed to the traditionally thought secondary role of epiphenomena. This implies that the spatial-temporal structures of the brain's endogenous fields are rich in information that directly convey brain health. Understanding the spatial-temporal structures of the endogenous fields under healthy and unhealthy conditions coupled with the technologies needed to sense and manage these fields opens a world of possibilities for the rational design of clinically accurate, wearable neurodevices to diagnose, therapeutically treat, and manage chronic neurological dysfunctions, mental disorders, and traumatic injuries. The World Health Organization reports that more than 1 billion people worldwide, irrespective of age, sex, education, or income, suffer because of neurological disorders. Devices of the type described here will provide clarity and relief to those individuals that have an impaired neurological system.

**Keywords:** axon, nerve fiber, electric near-field, nerve fiber crosstalk, regenerative repeater, node of Ranvier, endogenous field, exogenous field, brain function, brain structure, neurodegenerative diseases, linear antenna array, myelin sheath, wireless neurological networking, multiple sclerosis, autism spectrum disorder, Alzheimer's disease, chronic care, wearable device, rational design

## 1. Introduction

It is our hope that the avenue of investigation reported here leads to new modalities for brain mapping and the non-invasive diagnosis and treatment of neurological and mental disorders. We recommend that future work in brain mapping include both the current neuroimaging methods and the spatial-temporal monitoring of electric near-field signatures. By doing this, we will better understand how brain structure leads to dynamic brain function and further reveal that the highly sophisticated brain network is comprised of a fixed anatomical network discovered by neuroimaging augmented by a near-field wireless network which is shown here to naturally occur at the most fundamental levels of the nervous system. The problem we treat is at the intersection of medicine, science, and electrical engineering, and further research into the ideas presented has the potential to not only meet a great many societal health needs, but also impact a variety of allied fields, including advanced materials, biosensors, fluid dynamics, near-field electromagnetics, nanotechnology, and information (Shannon) theory.

The defining framework for one of the fundamental problems in neuroscience was laid down in the seminal work presented in [1]. In relation to this work, we believe that understanding circuit functions not only encompasses the intrinsic properties of individual neurons and their synaptic connections, but also their connections to, and interactions with, other neurons. Several mechanisms for transmission of excitation from one nerve cell to another are discussed in [2]. In this work, we will discuss a new mechanism.

The plan of the chapter is as follows. Section 2 describes nerve bundles and tracts, reviews fundamental axon structure-function relationships for myelinated, or insulated, fibers, and provides the reader with a feeling for how densely fibers are packed in bundles and tracts. This section also reminds us of the concept of *tortuosity*. Although axonal tortuosity is a ubiquitous phenomenon, no mathematical models exist to describe it in the context of nerve bundles and tracts.

Section 3 casts a Node of Ranvier, a critical element of a myelinated fiber, in the role of a *dual channel (or dual port) device* having both electrochemical (EC) *and* electromagnetic (EM) signaling channels. The EC channel is well known and is generally governed by Hodgkin-Huxley dynamics; whereas, the EM channel described here is not well known and is enabled by electric near-field dynamics. In the case of the EM channel, electric near-field characteristics for approximately linear arrays of Nodes of Ranvier consecutively located along myelinated nerve fibers are simulated and discussed, as is the much larger bandwidth and speed of the EM channel vis-à-vis the EC channel. This relatively rapid ability to traverse physically different regions of the brain can account for the statistically correlated concurrent activity observed during the execution of cognitive tasks which the anatomical links of the human connectome cannot currently all account for.

Indeed, a more comprehensive story of brain function can be constructed from the *fixed network* of the human connectome augmented by the *wireless network* described here. A study of this type opens many exciting avenues of research. Of interest are those that examine the spatial-temporal signatures of the electric near-fields, the *endogenous* (internal) fields for different central nervous system neuropathies, and the subsequent design of non-invasive or minimally invasive, clinically accurate medical devices capable of delivering rationally designed spatial-temporal *exogenous* (external) fields for neurological dysfunction mitigation. No such rationally designed, clinically accurate medical devices exist today. Section 4 provides a discussion that presents several ideas that we are certain will stimulate further the world research.

Section 5 describes the continuing research and development phases conducted in our laboratories for a non-invasive, wearable neurodevice that provides clinical grade information on cognitive and behavioral states of the brain and insight into impaired central nervous neurological network dynamics. We believe that this section is a model for those desiring to develop such medical devices using an integrated clinical *in vitro* wet laboratory/computer simulation *in silico* dry laboratory environment. The reader will note that Section 5 provides additional neurobiological support for the electrical engineering investigation found in Sections 2 and 3. Section 6 presents the conclusions of this work.

## 2. Nerve bundles/tracts

A nerve is a grossly visible anatomic structure and is a bundle or tract of axonal processes from many different neurons wrapped in connective tissue. Axons with larger diameters are generally insulated, or myelinated. The insulation is a highly intricate lamination of plasma membrane known as a myelin sheath and forms an efficient barrier to charge leakage, signal loss, and interference.

The myelin sheath is discontinuous along the length of the axon. The "gaps" are known as the Nodes of Ranvier where the axonal membrane is exposed (uninsulated) to the extracellular space. The Nodes of Ranvier act like digital repeater stations in communications networks. These digital repeater stations contain a high density of voltage-gated Na + channels that regenerate action potentials, if incoming current exceeds a threshold and incoming rate does not violate certain refractory conditions. In the following, we briefly review some properties of myelinated fibers.

### 2.1. Axon structure-function in myelinated Fibers

Early work [3] discusses axon structure-function relationships. **Figure 1** defines axon structural parameters and illustrates the recent analogy with communications networks.

Using current flow arguments, we arrive at the following proportionality:

$$\frac{l}{D} \propto \left(\frac{d}{D}\right)\left[\ln\left(\frac{D}{d}\right)\right]^{1/2} \tag{1}$$

where comparison with experimental data shows that the ratio $g = d/D$ varies in a relatively tight range about $g^* = 0.6$. This highly interesting behavior gives rise to a nearly linear variation of internodal length (average distance between Nodes of Ranvier) with fiber diameter [3]. We note that the right-hand side of (1) is proportional to the length constant which also attains a maximum at $g^*$. The length constant is a mathematical constant used to quantify the distance that a graded electric potential travels along a fiber via passive electrical conduction. The greater the value of the length constant, the farther the potential will travel. Note that a large length constant can contribute to spatial summation, the electrical addition of one potential with potentials from adjacent areas of a cell.

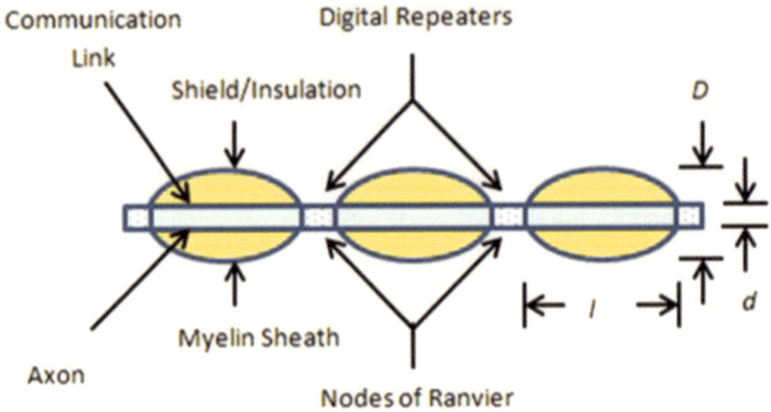


**Figure 1.** Elements of a Myelinated axon.

The situation with fiber conduction velocity is more complex, as pointed out in Ref. [4]. Conduction velocity is related to the delay between the times at which adjacent nodes reach threshold. There is an optimal value of $g$ (and, therefore, myelin thickness) at which conduction velocity is maximized. It appears that nerve fibers enjoy a structural relationship which ensures that currents rapidly reach repeaters and exceed the threshold necessary for regeneration through a delicate balance of relatively costly (in terms of energy) Nodes of Ranvier and passive internode conduction.

### 2.2. Morphometric studies of nerve bundles/tracts

We provide two distinctly different examples. The human trigeminal nerve is composed of several bundles of primarily myelinated fibers in the motor and sensory roots. The average transverse sectional sensory (motor) nerve area is 2.147 (0.295) $mm^2$, into which 51.862 (5.268) axons are packed, each with an average cross-sectional area of 0.896 (2.526) $\mu m^2$ [5]. The diameter of the optic nerve increases from about 1.6 mm within the eye to 3.5 mm in the orbit to 4.5 mm within the cranial space. Within the optic nerve, there are from 0.77 to 1.7 million fibers [6].

To appreciate the complexity and elegance of nerves, we consider what are known as *packing problems*, a class of spatial optimization problems. The model we employ is that of non-overlapping congruent circular cylinders (a rough model for fibers) of aspect (fiber length/fiber diameter) much greater than unity. This would appear to be a poor model for trigeminal nerves, where the packing density for sensor (motor) nerves is 0.022 (0.045), and a much better model for the optic nerve, where the average packing density is 0.846, close to the maximum packing density of $\pi/\sqrt{12} = 0.907$.

The reason that we differentiate the applicability of the packing model in the two cases is because optic nerve fibers tend to run approximately parallel within very small spaces and, thus, achieve a high packing density; whereas, trigeminal fibers generally do not run parallel, and require considerably more nerve volume for their runs. The difference

http://dx.doi.org/10.5772/intechopen.71945

may be described by the term *tortuosity*, which refers to the twisting, turning property of curves [7]. When we couple the concept of tortuosity to the quantum nature of the reactive near-field electromagnetic environment to be described in the sequel, it leads us to ponder the question of whether the ground-breaking results reported in [8] on entanglement and communications-assisted entanglement provide an approach to the study of axonal interaction. This is a thought to store away for future research. In this work, we stay on classical ground.

We refer to the interaction of tightly packed nerve fibers in a tract as *crosstalk*. The concept of crosstalk is well known in the discipline of electrical engineering, where it is characterized as any phenomenon by which a signal transmitted on one circuit or channel of a transmission system creates an *undesired* effect in another circuit or channel. There is a wealth of information in the electrical engineering literature on ways of mitigating the undesired effect of crosstalk. The situation may not be so clear in the current case of neurobiology, where the crosstalk may not necessarily be undesired. In fact, if we strictly apply the thinking that structure leads to function, then the fact that the Nodes of Ranvier are uninsulated, a structure which directly leads to crosstalk, we tentatively conclude that the crosstalk may serve a useful function [9].

It is instructive to provide a few remarks on the techniques for reducing crosstalk. Crosstalk in the electrical engineering context refers to interactions among wires carrying currents. With the full understanding that the nerve fiber is not a wire, we do know that it does exhibit activity which generates potential differences, current flow and electromagnetic fields that can influence the surrounding environment. Best practices in electrical engineering indicates that the following techniques can reduce crosstalk (analogies to nerve fibers are in parentheses):

- Widen spacing between signal lines (reduce packing density of fibers).
- Position signal lines as close to ground plane as possible (surround fibers by substantial amounts of extracellular material).
- Route signal lines on different layers orthogonal to each other (we see this in the cortex).
- If necessary, route signal lines in parallel, but minimize run lengths (as previously noted in the optic nerve).

Anatomy within the nervous system reflects a great variety of structural practice. In many cases, we see high fiber packing density and relatively little extracellular space, conditions which promote crosstalk. It is important to realize that, in the neurobiological context, crosstalk may be beneficial and not the problematic phenomenon it is viewed as in the electrical engineering context. Depending on where in the central nervous system we focus, we note that crosstalk may enable communication, cooperation, or competition among nerve fibers[1].

[1]The reader may wish to view the TEDx talk at https://www.youtube.com/watch?v=Jg50wEHqpas on the "Grandest Social Network" to obtain an additional perspective on how the vast population of 100 billion nerve fibers in the human brain interact in a manner similar to the way in which humans interact in much smaller social networks like Facebook.

## 3. The Node of Ranvier—A dual channel EC/EM device

In [10], we see that each Node of Ranvier serves two functions which may be differentiated by propagation process:

1. An element of a longitudinal EC conduction digital repeater string, as discussed earlier. This is the classical function discovered by Hodgkin and Huxley.
2. An element which provides bidirectional coupling into a transverse EM near-field channel. This is the new function described here. Note that we use the acronym EM, although the mechanism is solely an electric field.

We focus on item 2 above. Coupling, or interaction among nerve fibers, is accomplished by one or more consecutive Nodes of Ranvier which serve as elements of an approximately *linear antenna array*. **Figure 2** illustrates the dual channel nature associated with each Node of Ranvier.

### 3.1. Reactive near-field characteristics

To model the transverse EM propagation, we start at basic principles which describe the opening and closing of ion pores (channels) at a Node of Ranvier. Each pore opens and closes as what is called in communications systems a stochastic Pulse Position Modulation/Pulse Width Modulation (PPM/PWM) signal, i.e., the opening of a pore is rather abrupt and occurs at a stochastic position in time, the duration of pore opening is stochastic, and the closing of a pore is also rather abrupt.

Pulse position and pulse duration are assumed to be statistically independent uniformly distributed random variables with ranges established by clinical measurements. The time and frequency domain characteristics of an ensemble of such signals is shown in **Figures 3** and **4**, respectively.

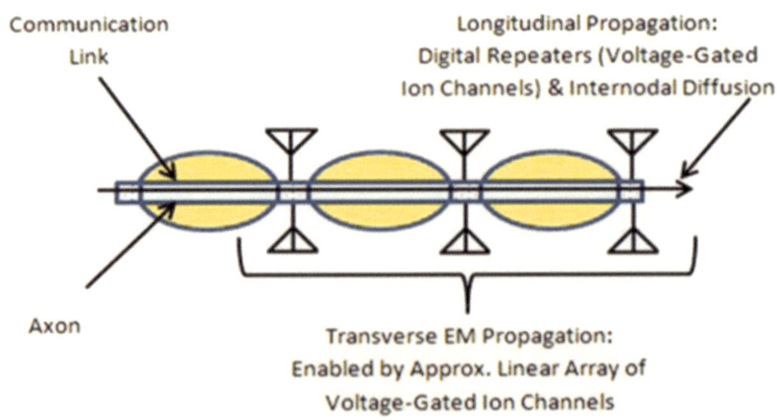


**Figure 2.** Axonal EC longitudinal Classical Channel and EM Transverse Channel (small triangular shapes positioned at nodes of Ranvier are symbols for antenna elements).

http://dx.doi.org/10.5772/intechopen.71945

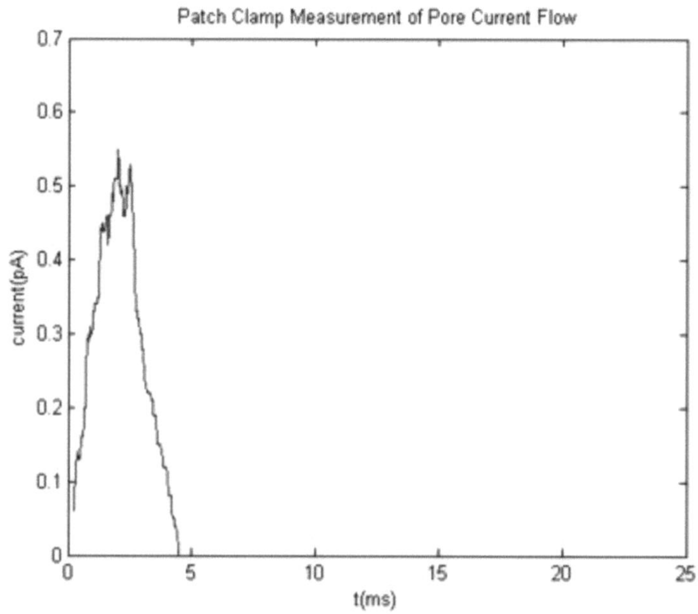


**Figure 3.** Typical patch clamp measurement of pore current flow.

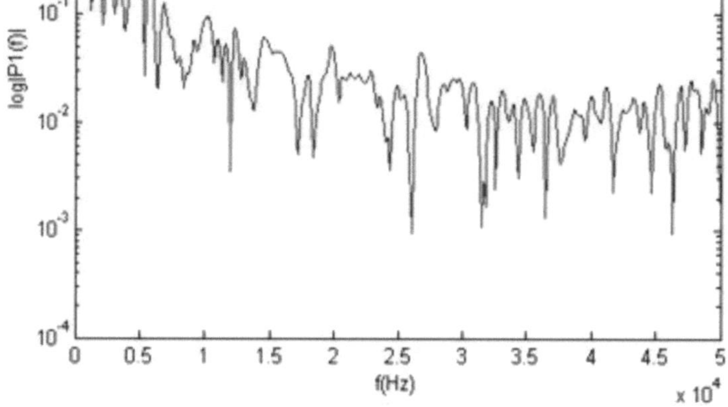


**Figure 4.** Fourier transform of an average of 100 patch clamp measurements.

A Node of Ranvier consists of several thousand pores per square micrometer of membrane each with diameter ≅ 0.4 nm disposed on an annulus with width ≅1–2 μm and diameter that of the nerve fiber. If we assume that the pore signals as modeled above are statistically independent, it is then possible to obtain a comprehensive model for the transverse EM propagation at a Node of Ranvier due to the ensemble of pores, or voltage-gated ion channels.

It is important to note that the electric field generated as above is not a single frequency field, but a field consisting of many frequencies associated with the characteristics of an ensemble of pores rather abruptly opening and closing. Looking at the Fourier transform of **Figure 4** and adopting a criterion that the significant band of frequencies are those that are no more than 30 dB down from the maximum, we see that the significant band of frequencies goes up to at least 50 kHz. This is a startling result, as research on nerve fibers is commonly conducted in the much lower 1 or 2 kHz regime associated with the time domain characteristics of the EC channel action potentials. We now see that the nerve fiber operates in two widely different frequency regions, EC at low frequencies and EM at much higher frequencies.

The need to treat a wide band of frequencies in the EM case distinguishes this problem from the many problems and solutions in electromagnetics that assume a single frequency. The offshoot of this is that advanced methods that start with dyadic Green's functions and can accommodate a wide band of frequencies at the outset are used. In this work, we make certain assumptions that reasonably accurately reflect first-order effects to present some new results in as simplistic terms as possible. In any event, we note that the expected range of frequencies leads to wavelengths $\lambda$ that are long relative to physical dimensions associated with axons and nerve bundles and tracts; therefore, EM interaction among Nodes of Ranvier for closely packed axons within a bundle or tract, or even among Nodes of Ranvier for physically distant tracts as occurs in the brain, is in the *reactive near-field*. For wave number $k = 2\pi/\lambda$ and interaction distance $r$, the near-field condition satisfies $k \cdot r \ll 1$.

As noted above, precise wide band near-field prediction of the electric field is challenging and few accurate tools are available. In the near-field region, in contrast to the far-field, absorption of radiation *does* affect the load on the source Node of Ranvier, electric and magnetic fields can exist independently of each other, and both phase and group speeds can be superluminal [11]. The speeds of potential *Shannon information* transfer from one axon to another over these transverse EM channels are slower and depend on EC channel speed.

Focusing on a single Node of Ranvier, and keeping things simple for this work, an action potential regenerates because of the transient ion flows passing through voltage-gated ion channels. Maxwell's equations show that such transient ion flows emit EM waves in the form of electrical pulses in a manner approximately like that of a pulsating dipole antenna [12]. This was the approach taken in [13, 14].

### 3.2. Antenna array of pulsating dipole elements

As shown in **Figure 2**, transverse EM propagation is enabled by several consecutive Nodes of Ranvier, modeled as pulsating dipole antennas. Each pulsating dipole is, in turn, derived from many filamentary currents associated with the voltage-gated ion channels. The rate and/or coding of action potentials traveling longitudinally determine whether all consecutive nodes are active or not at a given instant of time. If several nodes are active, we coherently add their individual fields [15]. Should some consecutive nodes be active and others inactive, we refer to those that are not active as *parasitic* and the resulting array of nodes as *sparse*. Note that the pore opening times described earlier at consecutive Nodes of Ranvier are not statistically independent.

It is important to note that the rate and/or coding patterns of action potentials converts into EM near-field *spatial-temporal patterns* which impact the extracellular space of the fiber itself, and other fibers within nerve bundles and tracts. In the brain, the EM near-fields easily have reach across the span of the brain and may help to account for the activation of different regions of the brain during the execution of cognitive tasks. This type of activity is entirely missed by the fixed network studies undertaken in the Human Connectome project.

As we discuss in Section 5, the electric near-field levels measured using patch clamp techniques range from $9 \times 10^{8}$ V/m at the very surface of the node of Ranvier to approximately $5 \times 10^{3}$ V/m at about 1.767 millimeters from the surface. This is a field level that is measurable and biologically significant in that it can impact neighboring nerve fibers. As a rough comparison, the average electric field strength from wiring and appliances can range from 5 to 20 V/m, but is often less than 10 V/m. Let us consider near-field communication (NFC). NFC is a new short-range, standards-based wireless connectivity technology, that uses magnetic field induction to enable communication, such as Apple Pay®, between devices. Even though NFC systems operate using magnetic fields vis- à -vis electric fields, we can look at the equivalent power level, which is roughly 0.55 V/m for an NFC system. At this level, which is lower than axon interaction levels, a complex transaction between proximate devices can be completed. Further, several consecutive active Nodes of Ranvier increase the aperture and, consequently, both the reception and transmission gains and, therefore, the electric field levels, relative to a single Node of Ranvier. As discussed in Section 2, some fibers run more parallel than others because of tortuosity and other evolutionary factors; thus, we refer to the arrays as being *approximately* linear.

We have developed software for approximately linear array electric near-field modeling. The geometry for a single dipole of dimension $L \ll \lambda$ is shown in **Figure 5**. The electric field vector $\mathbf{E}(E_\theta, E_r)$ is inhomogeneous. These components are not assumed to be constant over near field nerve bundle/tract fiber membranes and extracellular structures.

When $I(z)$ is the current supplied to the dipole the electric field is given by,

$$E_\theta = -j\frac{LI(z)\,k^3 \sin\theta}{4\pi\omega\varepsilon}\left(\frac{1}{(kr)^3} + \frac{1}{(kr)^2} - \frac{1}{kr}\right) e^{-jkr} \tag{2}$$

$$E_r = -j\frac{2LI(z)\,k^3 \cos\theta}{4\pi\omega\varepsilon}\left(\frac{1}{(kr)^3} + \frac{1}{(kr)^2}\right) e^{-jkr} \tag{3}$$

where $z = r\cos\theta$, $\omega$ is the angular frequency, and $\varepsilon$ is the (variable) complex permittivity of the near field bundle or tract membranes and extracellular structures. Examples of values for biological tissue may be found in Ref. [16].

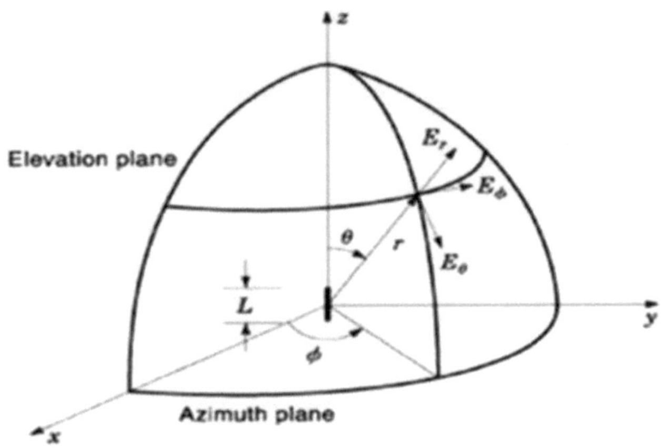


**Figure 5.** Electric field geometry and components.

These "biological antennae" are *electrically short*, and self and mutual impedance (coupling) effects cannot be disregarded. Note that coupling between Nodes of Ranvier is generally decreased in the tortuous case, roughly like the way crosstalk is reduced for twisted pair wires. We associate an $(N \times N)$-dimensional generalized impedance matrix $\mathbf{Z} = \mathbf{R} + j\mathbf{X}$ with an array of $N$ Nodes of Ranvier, where the real part $\mathbf{R}$ is resistance and the imaginary part $\mathbf{X}$ is capacitive reactance.

Electrically short antennas have low radiation resistance into which it is difficult to couple significant power. On the other hand, the energy transferred to the antenna, the extracellular region, and other fibers by the reactive power flow is stored mostly in the reactive near-field. Simulations indicate that the average radiation resistance for diagonal elements of $\mathbf{R}$ is $\cong 0.02\Omega$, indicating low EM radiation efficiency, and the capacitive reactance for diagonal elements of $\mathbf{X}$ is $\cong -40\mathrm{k}\Omega$, indicating a capacitance at a frequency of 10 kHz that is about that of closely packed fiber membrane capacitances. The fundamental research in impedance matching found in [17] helps to understand the nature of reactive power flow.

### 3.3. Simulation of the electric near-field

As previously mentioned, in a tightly packed bundle or tract of axons, the dimensions are such that axons are immersed in a reactive near-field. We choose an array length of $10l \cong 10$mm, where $l$ is internode length. Our simulation therefore includes 10 Nodes of Ranvier, equally distributed over $x \in [-5, 5]$ mm, $y = 0$.

The electric field is calculated using (2) and (3) for a frequency of 10 kHz on an $(x, y)$ axis planar rectangular surface located in the extracellular space between fibers [18]. No tortuosity is included in the simulations presented here. **Figure 6** illustrates a snapshot of the electric near-field magnitude and applies to the case where all 10 Nodes of Ranvier are concurrently active. We have also examined the phase of the near-field, the role of which is not well understood, but believed to be important in developing time synchronicity among groups of axons. This information is presented in Section 5.

**Figure 7** illustrates another snapshot of the electric near-field. **Figure 7** applies to the case where 5 of the 10 Nodes of Ranvier are alternately active. As mentioned earlier, we refer to such an array of nodes as a *sparse* array. Note that **Figures 6** and **7** display substantial and clinically measurable differences in electric field signature.

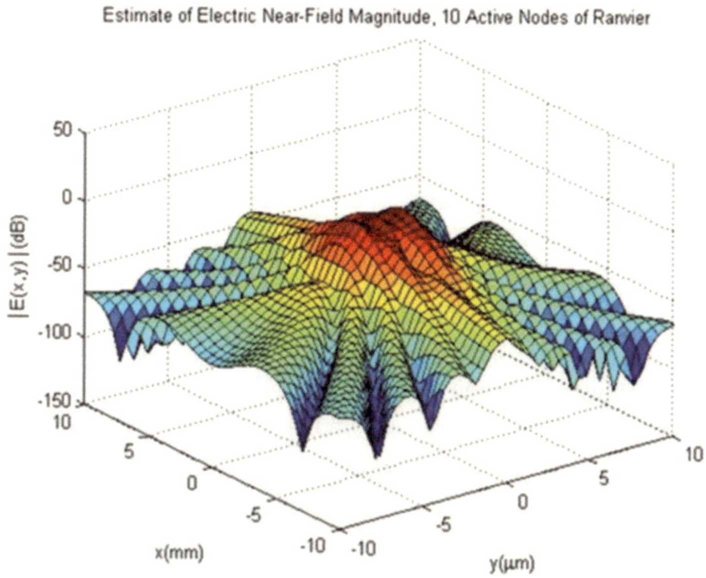


**Figure 6.** Extracellular region electric near-field magnitude, full Array.

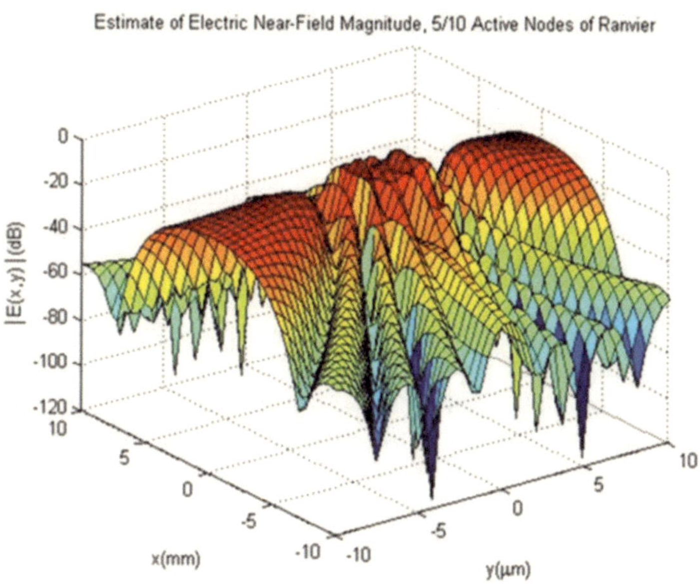


**Figure 7.** Extracellular region electric near-field magnitude, sparse Array.

## 4. Discussion of nerve fiber crosstalk modeling

Regarding the EM near-field model, if the environment consisting of tightly bundled fibers and extracellular material has any inductive properties, these can resonate out the capacitive reactance and provide a matching effect which can improve power transfer [17] for nerve fiber crosstalk. We also note that, in this case, the overall antenna array becomes more tuned or *narrowband*. Should this be the situation, the array of Nodes of Ranvier will become more sensitive to the disposition of any scattering objects in the near-field. Whether this is seen depends on the mutual impedances between the antenna and the scattering objects. Understanding these phenomena requires accurate near-field estimation methods, non-homogeneous extracellular matrix measurement and modeling, and improved understanding of electrical phenomena at biointerfaces. The random field model described in [19] is relevant here, as is the model presented in [20].

We leave the reader with several quite remarkable and game-changing thoughts regarding the potential functionality of the EM near-field. As we have been taught, the human nervous system provides extremely energy efficient, highly complex realization and control of how we sense and think. For machines designed by humans, the ideas of energy efficiency and complexity tend to be at odds, thus the question of how the nervous system *really* works has received intense scrutiny for decades. We believe that this chapter takes a first step in this understanding.

The electric near-field described in this work is generated in an energy efficient manner as the *natural by-product* of action potential generation by millions of nerve fibers. The electric near-field bathes regions of the central nervous system like the brain, optic nerves, and spinal cords, thereby enabling a wireless network that has the capability, for example, to interconnect different physical regions of the brain at instants of time as is required for the execution of many cognitive tasks. This realization leads to a change in thinking that augments the Human Connectome,[2] a fixed anatomical network, with a relatively rapid high-capacity wireless network. **Figure 8** explains this augmentation in simple terms.

On the left-half of **Figure 8** is the Human Connectome shown as a tractographic map of the nerve fiber connections in the human brain. This mapping of anatomical connections is the *fixed network* of the brain. An example of advanced work contributing to the Human Connectome is [21] which has led to a new anatomical map of the cortex based on neuroimaging. It is important to note that neuroimaging only provides indirect measurements of brain activity and that no additional modalities of measurement such as electric near-field measurements have been made in the Human Connectome project studies. On the right-half of the **Figure 8** are magnifications showing neuronal groups in two regions of the brain. There is an electric near-field between these neuronal groups, a point illustrated by the red spherical waves. These near-field waves communicate action potential activity and fine details of their regions in a bi-directional manner. This creates a wireless connection,

[2]See the Human Connectome Project, www.humanconnectomeproject.org.

http://dx.doi.org/10.5772/intechopen.71945

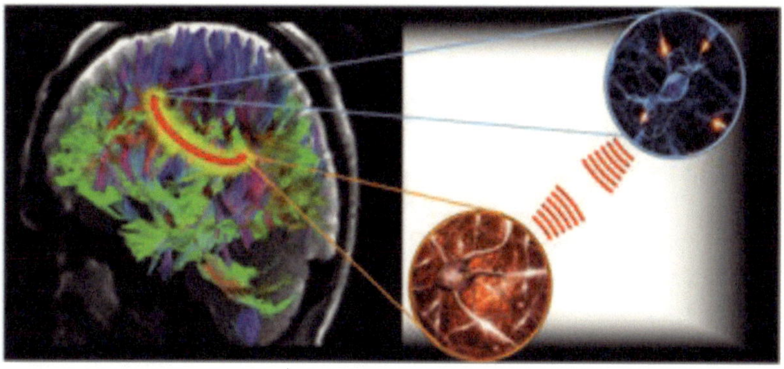

**Figure 8.** Electric near-field wireless network augmentation of the human Connectome.

as shown by the red loop across these brain regions in the left-half of the figure. Thus, the brain consists of a fixed network augmented by a wireless network which helps to explain how physically disparate regions of the brain connect so quickly during the execution of certain cognitive tasks.

Furthermore, as additional impetus for conducting electrical near-field measurements along with neuroimaging, neurological diseases, such as Multiple Sclerosis, Parkinson's, and Autism Spectrum Disorder, are associated with dysfunctions that are like those experienced in conventional wireless networks and electrical near-field measurements will detect neurological dysfunctions. For example, in Multiple Sclerosis, the physical change is a deterioration in the myelin sheath which insulates the nerve fiber. Damage to the myelin sheath can slow and even eliminate nerve conduction, in which case, Nodes of Ranvier do not regenerate action potentials and electric near-fields have levels that are significantly reduced, or do not occur. Our studies indicate that the electric near-field spatial-temporal signatures of healthy nerve fibers and demyelinated nerve fibers are quite different, with the radiation characteristics generally being more diffuse across both internodal and Node of Ranvier regions depending on the nature and extent of the lesions. This also suggests the possibility of a novel diagnostic tool to accompany the usual nerve conduction studies and electromyography used to assess axonal conduction speeds. This is discussed more precisely in Section 5.

Pursuing these ideas will lead to a new understanding of brain function as enabled by a highly intricate combination of fixed and wireless network connectivity [22]. This vision of brain function is indeed exciting, particularly when we recall the previous discussion that describes the EM near-field spatial directivity in terms of the data (action potentials) propagating down nerve fibers. One implication is that synchronous activity in groups of nerve fibers, which is a well-known clinically observable phenomenon, can spatially direct significant, frequency-rich electric fields toward certain target brain regions, thereby assuring that point-to-point connections are established. Thus, physically disparate brain destination regions are activated as a function of synchronous source region activity. This is a *data-driven* form of spatial connectivity.

## 5. Development of a wearable neurodevice

This section describes the research and development phases conducted in our laboratories for a non-invasive, wearable neurodevice that provides clinical grade information on cognitive and behavioral states of the brain and insight into impaired central nervous system neurological network dynamics. This medical device will be a game changer for the non-invasive diagnosis and treatment of a variety of neurological dysfunctions, including Multiple Sclerosis, Autism Spectrum Disorder, and Alzheimer's Disease; mental disorders; and traumatic brain and spinal cord injuries.

The material in this section provides neurobiological support for the engineering investigation presented in Section 3. To accomplish this, we model and clinically validate the origins, levels, and spatial-temporal structures of the endogenous fields under healthy and non-healthy conditions. As we have seen, the endogenous fields of the brain are characterized as low frequency electric near-fields (the electric field version of well-known NFC magnetic field technology). We believe that certain scientific and engineering challenges can be overcome because we approach this problem from a bioengineering perspective, with a team having a strong background in wireless communications; the design of "electrically short" antenna arrays; low frequency, highly sensitive receivers; and electric field transmission systems capable of a rich variety of modulation formats, along with partnerships with research laboratories, institutes and clinics dedicated to the understanding and therapeutic treatment of neurological disorders.

The overarching goals that we have set are:

1. Benchmark the Wearable Neurodevice as being clinically differentiated in both diagnostic and therapeutic efficacy and accuracy vis-à-vis fitness and lifestyle oriented wearables.[3]

2. Significantly impact chronic care management for patients with certain neurological and mental health dysfunctions.

3. Fill the enormous care gap for those with chronic neurological dysfunctions, where patients are not getting the long-term support they need and improve health outcomes for both doctors and patients.

4. Ensure patient compliance with physician treatment plans involving the Wearable Neurodevice by building useful levels of feedback into the device software.

5. As medical wearables are integrated into the chronic disease healthcare space, the biggest effect on the market will come from physicians who prescribe a Wearable Neurodevice such as that described here. Such a device will make a real difference in the lives of countless numbers of individuals afflicted with neurological dysfunctions and their families.

[3]To gauge reliability, we have run a clinical trial or benchmark against brain imaging modalities and neuromodulation devices that are the best known. Of these, there are only a small number that are US FDA-cleared and clinically-tried; therefore, the field is relatively open to the development of a device that becomes the benchmark for future device development.

### 5.1. Research objectives

The research objectives are as follows:

1. Conduct computer simulation of brain fiber tracts, or neurological networks, consisting of thousands of myelinated nerve fibers (axons) packed tightly together as found in several regions of the brain.

2. Establish the levels and spatial-temporal structure of the electric near-fields generated during typical physiological activity states that can be pharmacologically evoked (so that simulation and clinical data can be more easily compared).

3. Conduct a set of simulation studies for healthy conditions and for selected neurological dysfunctions. Each dysfunction will be associated with a different nerve fiber disease model, e.g., Multiple Sclerosis presents as a demyelination of the nerve fibers; this results in a change in the level and spatial-temporal structure of the resulting endogenous field.

4. Use the simulation studies, refined by the clinical data, to establish a specification for the near-field system and low frequency transceiver used to passively sense the brain's endogenous fields (diagnostic mode) and actively modulate them (therapeutic treatment mode).

5. Accomplishing these objectives requires the use of clinical data to refine simulation models. We have clinical data available to us from our partners. A carefully executed simulation study will reduce the time and risk associated with the development of a device approved for use. Future work includes the study of recommended feedback modes, personalization, and protocols for integration into existing legacy and proposed next generation healthcare systems.

### 5.2. Significance of the research and development

We mention three neurological dysfunctions in this chapter, because we feel that we can bring a new perspective to their diagnosis and therapeutic treatment. These diseases and their current prognoses are:

1. Multiple sclerosis (MS)—A chronic disease in which the immune system eats away at the protective coating (myelin) of nerve fibers. This disease requires a medical diagnosis. It appears that this disease cannot be cured, but treatment in the way of immunosuppressant drugs may help. These drugs have significant side effects. This disease primarily affects those in age between 14 and 60+ years.

2. Autism spectrum disorder—A chronic, serious developmental disorder that impairs the ability to communicate and interact with others. This disease requires a medical diagnosis. It appears that this disease cannot be cured, but treatment in the way of antipsychotic drugs may help. These drugs have significant side effects. This disease primarily affects those in age between 3 and 60 years.

3. Alzheimer's disease (*senile dementia*)—A chronic, progressive disease that impairs memory and other important mental functions. This disease requires a medical diagnosis. It appears

that this disease cannot be cured, but treatment in the way of cognition-enhancing medications may help. These drugs have significant side effects. This disease primarily affects those in age from 41 to 60+ years.

There are over 4 million individuals diagnosed with just these three neurological conditions each year in the US. Individuals with these diseases suffer throughout the diagnosis process, as it is many times painfully invasive, and they suffer throughout the treatment process, as the best drugs available have long lists of side effects. The significance of this research is the firm belief that a rationally designed, clinically accurate, wearable neurodevice can mitigate their suffering and allow them to take a greater role in their disease management.

We do not pretend to have cures for these neurological dysfunctions; however, we do claim to bring a new perspective to the table. This new perspective represents a paradigm shift in current thinking. Our biologically realistic preliminary *in-silico* disease models illustrate that each of the three neurological dysfunctions mentioned above is a specific impairment of the neurological networks of the brain that is reflected in specific spatial-temporal characteristics of the brain's electric near-fields. We provide a brief description and use an example to further underline the significance of the proposed research.

The most fundamental element of the neurological network is the nerve fiber, or axon. In vertebrates, many CNS nerve fibers have Oligodendrocyte derived myelin sheaths, multi-lamellar structures that wrap around the fibers, insulate them, and allow rapid information transmission over long distances with minimal energy as mentioned in Section 3. The Oligodendrocyte cells also actively promote formation of distinct membrane domains on the fiber that are not insulated. The distinct membranes of interest to us are the Nodes of Ranvier, short gaps between adjacent myelin segments as shown in **Figure 9**.

Voltage-gated ion channels are densely packed in these Nodes of Ranvier regions; key among these molecules are the Nav sodium ($Na^+$) channels that regenerate action potentials. Research shows that the bi-directional flow of ions through the Nav channels gives rise to electric

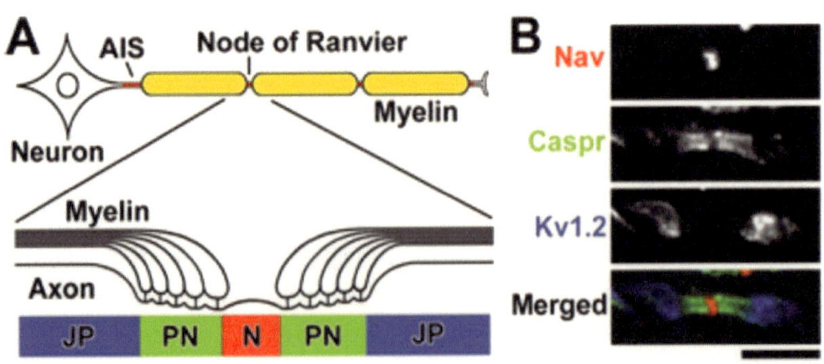


**Figure 9.** Molecular composition at nodes of Ranvier. (A) Diagram illustrating the structures of the myelinated nerve fiber and axonal subdomains: Axon initial segments (AIS), nodes of Ranvier (N), paranode (PN), and juxtaparanodes (JP). (B) Longitudinal sections of mouse optic nerve section immunostained with antibodies to Nav channels (nodal marker, red), Caspr (paranodal marker, green), and Kv1.2 channels (juxtaparanodal marker, blue) [23].

near-fields present in the extracellular space outside the uninsulated Nodes of Ranvier. These electric near-fields can be measured and have distinctly different spatial-temporal signatures when the Nodes of Ranvier function in healthy and unhealthy manners.

Since nodal Nav channel clusters are critical for action potential transmission, it is not surprising that changes and impairments in Nav channels are closely correlated with neurological symptoms. Furthermore, since the formation and maintenance of nodes depend on Oligodendrocyte interactions, defects in this interaction and axonal damage can disrupt nodal Nav channel clusters and consequently cause nerve conduction failure. There is a growing body of evidence for a pathogenic role of nodal dysfunction and/or disruption during neurological diseases. Dysfunction or disruption of Nodes of Ranvier are a primary focus for understanding the pathophysiology of neurological diseases. Our work is based on this evidence and our preliminary models which illustrate that the electric near-fields generated at impaired Nodes of Ranvier have spatial-temporal signatures which differ considerably from healthy profiles and that these signatures can be accurately classified as to the specific type of neurological dysfunction.

### 5.3. Research plan and technical approach

To fully understand our approach and the associated scientific and engineering challenges, we provide a brief introduction to the way various measurements are currently made. The study of neurological network connectivity, physiology, and pathology, is carried out using both invasive *in vivo* and *in vitro* methods and non-invasive modalities [24]. Invasive modalities range from sharp-glass and patch-clamp electrodes to planar electrodes (MEA) and FET arrays to recently developed intracellular recording nano- and micro-devices. These modalities generally measure what is known as the local field potential (LFP), an electrophysiological signal generated by the summed electric current flowing from a relatively localized population of neurons within a volume of neuronal tissue. Voltage is produced across the local extracellular space by action potentials in neurons in the volume and varies primarily because of synaptic activity.

What we would call "semi-invasive" modalities include optical imaging and stimulation technologies using either fluorescent indicators or genetically encoded molecular probes and the electrocorticogram (EcoG). Perhaps the most common non-invasive modality is the electroencephalogram (EEG). LFP signals differ from EEG and EcoG signals, principally because LFP signals are recorded in depth, from within deep brain structures, as opposed to at the surface of the scalp using macro-electrodes and at the surface of the brain using large subdural electrodes, respectively.

There are three advantages and considerations from the standpoints of diagnosis and therapeutic treatment of invasively measuring LPF signals vis-à-vis EEG signals:

1. They are representative of the activity of relatively localized populations of neurons [25], as opposed to characterizing the average activity of much larger populations [26].

2. They are not subject to the extensive filtering, diffusion, and distortion that EEG signals are because of propagation through many layers of strongly heterogeneous media [27].

3. It is believed that LFP measurements are correlated with normal and pathological excitable cell tissue operation [28–31].

From this discussion, we see that it would be desirable to develop a means of measuring LFP signals or other, even more relevant localized, non-distorted indicators, in a *non-invasive* manner for diagnostic purposes. Accomplishing this will require innovative technology for passively accessing localized populations of neurons from the level of the scalp.

**Research Objective 1:** Conduct computer simulation of brain fiber tracts, or neurological networks, consisting of thousands of myelinated nerve fibers (axons) packed tightly together as found in several regions of the brain.

**Research Plan:** There are two fundamental physical fields of the brain, the vector electric field $\mathbf{E}(\mathbf{r}, t)$(V/m) and the vector magnetic field $\mathbf{B}(\mathbf{r}, t)$(V-s/m$^2$). Our focus is on the vector electric field. The vector electric field can be Helmholtz decomposed into the gradient of a scalar potential $\Phi(\mathbf{r}, \mathrm{t})$ which can be measured with appropriate sensing technology at the spatial resolution of tissue fine structure [32, 33]. Our hypothesis is that membrane-related sources, specifically the Nodes of Ranvier, through their transmembrane currents and supporting systems of voltage-gated ion channels, act as electromagnetic field sources that contribute to the spatial-temporal structure of $\mathbf{E}(\mathbf{r}, t)$ and, consequently, $\Phi(\mathbf{r}, t)$. These fundamental fields then, in turn, mediate LFP expression, which is essentially a filtered, distorted, and spatially averaged version of $\Phi(\mathbf{r}, t)$. As described in recent work [34], in the neurological diseases involving myelinated nerve fibers (all those mentioned in this proposal and more), altered functions of nodal Nav channels and juxtaparanodal Kv channels lead to conduction failure. Indeed, the disruption of the molecular organization, altered ion channel expression, function, location, and/or density at the nodes of Ranvier are emerging as key players in the pathophysiology of neurological disorders [35].

We have developed our neurological network simulation software to accommodate thousands of myelinated nerve fibers tightly packed into fiber tracts. The simulation environment is COMSOL Multiphysics, a widely-accepted platform for coupled or Multiphysics phenomena executed within our CIRCE 6000 core computing cluster environment. Not only will this simulation be unique in its dimension (number of nerve fibers) and attention to fiber cross talk, a primary phenomenon in shaping electric field spatial-temporal structure [13], COMSOL Desktop will be used to ensure cross-disciplinary product development. Our software computing laboratory is fully integrated with the wet laboratory, so that *in vitro* model and parameter knowledge can be immediately transferred to the *in-silico* model development. This is a particularly important factor for ensuring that the biologically realistic nerve fiber functional deficit (disease) models needed for each type of neurological disorder are reflected in a biologically realistic manner in the *in-silico* model development.

In this work, we focus on deficit models associated with the Nodes of Ranvier. Dysfunction and/or disruption of the Nodes of Ranvier play significant roles in the development of neurological symptoms. Voltage-gated ion channel functions are disturbed by genetic mutations and by toxins. Autoimmunity against molecules at and near Nodes of Ranvier, myelin defects, and nerve fiber damage alter the localization and expression of ion channels and disrupt nerve fiber-Oligodendrocyte interactions. Our deficit models will be fully integrated with Hodgkin-Huxley models to account for how disease impacts both the electrophysiological passive and active regions of nerve fibers. **Figure 10** depicts the simulation for a disease model.

**Research Objective 2:** Establish the levels and spatial-temporal structure of the electric near-fields generated during typical physiological activity states that can be pharmacologically evoked (so that simulation and clinical data can be more easily compared).

**Research Plan:** With the nerve tract geometry set up, we treat each Node of Ranvier as the initiation site of an electric near-field[4] component. As described in Section 3, we mathematically model the electric near-field generated by the ensemble of voltage-gated ion channels distributed around a Node of Ranvier and then extend our development to the consecutive Nodes of Ranvier on a single nerve fiber. If we model each Node of Ranvier as a cylindrical electrical field source, it is then natural to model several consecutive Nodes of Ranvier as an approximately linear antenna with cylindrical elements. Note that the radiation characteristics of this linear antenna will depend on which Nodes of Ranvier are active or parasitic; therefore, the radiation characteristics are said to be data driven [13].

The resulting field from a small number of neighboring nerve fibers is what we referred to earlier as the endogenous field. This field's spatial-temporal signature is representative of the health of a localized population of nerve fibers in specific regions of the brain. Using the mathematical development, we then use COMSOL Desktop to create simulation software which we refine in parallel with ongoing wet lab studies. To test the validity of this approach, we have carried out a preliminary simulation of the electric near-field generated by a ring of ion channels in a Node of Ranvier. For a single ion channel (pore), clinical patch clamp measurements provide the information shown in **Figures 3** and **4**, e.g., pore current flow lasts several milliseconds and peaks at 0.5 pA.

It is important to note that the electric field generated by the pore ion flow is not a single frequency field, but a field consisting of many frequencies associated with the pore rather abruptly opening and closing. Adopting a criterion that the significant band of frequencies are those that are no more than 30 dB down from the maximum, we see that the significant band of frequencies goes up to at least 50 kHz. This is a startling result, as research on nerve fibers is commonly conducted below 1 kHz. The electric field is characterized as a multiple frequency, reactive near-field [13, 36]. Our preliminary analysis for a ring of ion pores at a node of Ranvier partially follows the theory presented in [37] and employs the geometry shown in **Figure 5**. A snapshot of the time-varying electric field is shown in **Figure 11**.

From **Figure 11** and the associated calculations, the electric near-field level is approximately $9 \times 10^8$ V/m at the very surface of the Node of Ranvier and approximately $5 \times 10^3$ V/m at about 1.767 millimeters from the surface. This is a field level that is biologically significant in that it can

[4]More precisely, we treat each voltage-gated ion channel (pore) as the initiation site of an electric near-field component. Voltage-gated ion channels are found in greatest density along myelinated nerve fibers at the nodes of Ranvier. For example, Na + channel density is ~2000 channels/μm at the nodes of Ranvier. They directionally propagate electrical signals. Voltage-gated ion-channels specific to sodium (Na), potassium (K), calcium (Ca), and chloride (Cl) ions have been identified. The opening and closing of the channels is best modeled as a stochastic process triggered by changing ion concentration, and hence charge gradient, between the sides of the cell membrane.

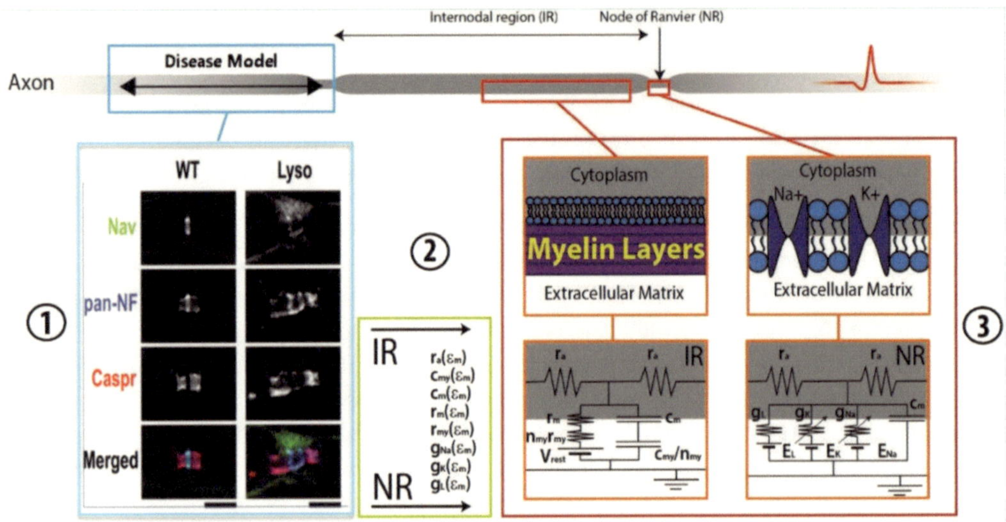


**Figure 10.** Schematic plan for the simulation. (1) the disease model. Shown is one example of the impact of demyelination in longitudinal sections of mouse nerves immunostained as indicated. Nerve fibers run horizontally. Scale bars = 10 microns. WT denotes wild-type; Lyso denotes lysolecithin. Note the anti-pan-NF antibodies display both nodal NF186 (strong signal colocalized with Nav channel staining) and paranodal NF155 (relatively weak signal colocalized with Caspr staining). Demyelination model induced by intraneural injection of lysolecithin (7 days after injection). Nodal cluster of Nav channel and NF186 is remarkably dispersed. (2) the coupling model reflects the disease model into the nerve fiber model as a set of altered parameters. (3) the nerve fiber model for both the Internodal regions (IR) and the nodes of Ranvier (NR) [38].

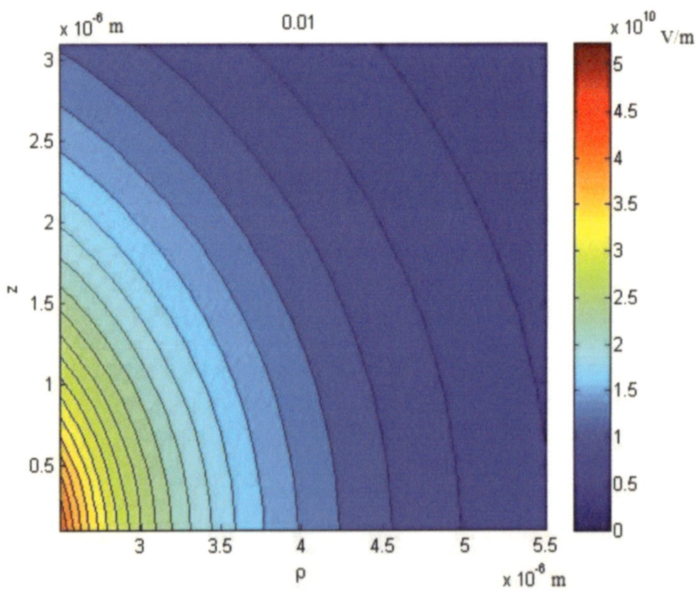


**Figure 11.** Snapshot of the time-varying electric near-field generated due to a ring of voltage-gated ion channels in a node of Ranvier. Video of time-varying field and clinical measurements available from the author.

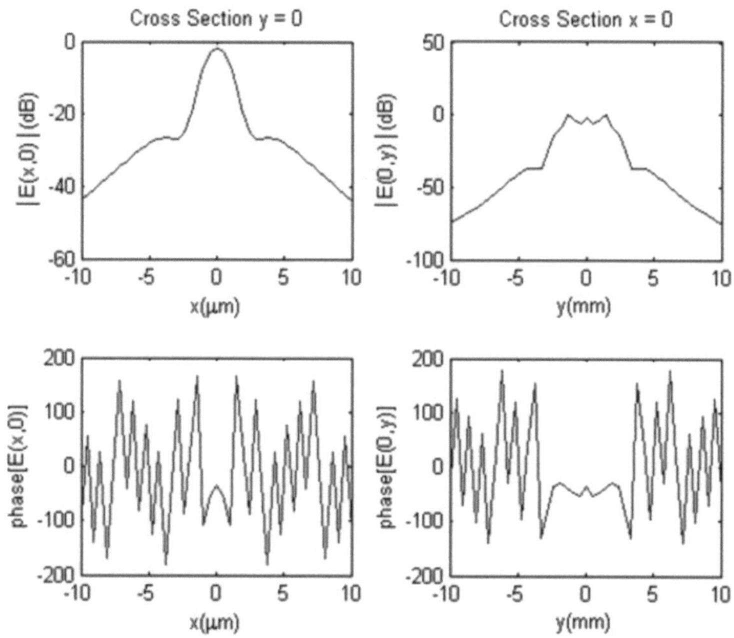


**Figure 12.** Cross-sectional plots of magnitude and phase of electric near-field at 10 kHz. Healthy nerve fiber with 10 active nodes of Ranvier.

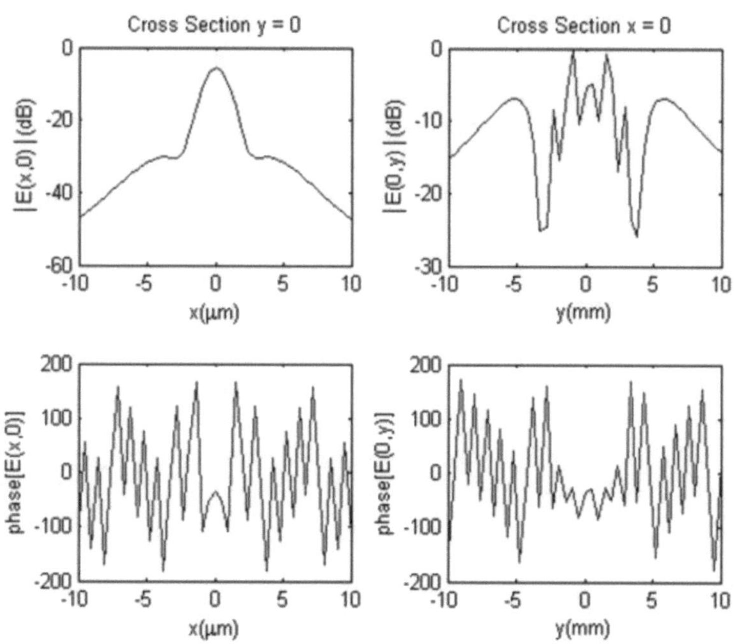


**Figure 13.** Cross-sectional plots of magnitude and phase of electric near-field at 10 kHz. Unhealthy nerve fiber with 50% nodal disruption. The unhealthy nerve fiber model is consistent with a progression of multiple sclerosis.

impact neighboring nerve fibers. This preliminary study indicates that this objective has merit and establishes the frequency range and range of levels expected for the electric near-field.

**Research Objective 3:** Conduct a set of simulation studies for healthy conditions and for selected neurological dysfunctions. Each dysfunction is associated with a different nerve fiber disease model, e.g., Multiple Sclerosis presents as a demyelination of the nerve fibers and as a change in the level and spatial-temporal structure of the resulting endogenous field.

**Research Plan:** Our selected neurological dysfunctions include Multiple Sclerosis (MS), Autism Spectrum Disorder (ASD), and Alzheimer's Disease, as mentioned earlier. These are dysfunctions that can be traced primarily to either the balance of excitatory and inhibitory nerve fibers or nodal demyelination in the central nervous system. Each dysfunction can be associated with a different nerve fiber disease model as described in Research Objective 1. To meet this research objective, we are concerned with the separability of spatial-temporal electric near-field signatures, or patterns, associated with healthy and non-healthy conditions. Note that the separability of patterns is exhibited by measurable differences in electric near-field signatures, which, in turn, are associated with time and frequency dynamics of localized, small nerve fiber populations. Here, frequency (up to 50 kHz, as described earlier) is a valuable parameter but unavailable to the EEG modality.[5]

Consider an experiment to demonstrate the separability of spatial-temporal electric near-field signatures when Nodes of Ranvier are disrupted [39–42]. For more detailed information about disruption of axon-Oligodendrocyte interaction in autoimmune diseases, see [43, 44]. In a tightly packed tract of nerve fibers, the dimensions are such that axons are mutually immersed in a reactive near-field. To examine the near-field characteristics, we choose an array length of $10\ell \approx 10$ mm, where $\ell$ is internodal length. Our simulation includes 10 nodes of Ranvier, equally distributed over $x \in [-5, 5]$ mm, $y = 0$.

The electric near-field is calculated for a frequency of 10 kHz on an $(x, y)$ axis planar rectangular surface located in the extracellular space between fibers [18]. **Figure 12** and **13** shows two cross sections of the magnitude and phase of the electric near-field for a healthy fiber and for a fiber where 50% of the nodes of Ranvier are disrupted, respectively. Note that the healthy and unhealthy nerve fibers display significant differences in electric field magnitude and phase characteristics and spatial extent.

Clearly, diagnosis and therapeutic treatment requires both position accuracy and spatial resolution within the brain. Our vision for this is a cap of flexible electronics sensors with time-delay electronics under control of a smart phone application that can facilitate beam formation to allow for passive listening and active therapeutic treatment of specific regions within the brain.

[5]Among the different brain imaging techniques, the EEG is classically considered as having an excellent temporal resolution, but a poor spatial one. We argue that the actual temporal resolution of conventional (scalp potentials) EEG is overestimated, and that volume conduction, the main cause of the poor spatial resolution of EEG, also distorts the recovered time course of the underlying sources at scalp level, and hence degrades the actual temporal resolution of EEG. A few facts about the temporal and spatial resolution of EEG: millisecond temporal resolution; localization of neural generators is complicated and usually not carried out; different tissues and the skull differ in their conductivity, and electric potentials do not pass through these structures undistorted; localization requires realistic head models. This last point is important and useful regarding the proposed wearable neurodevice.

http://dx.doi.org/10.5772/intechopen.71945

Since we are dealing with near-field phenomena, resolution significantly better than $\lambda/2$ is possible by accessing the evanescent modes of the spatial spectrum. We believe that utilizing near-field properties plus a head model will provide unprecedented spatial positioning capability.

It is important to note that in addition to neurological disease, traumatic injuries in the CNS may involve disruption of nodes. In an experimental model for spinal cord compression, exposure of juxtaparanodal Kv1.2 channels with accompanying myelin retraction at the nodes contributed to the induction of conduction block [22, 45]. In an animal model of traumatic diffuse brain axonal injury produced by fluid percussion insult, calpain-mediated proteolysis of cytoskeletal and scaffolding proteins, ankyrin-G at nodes and $\alpha$II-spectrin at paranodes, was associated with nodal damage, suggesting a possible contribution of nodal disruption to the complex mechanisms of traumatic brain injury [46].

These findings suggest that the wearable neurodevice can be designed to also monitor and therapeutically treat individuals that suffer insults to the spinal cord and brain. The possibilities for a device of this type are extraordinary. The Wearable Neurodevice can be a game changer in the mental health space for both diagnosis and therapeutic treatment [47, 48].

**Research Objective 4:** Use the simulation studies, refined by the clinical data, to establish a specification for the near-field antenna array system and low frequency transceiver used to passively sense the brain's endogenous fields and to actively modulate them.

**Research Plan:** The progress toward this research objective will evolve as we continue to experiment and explore and explain new concepts; however, to diagnose and therapeutically treat neurological dysfunctions of the type described in this proposal, we need a wearable neurodevice that meets exceptionally rigorous specifications. Our only interest is in non-invasive methods and technologies. We have worked on an initial set of specifications which we will detail here and refine as research progresses. The specifications include, but are not limited to, the following:

1. A passive listening capability and an active neuromodulation capability that allows recording and stimulation of small populations of neuronal targets.

2. Receiving sensitivity outside the human body consistent with neuron physiological parameters, e.g.,

    a. Transmembrane potentials in the range [−80,30] mV

    b. Subthreshold potentials in the range [±0.5,10] mV with rapid rise times (<1 ms) and slow decay times (100–1000 ms).

    c. Membrane oscillations in the range of ±5 mV at frequencies of 1–50 Hz.

3. Passive/active spatial resolution goal of 1 cubic nanometer for frequencies in the range 10 Hz to 10 kHz.

These specifications represent an unprecedented and extraordinarily exciting challenge to the fields of science and engineering.

## 6. Conclusions

Several conclusions can be drawn from this work:

(1) Neuroimaging principally drives current models of neurological networks through anatomical discovery, and it is known that the models are deficient in explaining all aspects of cognition and dysfunction. This deficiency is particularly acute in the cases of brain enhancement, Multiple Sclerosis and Autism Spectrum Disorder, examples where we believe that the wireless neurological networking described here plays a primary role.

(2) We have presented a wireless neurological networking model that augments current models and is based on conventional electrical engineering principles. The fundamental principle that is at play is that of crosstalk between nerve fibers. *In silico* and initial *in vivo* experiments carried out at the University of South Florida Bioengineering Laboratories and the Global Center for Neurological Networks indicates that this crosstalk may enable communication, cooperation, and/or competition among nerve fibers and that the electric near-field level is approximately $5{\times}10^{10}$ V/m at the very surface of an active Node of Ranvier. This level is orders of magnitude higher than the equivalent electric field level for the NFC Apple Pay® system and other contactless smart chips and cards which clearly accomplish complex information transfers. The electric near-field level increases when several consecutive active Nodes of Ranvier are active and when neural oscillations and nerve fiber synchronicity occur.

(3) As a matter of interest, the wireless neurological networking model can be cast in the form of a massive biological multiple input multiple output (MIMO) network [49] that permeates the brain and nerve bundles/tracts and explains cognition and dysfunction in terms of local and global effects and dynamics. Defining this model in detail is the target of future research.

Note that the spatial-temporal structure of the electric near-field depends on whether healthy or unhealthy, for example, demyelinated, conditions are present and may contribute to these conditions. Healthy conditions are associated with only the unshielded Nodes of Ranvier having the capacity to create electric near-fields in extracellular space; whereas, unhealthy conditions as clinically understood in Multiple Sclerosis, for example, present as demyelinated fibers which provide many more opportunities between Nodes of Ranvier for leakage and electric near-field activity as voltage gated ion channels migrate to these positions [50].

In short, the electric near-field spatial-temporal signatures are different for healthy and unhealthy conditions, as we have verified by simulation. Understanding and measuring electric near-field signatures could serve as a diagnostic tool, on one hand, and as an endogenous field, on the other hand, which could presumably be manipulated for therapeutic purposes by the application of exogenous neurostimulation devices. At present, there exists no rational design for neurostimulation devices and emitted waveforms based on endogenous field structure. This work represents one of the first steps toward such a rational design of non-invasive devices.

## Author details

Salvatore Domenic Morgera

Address all correspondence to: sdmorgera@usf.edu

University of South Florida and Global Center for Neurological Networks, Tampa, Florida, USA

## References

[1] Marder E. Variability, compensation, and modulation of neurons and circuits. Proceedings of the National Academy of Sciences of the United States of America. 2011;**108**(3):15542-15548. DOI: 10.1073/pnas.1010674108

[2] DeBanne D, Campanac E, Bialowas A, Carlier E, Alcaraz G. Axon physiology. Physiological Reviews. 2011;**91**:555-602. DOI: 10.1152/physrev.00048.2009

[3] Rushton WAH. A theory of the effects of fiber size in medullated nerve. The Journal of Physiology. 1951;**115**:101-122

[4] Waxman SG. Determinants of conduction velocity in myelinated nerve fibers. Muscle & Nerve. 1980;**3**(2):141-150. DOI: 10.1002/mus.880030207

[5] Ezure H, Goto N, Nonaka N, Goto J, Tani H. Morphometric analysis of the human trigeminal nerve. Okajimas folia anatomica Japonica. 2001;**78**(2-3):49-54

[6] Jonas JB, Schmidt AM, Muller-Bergh JA, Schlotzer-Schredthardt UM, Naumann GO. Human optic nerve fiber count and optic disc size. Investigative Ophthalmology & Visual Science. 1992;**33**(6):2012-2018. PMID: 1582806

[7] Dougherty D, Johnson MJ. Clinical validation of three-dimensional tortuosity metrics based on the minimum curvature of approximating polynomial splines. Medical Engineering & Physics. 2008;**30**(2):190-198. DOI: 10.1016/j.medengphy.2007.02.009

[8] Bennett CH. Theory of entanglement and entanglement-assisted communication. Bulletin of American Physical Society. 2011;**56**. DOI: 10.1109/TIT.2010.2053903. http://meetings.aps.org/Meeting/MAR11/Event/139600

[9] Morgera SD. The axonal spatial-temporal reactive assembly (ASTRA): Part II. In: IEEE EMBS Micro & Nanotechnology in Medicine; Dec 8-12. Oahu, Hawaii: IEEE; 2014

[10] Morgera SD. A stochastic multi-axon model for guiding a nanomedicine strategy for multiple sclerosis: Part I. In: IEEE EMBS Micro & Nanotechnology in Medicine; Dec 3-7. Maui, Hawaii: IEEE; 2012

[11] Walker WD. Superluminal near-field dipole electromagnetic fields. In: Vigier III Symposium, Gravitation and Cosmology. Berkely CA; 2000. pp. 21-25

[12] Hales CG, Grayden DB, Quiney H. The electric field system of a macular ion channel plaque. In: 33rd Ann Intern Conf IEEE EMBS. Boston, MA: IEEE; 2011. pp. 294-297

[13] Morgera SD. Reactive near field electromagnetic axonal communication channels and their role in neurodegenerative diseases. In: IEEE EMBS Eng Med Bio Conf; Aug 25-29. Milano: IEEE. pp. 2307-2310

[14] Morgera SD. Near field axonal communication networks and their role in neurodegenerative diseases. In: Bio Eng Soc BMES 21st Ann Mtg; Oct 7-10. Tampa, FL; 2015

[15] Hill DA. Theory of near-Field Phased Arrays for Electromagnetic Susceptibility Testing. Washington DC: National Bureau of Standards, US Govt Printing Office; Feb. 1984

[16] Gabriel C, Gabriel S, Corthout E. The dielectric properties of biological tissues: I. Literature survey. Physics in Medicine and Biology. 1996;**41**:2231-2249. PMID: 8938024

[17] Fano R. Theoretical Limitations on the Broadband Matching of Arbitrary Impedances. Cambridge, MA: Technical Report No 41: MIT Research Laboratory of Electronics; 1948

[18] Wang JJH. An examination of the theory and practice of planar near-field measurement. IEEE Transactions on Antennas and Propagation. 1988;**36**(6):746-753. DOI: 10.1109/8.1176

[19] Romero PA, Arnold FH. Random field model reveals structure of the protein recombinational landscape. PLoS Computational Biology. 2012;**8**(10):E1002713. DOI: 10.1371/journal.pcbi.1002713

[20] Morgera SD, Forbes ZP. Long correlation random image field models. In: 1988 Conf Acoust, Speech and Sig Proc. Vol. 2. IEEE; 1988. pp. 1036-1039

[21] Glasser MF, Coalson TS, Robinson TC, Hacker CD, Harwell J, Yacoub E, et al. A multimodal parcellation of human cerebral cortex. Nature. 2016;**531**(7615):171-178. DOI: 10.1038/nature18933

[22] Morgera SD. The fixed and wireless cooperative networks of the brain. In: Soc Brain Mapping & Therapeutics 13th Ann World Cong; April 8-10; Miami, FL. 2016. (Invited Speaker)

[23] Susuki K. Node of Ranvier disruption as a cause of neurological diseases. ASN NEURO. 2013;**5**(3):art:e00118. DOI: 10.1042/AN20130025

[24] Morgera SD. Electric field imaging of the brain. In: IEEE Intern Symp Brain Imaging; April 18-21. Melbourne: IEEE; 2017

[25] Katzner S, Nauhaus I, Benucci A, Bonin V, Ringach DL, Carandini M. Local origin of field potentials in visual cortex. New York, NY: Neuron. 2009;**61**:35-41. DOI: 10.1016/j.neuron.2008.11.016

[26] Niedermeyer E, Lopes Da Silva FH, editors. Electroencephalography: Basic Principles, Clinical Applications, and Related Fields. Boca Raton, FL: Lippincott Williams & Wilkins, 2005. ISBN-13: 978-0781751261

[27] Varsavsky A, Mareels I, Cook M. Epileptic Seizures and the EEG: Measurement, Models, Detection and Prediction. Taylor & Francis; 2011. ISBN 9781439812006

[28] Einvoll GT, Kayser C, Logothetis NK, Panzeri S. Modelling and analysis of local field potentials for studying the function of cortical circuits. Nature Reviews. Neuroscience. 2013;**14**:770-785. DOI: 10.1038/nrn3599

[29] Friston KJ, Bastos AM, Pinotsis D, Litvak V. LPF and oscillations - what do they tell us? Current Opinion in Neurobiology. 2014;**31**:1-6. DOI: 10.1016/j.conb.2014.05.004

[30] Riemann MW, Anastassiou CA, Perin R, Hill SL, Makram H, Koch C. A biophysically detailed model of neocortical local field potentials predicts the critical role of active membrane currents. Neuron. 2013;**79**:375-390. DOI: 10.1016/j.neuron.2013.05.023

[31] Hales C, Pockett S. The relationship between local field potentials (LFPs) and the electromagnetic fields that give rise to them. New York, NY: Frontiers in Systems Neuroscience. New York, NY. 2014;**12**(8):233. DOI: 10.3389/fnsys.2014.00233

[32] Malmivuo J, Plonsey R. Bioelectromagnetism: Principles and Applications of Bioelectric and Biomagnetic Fields. Oxford Unversity Press; 1995. ISBN: 9780195058239

[33] Jackson JD. Classical Electrodynamics. New York, NY: Wiley; 1999. ISBN: 978-0-471-30932-1

[34] Susuki K, Chang K-J, Zollinger DR, Liu Y, Ogawa Y, Eshed-Eisenbach Y, et al. Three mechanisms assemble central nervous system nodes of Ranvier. Neuron. 2013;**78**:469-482. DOI: 10.1016/j.neuron.2013.03.005

[35] Buffington SA, Rasband MN. The axon initial segment in nervous system disease and injury. The European Journal of Neuroscience. 2011;**34**:1609-1619. DOI: 10.1111/j.1460-9568.2011.07875.x

[36] Girard C, Joachim C, Gauthier S. The physics of the near-field. Singapore: Reports on Progress in Physics; 2000;**63**:893-938

[37] Horch KW, Dhillon GS. Neuroprosthetics: Theory and Practice. Vol. 2. World Scientific; 2004. ISBN: 978-981-238-022-7

[38] García-Grajales JA, Rucabado G, García-Dopico A, Peña J-M, Jérusalem A. Neurite, a finite difference large scale parallel program for the simulation of electrical signal propagation in neurites under mechanical loading. PLoS ONE. 2015;**10**(2):e0116532. DOI: 10.1371/journal.pone.0116532

[39] Wolswijk G, Balesar R. Changes in the expression and localization of the paranodal protein Caspr on axons in chronic multiple sclerosis. Brain. 2003;**126**:1638-1649. DOI: 10.1093/brain/awg151

[40] Craner MJ, Newcombe J, Black JA, Hartle C, Cuzner M, Waxman SG. Molecular changes in neurons in mutiple sclerosis: Altered axonal expression of Nav 1.2 and Nav 1.6 sodium channels and Na+/Ca2+ exchanger. Proceedings of the National Academy of Sciences of the United States of America. 2004;**101**:8168-8173. DOI: 10.1073/pnas.0402765101

[41] Howell OW, Palser A, Polito A, Melrose S, Zonta B, Scheiermann C, et al. Disruption of neurofascin localization reveals early changes preceding demyelination and remyelination in multiple sclerosis. Brain. 2006;**129**:3173-3185. DOI: 10.1093/brain/awl290

[42] Coman I, Aigrot MS, Seilhean D, Reynolds R, Girault JA, Zale B, et al. Nodal, paranodal and juxtaparanodal axonal proteins during demyelination and remyelination in multiple sclerosis. Brain. 2006;**129**:3186-3195. DOI: 10.1093/brain/awl144

[43] Derfuss T, Linington C, Hohlfeld R, Meinl E. Axo-glial antigens as targets in multiple sclerosis: Implications for axonal and grey matter injury. Journal of Molecular Medicine (Berlin). 2010;**88**:753-761

[44] Desmazi'eres A, Sol-Foulon N, Lubetzki C. Changes at the nodal and perinodal axonal domains: A basis for multiple sclerosis pathology? Multiple Sclerosis. 2012;**18**:133-137

[45] Ouyang H, Sun W, Fu Y, Li J, Cheng JX, Nauman E. Compression induces acute demyelination and potassium channel exposure in spinal cord. Journal of Neurotrauma. 2010;**27**:1109-1120. DOI: 10.1089/neu.2010.1271

[46] Reeves TM, Greer JE, VAnderveer AS, Phillips LL. Proteolysis of submembrane cytoskeletal proteins ankyrin-G and αII-spectrin following diffuse brain injury: A role in white matter vulnerability at nodes of Ranvier. Brain Pathology. 2010;**20**:1055-1068. DOI: 10.1111/j.1750-3639.2010.00412.x

[47] Frolich F, McCormick DA. Endogenous electric fields may guide neocortical network activity. Neuron. 2010;**67**(1):129-143. DOI: 10.1016/j.neuron.2010.06.005

[48] Frolich F. Endogenous and exogenous electric fields as modifiers of brain activity: Rational design of noninvasive brain stimulation with transcranial alternating current stimulation. Dialogues in Clinical Neuroscience. 2014;**16**(1):93-102, PMCID: PMC3984895

[49] Ngo HQ, Larsson EG, Marzetta T. Energy and spectral efficiency of very large multiuser MIMO systems. IEEE Transactions on Communications. 2013;**61**(4):1436-1449

[50] Black JA, Waxman SG. Noncanonical roles of voltage-gated sodium channels. Neuron. 2013;**80**(2):280-291. DOI: 10.1016/j.neuron.2013.09.012

Chapter 13

# Inverse Problem for an Electrical Dipole and the Lightning Location Passive Monitoring System

Anatoly V. Panyukov

Additional information is available at the end of the chapter

http://dx.doi.org/10.5772/intechopen.74264

**Abstract**

We solve the problem of the locating parameters, identifying equivalent dipole electromagnetic radiation source through measured horizontal magnetic and vertical electric components at some point of the infinite conducting ground. Methods based on analysis of measured signals are suggested. The problem under consideration, like any inverse problem of mathematical physics, is ill-conditioned. The consequences of this are the high sensitivity of the algorithm to the errors in the source data and calculation errors. All these circumstances do not allow to estimate the accuracy and reliability of the results obtained with the help of single-scale algorithms. The considered problem is contained in a complex of mathematical models of the practically important problem of forecasting the development of thunderstorm foci. Lightning meteorology focuses on investigating the lightning activities in different types of convective weather systems and the relationship of lightning to the dynamic and microphysical processes in thunderstorms. With the development and application of advanced lightning detection and location technologies, lightning meteorology has been developed into an important interdisciplinary between atmospheric electricity and meteorology. This paper reviews (1) methods to identify the dipole location and (2) possibilities to analyze the pre-radiation of thunderstorm clouds by the passive methods.

**Keywords:** algorithm, location finding, mathematical modeling, passive monitoring of the Earth's electromagnetic field, electric dipole

## 1. Introduction

The problem of identifying the position parameters of an arbitrary-oriented electric dipole over a plane with infinite conductivity from its electromagnetic field induced at the observation point is considered. The considered problem is contained in a complex of mathematical models of the practically important problem of forecasting the development of thunderstorm foci. To estimate the location of the electromagnetic radiation source (EMR) based on the results of a single-point observation of the electromagnetic field induced by it, up to 1990, a number of devices were developed, based on the use of a vertical dipole as a model of an EMR source and physically realizable analog algorithms. In the next decade, the development of the use of an arbitrarily oriented dipole as a model of the EMR source and digital processing of observed signals was developed [1–3].

The effective algorithm for the single-point distance determination to a pulsed EMR source was proposed and investigated in the papers [4, 5], and pointed out to the emergence of irremovable uncertainty in the location of the EMR source by a one-point method caused by the difference in the orientation of the dipole from the vertical. The use of two or more observation points that do not belong to the same straight line makes it possible to determine the position parameters of the equivalent dipole [4, 6, 7].

A review of the status of passive storm monitoring systems by the end of 2003 and the demonstration of the use of lightning-position detection systems for the passive radar of hazardous meteorological phenomena are presented in [1, 2]. Modern methods for analyzing the field, allowing to determine the parameters of the source of EMP, characterizing its location and orientation are presented in the works [4, 5, 8] within the ISTC project #1822 was developed. As a result of the conducted field tests of this sample, during May–August 2004 more than 2.5 million atmospherics were recorded. Of these, not more than 10% were classified as radiation from a lightning discharge, the rest were classified as pre-threat radiation. Thus, the proportion of inter-cloud and intra-cloud discharges relative to the cloud-ground discharges turned out to be much higher than those noted for the work [9].

The registration of ominous cloud radiation (i.e., before the first lightning flash) by a single-point lightning protection system, for the purposes of forecasting thunderstorm development, was provided by the expansion of the dynamic range of receiving equipment and the further development of mathematical and software. At present, active radar facilities using a comparison of sounding and reflected signals are used to analyze the pre-threat state [9–11]. The obvious drawbacks of this approach are: (1) the high cost and the presence of the human factor; (2) the low probability of detecting individual discharges; (3) the lack of ecological compatibility due to the application of microwave radiation and its effect on the clouds. Passive methods of analysis of the pre-threat state, based on the analysis of the intrinsic radiation of clouds, exclude the noted shortcomings of active methods.

Previously, in passive thunderstorm monitoring systems, pre-threat radiation was either filtered out, or it was worked out incorrectly by single-point systems of thunderstorm location [12]. The database of ISTC project #1822 [8] field trials provides a large experimental material for testing the adequacy of thunderstorm models and testing of the developed software and devices. In general, the results of the project demonstrated the possibility and necessity of creating a new generation of thunderstorm detection systems and expanding the range of tasks they solve. This leads not only to a significant revision of the requirements for their technical characteristics, but also to the development of new mathematical models and algorithms for analyzing thunderstorm phenomena, their tracing and display, as well as archiving and their use by specialists from different subject areas.

http://dx.doi.org/10.5772/intechopen.74264

## 2. Statement of the dipole location problem

### 2.1. Direct problem

Born and Wolf [13] give the expressions for the electromagnetic field of a dipole $\mathbf{P}_0 = p(t)\mathbf{n}_0\delta(\mathbf{r}_0)$ (moment $p$, direction $\mathbf{n}_0$, location $\mathbf{r}_0$, δ(*) is the Dirac delta function) in vacuum

$$E(P_0) = \frac{1}{4\pi\varepsilon_0}\left[\frac{3p(t)}{r^3} + \frac{3p'(t)}{cr^2} + \frac{p''(t)}{c^2r}\right](n_0, e_0)\, e_0 - \frac{1}{4\pi\varepsilon_0}\left[\frac{p(t)}{r^3} + \frac{p'(t)}{cr^2} + \frac{p''(t)}{c^2r}\right]n_0, \tag{1}$$

$$H(P_0) = \frac{1}{4\pi}\left[\frac{p'(t)}{r^2} + \frac{p''(t)}{r}\right](n_0 \times e_0) \tag{2}$$

here $\mathbf{e}_0$ is the direction from the dipole to the observer and the single and double primes denote time derivatives. For distances up to 300 km, a conducting plane surface is used to model the ground. The dipole in the half-space bounded by the infinitely conducting plane results in a field (**E, H**) consisting of the field (**E(P$_0$), H(P$_0$)**) of the source $\mathbf{P}_0$, and the reflected field; the latter can be represented by the field (**E(P$_1$), H(P$_1$)**) of an imaginary dipole $\mathbf{P}_1 = p(t)\mathbf{n}_1\delta(\mathbf{r}_1)$, which is the mirror image of the dipole $\mathbf{P}_1$ therefore we have $\mathbf{E} = \mathbf{E}(\mathbf{P}_0) + \mathbf{E}(\mathbf{P}_1)$, $\mathbf{H} = \mathbf{H}(\mathbf{P}_0) + \mathbf{H}(P_1)$ (see **Figure 1**).

Using the Cartesian system of coordinates with origin at the observation point O and the Oz axis being the normal of the bounding plane, we present the electric (E) and magnetic (H) components of the field in the coordinate form [4]:

$$E_x(t) = E_y(t) = H_z(t) = 0, \quad E_z(t) = [q(t) + q'(t)]\frac{3u - 2v}{\alpha} + q''(t)\frac{u}{\alpha^2}, \tag{3}$$

$$H_x(t) = \sin\varphi\left(q'(t)\frac{1}{\alpha} + q''(t)\frac{1}{\alpha^2}\right), \quad H_y(t) = \cos\varphi\left(q'(t)\frac{1}{\alpha} + q''(t)\frac{1}{\alpha^2}\right), \tag{4}$$

here $\alpha = c/r$ is the inverse of the time of wave propagation from the source to the observer and the variables $u, v, q(t)$ are determined from the equalities

$$w\sin\varphi = \sin\theta\cos\theta_0\sin\psi - \cos\theta\sin\theta_0\sin\psi_0, \tag{5}$$

$$w\cos\varphi = \sin\theta\cos\theta_0\cos\psi - \cos\theta\sin\theta_0\cos\psi_0, \tag{6}$$

$$u = \sin\theta\cos(\varphi - \psi), \quad v = \frac{\sin(\varphi - \psi_0)}{\sin\theta\,\sin(\psi - \psi_0)}, \quad q(t) = w\alpha^2 p(t)/\left(2\pi c^2 r\right). \tag{7}$$

Other parameters are given in **Figure 1**.

### 2.2. Inverse problem

Components $E_z$, $H_x$, and $H_y$ can be changed with the help of the antenna system consisting of a vertical electrical antenna and a pair of mutually orthogonal frame magnetic antennas, and

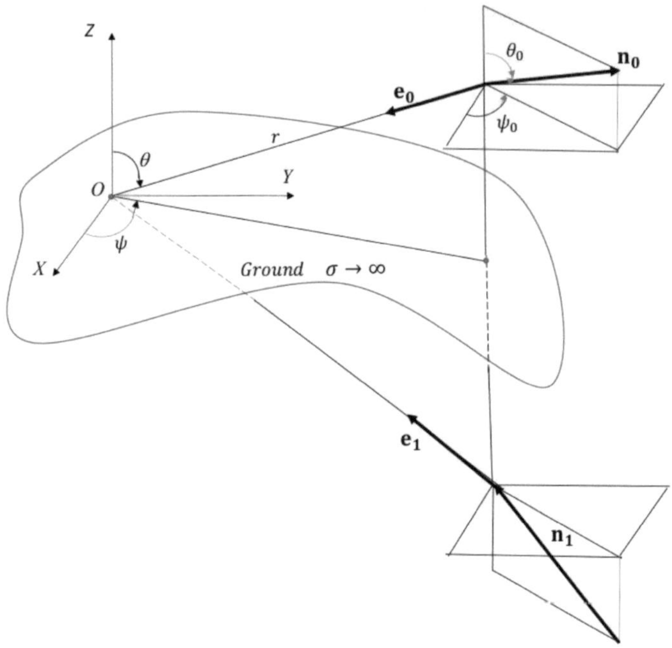


**Figure 1.** Geometrical parameters of the mathematical problem.

hence used as initial data for the task of evaluating coordinates $(r, \psi, \theta)$ for the dipole position and $(\psi_0, \theta_0)$ for the dipole orientation.

The inverse problem has a few specific aspects that differs it from the direct problem. If the source model is known, one often solves the inverse problem by searching the values of the parameters of the model, which give the best explanation of the measured data. Such a search is usually performed by an optimization algorithm which varies the values of the parameters until a necessary agreement between the measured and the model predicted fields is achieved. The search algorithms may get stuck in a local minimum and the solution will not be determined correctly. The use of a global optimization technique usually diminishes the problem of local minima but increase computational complexity.

For our case, it is possible to directly invert the values of the measured field into the values of the parameters describing the source location. In the present article, we use an explicit identification to infer the parameters of an electric dipole source situated over a perfectly conducting plane.

## 3. Calculation of the parameter $\varphi$. Exclusion of frames with interference of signals from different sources

Parameter $\varphi$, hereinafter called pseudo-bearing. For a vertical dipole (polar angle $\theta_0 = 0$ we have $\varphi = \psi, w = \sin\theta$). For an inclined dipole, the pseudo-bearing $\varphi$ is differs from actual direction $\psi$, this difference depends on the orientation angles $\theta_0$ and $\psi_0$. For a horizontal dipole, we have $\varphi = \psi_0 + \pi$ and $w = \cos\theta$, i.e., the pseudo-bearing is independent of the actual direction $\psi$ and is determined by only the dipole orientation $\psi_0$. That is why the well-known direction finders [1–3] correctly record only cloud-ground discharges and give false estimates of bearing on inter-cloud and intra-cloud discharges. In fact, these methods evaluate pseudo-bearing $\varphi$.

To calculate the pseudo-bearing we use the following formulas:

$$\cos 2\varphi = \frac{h_{xx} - h_{yy}}{\sqrt{\left(h_{xx} - h_{yy}\right)^2 + 4h_{xy}^2}}, \quad \sin 2\varphi = \frac{h_{xx} - h_{yy}}{\sqrt{\left(h_{xx} - h_{yy}\right)^2 + 4h_{xy}^2}}, \tag{8}$$

$$\overline{\varphi} = \operatorname{atan}\sqrt{\frac{1 - \cos 2\varphi}{1 + \cos 2\varphi}} \operatorname{sign} h_{xy}, \tag{9}$$

here $h_{xx} = \int_{T_1}^{T_2} [H_x(t)]^2 dt$, $h_{yy} = \int_{T_1}^{T_2} \left[H_y(t)\right]^2 dt$, $h_{xy} = \int_{T_1}^{T_2} H_x(t) H_y(t) dt$, $[T_1, T_2]$ is signal observation interval.

Obviously, the signals observed in the antenna system can represent interference of signals from several sources. For the correct work of the algorithms described in the work, it is necessary to exclude frames with the presence of interference. Let us consider a possible way of detecting interference.

Interference leads to the appearance of a trend in the array of instantaneous error values

$$\Delta(\overline{\varphi}, t_i) = H_x(t_i) \cos\overline{\varphi} - H_y(t_i) \sin\overline{\varphi}, \quad t_i \in [T_1, T_2], i = 0, 1, 2, \ldots, \frac{T_2 - T_1}{\Delta t} = N. \tag{10}$$

Really let be $H_x(t) = \sin\varphi_1 S_1(t) + \sin\varphi_2 S_2(t) + N_x(t)$, $H_y(t) = \cos\varphi_1 S_1(t) + \cos\varphi_2 S_2(t) + N_y(t)$, $N_x$, $N_y$ are uncorrelated white noise in the antenna channels having the same intensity $N$. Then $\Delta(\overline{\varphi}, t) = S_1(t) \sin\left(\overline{\varphi} - \varphi_1\right) + S_2(t) \sin\left(\overline{\varphi} - \varphi_2\right) + N_x(t) \sin\varphi - N_y(t) \cos\varphi$.

It is easy to see that for $\varphi_1 \neq \varphi_2$ function $\Delta(\overline{\varphi}, t)$ for any $\overline{\varphi}$ has a trend due to the difference in the functions $S_1(t)$, $S_2(t)$ of the radiation sources. On the contrary, for $\overline{\varphi} = \varphi_1 = \varphi_2$, the function $\Delta(\overline{\varphi}, t)$ has no such trend. Effect of interference make autocorrelation in function $\Delta(\overline{\varphi}, t)$. Thus, the fact of autocorrelation in the error function $\Delta(\overline{\varphi}, t)$ is criterion for the presence of interference.

Durbin-Watson test is often used to determine the presence of autocorrelation. This criterion is usually used to establish the fact of the presence of an autocorrelation dependence of the first order in the error series, i.e., between its neighboring values $\Delta(\overline{\varphi}, t_i)$ and $\Delta(\overline{\varphi}, t_{i+1})$, $t_i, t_{i+1} \in [T_1, T_2], i = 0, 1, 2, \ldots, \frac{T_2 - T_1}{\Delta t} = N$, $\Delta t$ is sampling step. Usually the neighboring error values are related by a stronger dependence than other values. Therefore, the absence of an autocorrelation dependence of the first order makes it possible to state quite confidently that there are no autocorrelation relationships in the errors.

Durbin-Watson statistics is calculated by the following formula:

$$d = \frac{\sum_{i=0}^{N-1} \left(\Delta(\overline{\varphi}, t_{i+1}) - \Delta(\overline{\varphi}, t_i)\right)^2}{\sum_{i=0}^{N-1} \left(\Delta(\overline{\varphi}, t_i)\right)^2}. \quad (11)$$

We have $0 \le d \le 4$. The values of $d = 0$ and $d = 4$ correspond to the cases when there is a strict positive or negative linear relationship between the shifted series $\{\Delta(\overline{\varphi}, t_i)\}$ and $\{\Delta(\overline{\varphi}, t_{i+1})\}$, respectively. If $d = 2$, then the series is independent.

The method has been tested based on data generated as a result of three-month field testing of the lightning detection system in 2004 within the framework of the ISTC project #1822 [8] and containing 796,112 atmospheric spheres suitable for computation [8]. After calculating the value of the statistics $d$ for the entire set of signals from the signals (over the whole implementation), obtained because of the field tests of the lightning detection system and analysis of the results, the intermediate point values were chosen:

$$d_1 = 0.4, \quad d_2 = 0.7, \quad d_3 = 3.3, \quad d_4 = 3.6.$$

If $d \in [0, d_1] \cup [d_4, 4]$, then the hypothesis of the existence of first order autocorrelation in a series $\{\Delta(\overline{\varphi}, t_i)\}$ and $\{\Delta(\overline{\varphi}, t_{i+1})\}$ is assumed. If $d \in [d_2, d_3]$, then this hypothesis is rejected. If the value of the statistics $d$ is in other intervals, then an unambiguous answer about the existence of autocorrelation cannot be given.

The analysis of the detected signals using the Durbin-Watson test showed the following results (see **Table 1**).

Thus, we can confidently determine the coordinates of only 144,695 of the 796,112, based on the data of the entire implementation of each of the signals.

To eliminate this influence, you can select a part of the signal length (the implementation interval) at which interference appears less, or where there is none (there is no trend). Of course, the calculation of discharge characteristics over this implementation interval will be more accurate. In addition, the approach to isolating a "clean" signal can help in filtering the transients at the beginning of the signal reception.

| Value Range of Durbin-Watson statistics | Number of signals |
|---|---|
| [0, 0.4] | 572,005 |
| [0.4, 0.7] | 79,412 |
| [0.7, 3.3] | 144,695 |
| [3.3, 4] | 0 |

**Table 1.** The relationship between the Durbin-Watson statistics interval and the number of corresponding signals.

## 4. Calculation of the parameters *u*, *v*, *α*

Further, we assume the parameter φ to be computed, and interference is absent in the recorded signals $E_z$, $H_x$, and $H_y$. Let us further use the function $H(t) = H_x(t)\sin\varphi + H_y(t)\cos\varphi$.

### 4.1. Primal algorithm

An analytical solution of the problem to identify parameters $u, v, \alpha$ is given in [4]

$$\begin{cases} \alpha = \sqrt{(h_1e_2 - h_2e_1)/(h_2e_0 - h_1e_1)}, \\ u = \left(e_0e_2 - e_1{}^2\right)/(h_2e_0 - h_1e_1), \\ v = (3h_0u\,\alpha - e_0\alpha - g)(2h_0\alpha), \end{cases} \tag{12}$$

here

$$e_k = \int_0^\infty E_z^{(k)}(t)H^{(k)}(t)dt,\ \ h_k = \int_0^\infty \left(H^{(k)}(t)\right)^2 dt,\ \ g = \int_0^\infty E_z'(t)H(t)dt,\ \ k = 0, 1, 2. \tag{13}$$

Calculated parameters $u, v, \alpha, \varphi$ can be used to determine the distance to the dipole $r = c/\alpha$, as well as guaranteed interval estimates and probabilistic characteristics of angular coordinates of the dipole position.

Primal algorithm calculates parameters using formulas (3). It is assumed that the signals are represented by counts of instantaneous values with a given sampling rate. To calculate the coefficients (4), we use the known quadrature formulas.

An analysis of the sensitivity of the direct algorithm to the noise in the measured signals is given in [5]. It is shown that to achieve acceptable accuracy, pre-processing of input signals is necessary. It is found the dependence of mathematical expectation and variance of the distance determination error on the source spectrum and transfer function of the input filter. The estimates for the case corresponding to the averaged source of lightning discharges indicate the possibility of using the algorithm under study in systems for locating the lightning centers and allow us to develop the requirements for hardware of such systems.

### 4.2. Extremal algorithm

The primal algorithm was obtained in the assumption that the radiation source is the dipole, and the underlying surface—a plane with the infinite conductivity. Due to non-ideal models and noises in observed signals, we are forced to allow discrepancies in the resultant equations. Hence, intuition shows that most preferable is the least-squares principle. According to this principle, parameters of the model are determined from

$$I(\boldsymbol{x}) = \int\limits_{T_1}^{T_2} \left[x_1E_z''(t) + x_2E_z{}'(t) + x_3H''(t) + x_4H'(t) + x_5H(t)\right]^2 dt = (\boldsymbol{x}, \boldsymbol{Ax}) \to \max_{\boldsymbol{x}} \tag{14}$$

$$\mathbf{A} = \int_{T_1}^{T_2} \mathbf{S}\,\mathbf{S}^T\,dt,\ \ \mathbf{S} = \begin{pmatrix} kE_z''(t) & E_z{}'(t) & kH''(t) & H'(t) & k^{-1}H(t) \end{pmatrix}^T,$$

$k$ is the time constant for differentiating and integrating sections. This constant is introduced to make scales of coordinates for vector $\mathbf{x}$ consistent.

The extreme algorithm [14] for determining parameters $u$, v, and $\alpha$ is calculated as follows:

**Step 1.** Let $\lambda_0 \le \lambda_1 \le \lambda_2 \le \lambda_3 \le 4$ be eigenvalues of matrix A, $\mathbf{x}^{(i)}, i = 0, 1, 2, 3, 4$ are corresponding eigenvectors.

**Step 2.** Let $\Delta_{mn}$ be the determinant of the matrix derived from the matrix $\left(\mathbf{x}^{(2)} \quad \mathbf{x}^{(3)} \quad \mathbf{x}^{4}\right)$ through deleting lines having numbers $m$ and $n$.

**Step 3.** For $i = 1, 2$ do

$$\beta_i = \begin{cases} \dfrac{2\Delta_{25}}{-\Delta_{24} - \Delta_{15} + (-1)^i \sqrt{(\Delta_{24} + \Delta_{15})^2 - 4\Delta_{14}\Delta_{25}}} & \text{if } |\Delta_{25}| \ge |\Delta_{14}|, \\ \dfrac{-\Delta_{24} - \Delta_{15} - (-1)^i \sqrt{(\Delta_{24} + \Delta_{15})^2 - 4\Delta_{14}\Delta_{25}}}{2\Delta_{14}} & \text{otherwise,} \end{cases}$$

$$u_i = \frac{\beta_i \Delta_{13} + \Delta_{23}}{\Delta_{12}}, \quad w_i = \frac{-\Delta_{45}}{\beta_i \Delta_{24} + \Delta_{25}}$$

.

**Step 4.** Let $i^* = \arg\min_{i=1,2} \left[ \left(1, \beta_i, -u_i, -\beta_i w_i, -\beta_i^2 w_i^2\right) \mathbf{A} \left(1, \beta_i, -u_i, -\beta_i w_i, -\beta_i^2 w_i^2\right)^T \right]$,

**Step 5.** Let $\alpha = \frac{\beta_{i^*}}{k}, \quad u = u_{i^*}, \quad v = \frac{w_{i^*} - 3u}{2}$.

A calculation experiment was performed to analyze accuracy and stability of the extreme algorithm, as well as to compare its characteristics with those of the primal algorithm. Software of the algorithm is implemented with the observation of agreements for the instrumental environment developed under the ISTC project 1822 [8]. Signals $E_z(t), H(t)$ are presented as integral counts of rapid values in discrete time moments, step of time discretization is $\Delta t = 2\mu s$, number of quantification levels is $2^{16}$ . Input signals $E_z(t), H(t)$ are preliminarily processed by a smoothing filter. Usual difference and quadrature formulas with the accuracy of $O(\Delta t^2)$ were used for numerical differentiation and integration. The Givens rotation method is used to solve the complete symmetric problem of eigenvalues.

Extreme algorithm gives better accuracy if compared with the primal algorithms. Most significant difference in characteristics is observed for distances less than 50 km: error of the primal method sharply increases with the distance decrease, and the error of the extreme method is decreasing. Since with the decreasing distance, one has increased error in presentation of the real source as an electrical dipole, the experimental results demonstrate the extreme algorithm to be expedient at the distance $r \le 50$ km.

### 4.3. Parametrization of the algorithms

To solve the problem to identify the location parameters because of its poor condition is proposed to use a parameterized set of algorithms, and the final decision to accept the results of statistical analysis. Parameterizations of the primal and extremal methods are considered in paper [15].

The spectrum of dipole moment $p(t)$ lies in a rather narrow frequency range, so the use of a bandpass filter with suitable lower and upper cutoff frequencies will maximize the use of all useful information that the signal carries and weaken the effect of noise present in the signal. Since the position parameters and the spectrum of the radiation source are not known a set of estimates $\tilde{E} = \{(\alpha_i, u_i, v_i): \ i = 1, 2, ..., N\}$ for a family of bandpass filters with an amplitude-frequency characteristic

$$F_i(\omega) = \begin{cases} 1, & \text{if } \omega \in \left[\widehat{\Omega}_i, \check{\Omega}_i\right], \\ 0, & \text{otherwise}, \end{cases} \quad i = 1, 2, ..., N \tag{15}$$

is constructed. The elements of the set $\tilde{E}$ are considered as an implementation of a vector random variable, and for estimating the true position parameters $u, v,$ and $\alpha$ to apply methods of statistical robust estimation.

It would be well to exploit more methods for increasing of statistical significance of result estimation. Put computational experiments' results confirms efficiency the approach, but a wide variety of the measurement results do not lets to form up the well-provided measuring estimation of the dipole location parameters adequately. It is suggested that the filtration of the measurement results based on the cleaner functionals and combination of them for improving of the measuring estimation of the dipole location parameters. For construction of the cleaner functionals is the projection method [16]. This method is based on the projection of the inverse image of the vector-valued function $(E_z(t), H(t))$ onto the linear manifold $\mathcal{L}$ of solutions of the system of differential Eqs. (1) and (2). There is a solution $q(t)$ satisfying the system of differential Eqs. (1) and (2) under error-free values of the estimated parameters $u, v, \alpha$. Therefore, the inverse image of the pair of signals $(E_z(t), H(t))$ belongs to $\mathcal{L}$, and the length of the projection is maximal. This allows you for filtering to use the length of the projection as a utility function. In terms of Fourier transforms, the utility function has the form [16]

$$U(u, w, \alpha) = \left\| \frac{E(\omega)\left(-u\omega^2 + j\omega w\alpha + w\alpha^2\right) + H(\omega)\left(-\omega^2 + j\omega\alpha\right)}{\left|-u\omega^2 + j\omega w\alpha + w\alpha^2\right|^2 + \left|-\omega^2 + j\omega\alpha\right|^2} \right\|. \tag{16}$$

Function $U\,(u, w, \alpha)$ is continuous for $(u, w, \alpha) \neq 0$. Algorithm for calculating $U\,(u, w, \alpha)$ is completely stable for signals represented of 4096 readouts with the sampling time 2–$10^{-6}$ s, i.e., containing 2047 harmonics with the first-harmonic discretization frequency 122 Hz.

Set $E(U_0) = \left\{(u, w, \alpha) \in \tilde{E}: \ U(u, w, \alpha) \geq U_0\right\}$ contains most reliable estimates. Statistical analysis of the set of estimates makes it possible to estimate $(u, v, \alpha)$ probability density.

### 4.4. Spectral statistical method for location parameters identifying of a dipole electromagnetic radiation source

The spectral statistical method makes it possible to obtain more stable solutions at a lower computation cost compared with the previously developed parametric extremum method. The spectral statistical method algorithm can be naturally parallelized. The proposed method based on the analysis of the measured-signal spectra, allows one to get many estimates of the source location, choose the final estimate of the results of analysis of the entire totality of these estimates, and therefore, reach stability in determining the source location. To reduce the computational complexity, it is preferable to analyze individual harmonics rather than the entire frequency band. In this case, the algorithms for estimating the dipole location are substantially simplified [17].

As before, we will consider signals represented as a fast Fourier transformation $E(\omega), H(\omega)$ containing 2047 harmonics with the first-harmonic discretization frequency 122 Hz. Let us divide the set $\Omega = \{\omega_1, \omega_2, \ldots, \omega_{2047}\}$ of all harmonics into subsets

$$\Omega_1 = \{\omega_1, \omega_2, \ldots, \omega_{131}\}, \Omega_2 = \{\omega_{132}, \omega_{133}, \ldots, \omega_{400}\}$$

and

$$\Omega_3 = \{\omega_{401}, \omega_{402}, \ldots, \omega_{2047}\}.$$

**Spectral statistical algorithm** to find estimates $\widehat{u}, \widehat{v}, \widehat{\alpha}$ of the parameters $u, v, \alpha$ and their variance $\sigma\widehat{u}^2, \sigma\widehat{v}^2, \sigma\widehat{\alpha}^2$ is calculated as follows:

**Step 1.** For each $\omega \in \Omega$ calculate

$$f(\omega) = \mathrm{Re}\left[E(\omega)\overline{H(\omega)}\right], \quad g(\omega) = \mathrm{Im}\left[E(\omega)\overline{H(\omega)}\right], \quad h(\omega) = |H(\omega)|^2.$$

**Step 2.** Calculate the noise intensity

$$\sigma_N^2 = \frac{\sum_{\omega \in \Omega_3} |H(\omega)|^2}{|\Omega_3|}.$$

**Step 3.** For each $\omega \in \Omega_2$ calculate

$$\widehat{u(\omega)} = \left(1 + \frac{\sigma_N^2}{h(\omega)}\right)\frac{f(\omega)}{h(\omega)}.$$

**Step 4.** Calculate

$$\widehat{u} = \frac{\sum_{\omega \in \Omega_2}\left(h(\omega)\widehat{u(\omega)}\right)}{\sum_{\omega \in \Omega_2} h(\omega)}, \quad \sigma\widehat{u}^2 = \frac{\sum_{\omega \in \Omega_2}(f(\omega) - h(\omega)\widehat{u})}{\sum_{\omega \in \Omega_2} h(\omega)}.$$

**Step 5.** For each $\omega \in \Omega_1$ calculate

$$A(\omega) = \omega\sqrt{\frac{h(\omega)\widehat{u}}{f(\omega)} - 1}, \quad S(\omega) = \frac{\sigma_N^2}{h(\omega)}, \quad \widehat{\alpha(\omega)} = \sqrt{\frac{A^2(\omega) - S(\omega)\omega^2}{S(\omega) + 0.5 + \sqrt{\frac{S(\omega)A^2(\omega)}{\omega^2} + 0.25}}}.$$

**Step 6.** Calculate

$$\widehat{\alpha} = \frac{\sum_{\omega \in \Omega_1} \left[ \frac{\sqrt{h(\omega)}}{\omega^2} A(\omega) \widehat{\alpha(\omega)} \right]}{\sum_{\omega \in \Omega_1} \left[ \frac{\sqrt{h(\omega)}}{\omega^2} A(\omega) \right]}, \quad \sigma \widehat{\alpha}^2 = \frac{\sum_{\omega \in \Omega_1} \left[ \frac{\sqrt{h(\omega)}}{\omega^2} A(\omega) \left( \widehat{\alpha} - \widehat{\alpha(\omega)} \right)^2 \right]}{\sum_{\omega \in \Omega_1} \left[ \frac{\sqrt{h(\omega)}}{\omega^2} A(\omega) \right]}.$$

**Step 7.** For each $\omega \in \Omega_1$ calculate

$$W(\omega) = \frac{f(\omega)A(\omega) - \omega g(\omega)}{h(\omega)A(\omega)}.$$

**Step 8.** Calculate

$$\widehat{w} = \frac{\sum_{\omega \in \Omega_1} \left[ \sqrt{h(\omega)} W(\omega) \right]}{\sum_{\omega \in \Omega_1} \sqrt{h(\omega)}}, \quad \sigma \widehat{w}^2 = \frac{\sum_{\omega \in \Omega_1} \left[ \sqrt{h(\omega)} [W(\omega) - \widehat{v}]^2 \right]}{\sum_{\omega \in \Omega_1} \sqrt{h(\omega)}}.$$

The rationale for the effectiveness of the algorithm and computational experiments are given in [17].

## 5. Secondary processing of electromagnetic field monitoring results

The most important problem in work of the single-point systems of storm location is ineradicable error in the bearing determination of each source of radiation taken separately. Amount of this error depends on orientation of the lightning equivalent dipole, and the "pseudo-bearing" determined by the device for horizontal dipoles does not depend from the real bearing at all and is defined only by projection of orientation of the dipoles to the ground plane.

The given problem can be solved only by the analysis of the information from all set of discharges of the storm. At International conference ETC'2006 [18], it was suggested to display the stream of signals as a map of lightning discharge hit probability density in this or that point of terrestrial surface. The representation proposed increases probability of definition of true thunderstorm discharges location, and hence, raises accuracy of their fixing.

Let us describe the technique of this approach realization, the difficulties arising, and the variants of their decision in detail.

Let values $(r, \psi, \theta)$ be the polar coordinates of the lightning equivalent dipole for the measuring device coordinate system. As it is known [4, 18], Cartesian coordinates $x, y$ of the dipole meet the condition

$$x \cos \varphi + y \sin \varphi = r \sin \theta \cos (\varphi - \psi) = ru. \tag{17}$$

This equation includes known parameters: $r$ is distance from the dipole to the observation point, $u, v$ are the variables of the source dipole model (3)—and unknown Cartesian coordinates $x, y$ of the dipole location that is a feature of an uncertainty. The coordinates of the points of possible location of a dipole most remote from origin are the solutions of a set of equations

$$x\cos\varphi + y\sin\varphi = ru, \quad x^2 + y^2 = r^2. \tag{18}$$

They are equal

$$x_{1,2} = r\left(u\cos\varphi \mp \sqrt{1-u^2}\,\sin\varphi\right), \quad y_{1,2} = r\left(u\sin\varphi \pm \sqrt{1-u^2}\,\cos\varphi\right). \tag{19}$$

The coordinates $x, y$ of all probable points of a source location introduce on a plane $XOY$ a section of a straight line (10) between points $(x_1, y_1)$ and $(x_2, y_2)$. The set of probable location of a dipole have one-parameter representation

$$\mathcal{L}(r,u,v) = \left\{ \begin{array}{c} x(\lambda) = ru\cos\varphi + r(2\lambda-1)\sqrt{1-u^2}\,\sin\varphi \\ y(\lambda) = ru\sin\varphi + r(1-2\lambda)\sqrt{1-u^2}\,\cos\varphi, \\ z(\lambda) = \sqrt{r^2 - x^2(\lambda) - y^2(\lambda)} = 2r\sqrt{\lambda(1-\lambda)(1-u^2)} \end{array} \;\middle|\; 0 \le \lambda \le 1 \right\}. \tag{20}$$

A measure of the coordinates set $\mathcal{L}(r,u,v)$ equal

$$\mathrm{L} = \int_0^1 \sqrt{\left(\frac{\mathrm{dx}(\lambda)}{d\lambda}\right)^2 + \left(\frac{\mathrm{dy}(\lambda)}{d\lambda}\right)^2 + \left(\frac{\mathrm{dz}(\lambda)}{d\lambda}\right)^2}\, d\lambda = 4.844\mathrm{r}\sqrt{1-\mathrm{u}^2}.$$

The value of $L$ represents a measure of the uncertainty in the estimation of the Cartesian coordinates of the dipole location. Uncertainty is absent when $L = 0$. It is reached for a vertical dipole ($u = \pm 1$). It is possible value $u = 0$ for a horizontal dipole when $L$ reaches the maximum value.

A more practical measure of uncertainty is probability

$$\mathbf{P}[i,j,k]\{[(x(\lambda), y(\lambda), z(\lambda)] \in C[i,j,k]\}$$

of the dipole location in space cell

$$C_{[i,j,k]} = \left\{ (x,y,z) \middle| \begin{array}{c} i\Delta \le x \le (i+1)\Delta \\ j\Delta \le y \le (j+1)\Delta \\ k\Delta \le z \le (k+1)\Delta \end{array} \right\},$$

here $\Delta$ is cell size.

It is obvious that $\mathbf{P}_{[i,j,k]} > \mathbf{0}$ if $C_{[i,j,k]} \cap \mathcal{L}(r,u,v) \neq \varnothing$. To find cells in which the dipole is located with a nonzero probability, it is sufficient to find the minimal covering $C_{XOY}(r,u,v)$ of the projection of $\mathcal{L}(r,u,v)$ onto the plane $XOY$ by flat cells $C_{[i,j]}$ :

$$\left\lfloor \min\left\{\frac{r\,u\cos(\varphi-\arccos u)}{\Delta},\frac{r\,u\cos(\varphi+\arccos u)}{\Delta}\right\}\right\rfloor \le i$$
$$\le \left\lceil \max\left\{\frac{r\,u\cos(\varphi-\arccos u)}{\Delta},\frac{r\,u\cos(\varphi+\arccos u)}{\Delta}\right\}\right\rceil,$$

$$\left\lfloor \min\left\{\frac{r\,u\cos(\varphi-\arcsin u)}{\Delta},\frac{r\,u\cos(\varphi+\arcsin u)}{\Delta}\right\}\right\rfloor \le j$$
$$\le \left\lceil \max\left\{\frac{r\,u\cos(\varphi-\arcsin u)}{\Delta},\frac{r\,u\cos(\varphi+\arcsin u)}{\Delta}\right\}\right\rceil.$$

An example of coverage $C_{XOY}(r,u,v)$ is shown in the **Figure 2**.

Cell $C_{[i,j]} \in C_{XOY}(r,u,v)$ in accordance with equality (12) uniquely defines a space cell $C_{[i,j,k]}, k = \left\lceil \sqrt{\frac{r}{\Delta} - i^2 - j^2} \right\rceil$ with a nonzero probability of the dipole location.

To find this probability, we accept a completely plausible hypothesis about the uniform distribution of the probabilities of all possible values of the angle $\psi_0$ (see **Figure 1**). In this case the probability density for angle $\psi$ is [4]

$$f_\psi(\eta) = \frac{s}{2\widehat{\psi}[s^2\cos^2(\varphi-\eta) + \sin^2(\varphi-\eta)]}, \quad -\arccos u \le \eta - \varphi \le \arccos u \tag{21}$$

here

$$s = \left|\frac{1-uv}{uv}\right|, \quad \widehat{\psi} = \arctan\left(\frac{1}{s}\tan\arccos u\right).$$

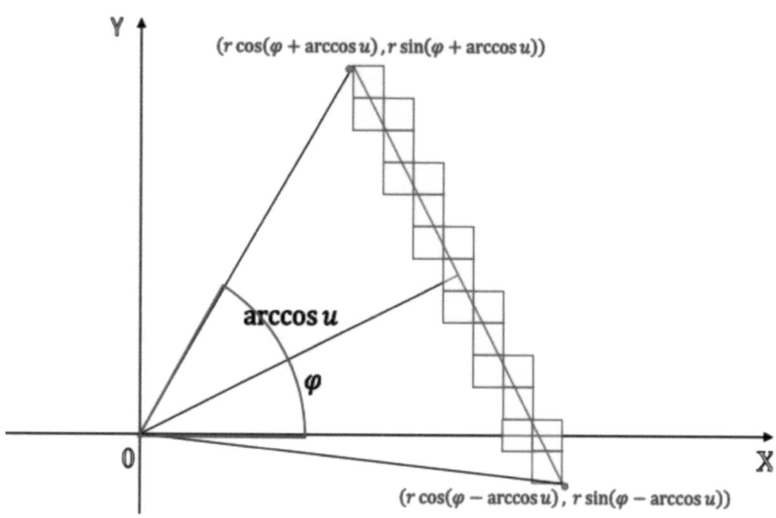


**Figure 2.** An example of coverage $C_{XOY}(r,u,v.)$

Thus, for $C_{[i,j]} \in C_{XOY}(r,u,v)$ one can take

$$\mathbf{P}_{[i,j,k]} = f_{\psi}(\eta),\ k = \left[\sqrt{\frac{r}{\Delta} - i^2 - j^2}\right],\quad \eta = \arctan\frac{j}{i}.$$

Let $R(T)$ be the set of registered signals in the time interval $T$. The integral estimate grade of radiation locus membership to cell $C_{[i,j,k]}$ over period $T$ let be

$$\text{grade}[i,j,k] = \frac{\sum_{d\in R(T)} \mathbf{P}_{[i,j,k]}(d)}{\max_{[l,m,n]} \sum_{d\in R(T)} \mathbf{P}_{[l,m,n]}(d)}.$$

It is suggested to display the stream of signals as a grade map. The representation proposed increases in probability of definition of true thunderstorm discharges location, and hence, raises accuracy of their fixing.

## 6. Using a network of sensors

Currently, mainly for monitoring thunderstorm activity, multi-point systems based on the application of the difference-ranging method and its variations to the results of monitoring the Earth's electromagnetic field in the VLF and VHF bands are used. Some lacks such an approach are: (1) insufficient reliability, which results from the need communication networks for the exchange of information between observation stations spaced over significant distances, and the need for time synchronization at the sub-microsecond level; (2) the inability to analyze the cloud radiation before the thunderstorm. Previously, this was the reason for the development of single-point systems.

Now, the rapid development of computer technologies and communication systems have made it possible to combine single-point lightning location finders into a system and to organize joint processing of signals from individual lightning range finder. Automating the collection of information from observation points helps different services to respond more exactly to changes in thunderstorm conditions.

The uncertainty in the dipole location is fundamentally unavoidable when single-point systems are using, however these errors can be eliminated by determining its generalized parameters at two or more points [4, 6, 7]. Thus, by integrating single-point systems of passive monitoring of thunderstorm activity into a single computer network, it is possible to increase the probability of detecting a lightning discharge and the accuracy of determining its coordinates. In addition, an increase in the degree of automation of information collection from observation posts makes it possible to receive data promptly and make adequate decisions in accordance with the current thunderstorm situation.

### 6.1. Direction-range method

Let $(x, y, z)$ be the Cartesian coordinates of the dipole location point. Let $(x_i, y_i)$ be Cartesian coordinates of $i$-th observation post, $i \in J$. Let $(r_i, \varphi_i, u_i)$ be the dipole location parameters of $i$-th observation post. Then the system of equations

$$(x_i - x)\cos\varphi_i + (y_i - y)\sin\varphi_i = r_i u_i, \qquad i \in J \tag{22}$$

holds. If the observation points do not lie on one line, then system (22) has full rank and can be solved by the least squares method

$$\begin{pmatrix} x \\ y \end{pmatrix} = \begin{pmatrix} \frac{1}{\Delta}\sum_{i\in J}\left(a_i \sum_{k\in J} \sin\varphi_k \sin(\varphi_k - \varphi_i)\right) \\ \frac{1}{\Delta}\sum_{i\in J}\left(a_i \sum_{k\in J} \cos\varphi_k \sin(\varphi_k - \varphi_i)\right) \end{pmatrix}, \tag{23}$$

$$\Delta = \sum_{1\le k<l\le |J|} \sin^2(\varphi_k - \varphi_i), \qquad a_i = x_i\cos\varphi_i + y_i\sin\varphi_i, \qquad i \in J.$$

The method uses for its calculations the values $\left(r_i = \frac{c}{\alpha_i},\ u_i, \varphi_i\right)$ measured by $i$-th autonomous lightning detection finder, and not the bearing estimation $\psi$, as a classical direction-finding method. This makes it insensitive to the presence of an anomalous component in the observed magnetic field. There is no requirement for high-precision synchronization for the direction and distance method, which is a characteristic of the difference-range method, and the possibility of estimating the position of the lightning discharge at each point eliminates collisions in identifying the correspondence of the recorded signals to specific lightning discharges. Finally, this method makes it possible to determine the three coordinates of the location of an equivalent dipole source.

The development of this method, which makes it possible to find the angular coordinates $(\theta_0, \psi_0)$ of the EMR dipole source, is given in [6, 7].

It should be noted that the use of the direction-ranging method involves knowledge of the ratio of the effective heights of magnetic and electric antennas and high requirements for the accuracy of the orientation of magnetic antennas. The determination of the ratio of effective antenna heights is a complex technical problem requiring a reference meter or a long time for the collection and processing of statistical data. Therefore, it is of interest to develop other methods for determining the location of a lightning discharge, which do not require knowledge of this relationship.

### 6.2. Range method

Range method for determining the coordinates of a dipole EMR source [19], in contrast to a direction- range finder, does not require a prior knowledge of the ratio of the effective antenna heights, and high accuracy of the orientation of magnetic antennas. The essence of the method is that the Cartesian coordinates $(x_i, y_i)$ of the $i$-th point of observation, $i \in J$. Cartesian coordinates $(x, y, z)$ of the dipole EMR source and ranges $r_i$ from the $i$-th observation points up to the dipole EMR source satisfy the equations system

$$r_i^2 = (x - x_i)^2 + (y - y_i)^2 + z^2, \qquad i \in J \tag{24}$$

reducible by the introduction of the variable

$$m = \frac{x_i^2 + y_i^2 + z_i^2}{C}$$

to a system of linear equations

$$mC - 2x_i x - 2y_i y = r_i^2 - x_i^2 - y_i^2, \qquad i \in J, \tag{25}$$

here constant C is the scaling factor for reducing the condition number of the equation system.

If the observation points do not lie on one line then the resulting system will have full rank, and its least-squares method, the solution is:

$$\begin{pmatrix} m \\ x \\ y \end{pmatrix} = \left(A^T A\right)^{-1} A^T B, \text{ where } A = \begin{pmatrix} C & -2x_1 & -2y_1 \\ C & -2x_2 & -2y_2 \\ \vdots & \vdots & \vdots \\ C & -2x_{|J|} & -2y_{|J|} \end{pmatrix}, \quad B = \begin{pmatrix} r_1^2 - x_1^2 - y_2^2 \\ r_2^2 - x_2^2 - y_2^2 \\ \vdots \\ r_{|J|}^2 - x_{|J|}^2 - y_{|J|}^2 \end{pmatrix}. \tag{26}$$

At fixed coordinates of all observation points, the matrix A is constant. Therefore, even at the design stage of the system it is easy to construct an optimal algorithm for calculating unknowns, since $\left(A^T A\right)^{-1} A^T$ can be counted once for the implementation of the system.

For example, let observation points be

$$(x_1, y_1) = (-R, O), (x_2, y_2) = (0, R), (x_3, y_3) = (R, O), (x_4, y_4) = (0, -R).$$

Then minimum value of the condition number equal 1 and is reached for $C = 2R\sqrt{2}$. The estimates of the dipole location coordinates are the following:

$$x = \frac{r_1^2 - r_3^2}{2R}, \qquad y = \frac{r_2^2 - r_4^2}{2R}. \tag{27}$$

It is easy to determine the coordinates both the dipole orientation and its location with Cartesian coordinates (27) and the distances $r_i$ to the observation points $i \in J$.

## 7. Conclusion

New generation of thunderstorm passive monitoring systems expand the range of problems they solve: development of new mathematical models and algorithms for analyzing thunderstorm phenomena, their tracing and display [20].

## Author details

Anatoly V. Panyukov

Address all correspondence to: anatoly.panyukov@gmail.com

Federal State Autonomous Educational Institution of Higher Education, "South Ural State University (National Research University)," Russian Federation

## References

[1] Kononov II, Yusupov IE, Kandaratskov NV. Analysis of one-point methods for lightning discharge passive location. Radiophysics and Quantum Electronics. 2014;**56**(11–12):875-888. DOI: 10.1007/s11141-9481-1

[2] Rakov AV, Rachidi F. Overview of reacent progress in lightning research and lightning protection. IEEE Transactions on Electromagnetic Compatibility. 2009;**51**(3):428-442

[3] Biagi CJ, Cummins KL, Kehoe KE, Krider EP. NLDN Performance in Southern Arizona, Texas, and Oklahoma in 2003–2004. Journal of Geophysical Research – Atmospheres. 2007;**112**(D5):D05208. DOI: 10.1029/2006JD007341

[4] Panyukov AV. Estimation of the location of an arbitrarily oriented dipole under single-point direction finding. Journal of Geophysical Research – Atmospheres. 1996;**101**(D10): 14977-14982. DOI: 10.1029/96JD00067

[5] Panyukov AV. Analysis of the error of a direct algorithm for determining the distance. Radiophysics and Quantum Electronics. 1999;**42**:239-250. DOI: 10.1007/BF02678847

[6] Popov M, He S. Identification of a transient electric dipole over a conducting half space using a simulated annealing algorithm. Journal of Geophysical Research – Atmospheres. 2000;**105**(D16):20821-20831. DOI: 10.1029/2000JD900261

[7] He S, Popov M, Romanov V. Explicit full identification of transient dipole source in the atmosphere from measurement of the electro magnetic fields at several points at ground level. Radio Science. 2000;**35**(1):107-117. DOI: 10.1029/1999RS002198

[8] The International Science and Technology Center. Project # 1822. Development and Investigation of the Single-Point System for Lightning Location in the Range of Super-Long Waves [Internet]. March 23, 2000 [Updated: November 03, 2005]. Available from: http://www.istc.int/ru/project/CA0166914DBF304CC3256A85003942D3 [Accessed: November 12, 2017]

[9] Hiushu Q, Dongxia L, Zhuling S. Recent advances in research of lightning meteorology. Journal of Meteorological Research. 2014;**28**(5):983-1001. DOI: 10.1007/s13351-014-3295-0

[10] Brown B, Bullock R, Gotway JH, Davis C, Ahijevych D, et al. New tools for evaluation of numerical weather prediction models using operationally relevant approaches. In: Battelspace Atmospheric and Cloud Impacts on Military Operations; 6–8 November; Chestnut Hill, Massachusetts. 2007

[11] Stagliano JJ, Spaight BV, Kerce JC. Lightning forecasting before the first strike. In: 4th Conference on Meteorological Applications of Lightning Data; 11–15 January; Phoenix, AZ: American Meteorological Society; 2009. pp. 1-9

[12] Taibin BZ. An approach to define parameters for localization of thunderstorms. IEEE Antennas and Propagation Magazine. 2006;**48**(2):48-54. DOI: 10.1109/MAP.2006.1650816

[13] Born M, Wolf E. Principles of Optics. 7th ed. Cambridge: Pergamon Press; 2003. 936p

[14] Panyukov AV, Buduev DV. An extremum algorithm for a passive method of determining the distance. Electrical Technology Russia. 2001;**2**:13-22. Available from: https://elibrary.ru/item.asp?id=13648530 [Accessed: November 12, 2017]

[15] Panyukov AV, Bogushov AK. Parametrizatsiya algoritmov identifikatsii elektricheskogo dipolya [Parameterization of algorithms of identification of an electric dipole]. Vestnik Yuzhno-Ural'skogo Gosudarstvennogo Universiteta. Seriya: Matematicheskoe Modelirovanie i Programmirovanie [Bulletin of South Ural State University. Series: Mathematical Modeling and Programming]. 2012;**18(277)**(12):32-43. Available from: http://mmp.vestnik.susu.ru/article/en/152 [Accessed: November 12, 2017]

[16] Panyukov AV. An extremal method for solving a parametric inverse problem for a system of linear functional equations. Russian Mathematics (Iz. VUZ). 2000;**44**(9):25-31. Available from: http://www.ams.org/mathscinet-getitem?mr=1801958 [Accessed: November 12, 2017]

[17] Panyukov AV, Bogushov AK. The spectral statistical method for determining the location parameters of a dipole source of electromagnetic radiation. Radiophysics and Quantum Electronics. 2016;**59**(4):278-288. DOI: 10.1007/s11141-016-9696-4

[18] Panyukov AV, Avramenko A. Increasing accuracy of the single-point thunderstorm location system (Conference Paper). In: Navickas A, Virbalis A, Lipnickas A, Sauhats AS, Azubalis V, editors. 1st International Conference on Electrical and Control Technologies, ECT 2006; 4–5 May 2006; Kaunas University of Technology. 2006

[19] Malov DN, Panyukov AV. Method for Determination of Location of Lightning Discharge and Multiple-Point System for Its Realization [Internet]. July 31, 2003. Available from: https://elibrary.ru/item.asp?id=17445487 [Accessed: December 21, 2017]

[20] Vaisala. ALARM Automated Lightning Alert and Risk Management System [Internet]. 2017 [Updated: 2017]. Available from: http://www.vaisala.fi/Vaisala%20Documents/Brochures%20and%20Datasheets/ALARM-Datasheet-B210321EN.pdf [Accessed: November 28, 2017]

Chapter 14

# Maxwell-Fredholm Equations

Juan Manuel Velázquez-Arcos,
Alejandro Pérez-Ricardez,
Ricardo Teodoro Páez-Hernández,
Samuel Alcántara Montes and
Jaime Granados-Samaniego

Additional information is available at the end of the chapter

http://dx.doi.org/10.5772/intechopen.76115

**Abstract**

With the aim to increase the knowledge of the broadcasting properties under circumstances like time reversal, change on refractive index, presence of random obstacles, and so on, we developed new type of hybrid equations named Maxwell-Fredholm equations. These new equations fuse the Maxwell equations' description of the electromagnetic fields with the Fourier transform of the Fredholm integral equations appropriate for a broadcasting process. Now we have a new tool, which resembles the Maxwell equations but including contributions from the Fredholm formulation like the resonant behaviour of the left-hand material conditions. To illustrate the usefulness of this new class of equations, we include an academic example that shows the deflection of an electromagnetic beam traveling among a highly anisotropic and left-handed behaviour media.

**Keywords:** Maxwell equations, Fredholm equations, left-hand materials conditions, evanescent waves, broadcasting

## 1. Introduction

In the previous works [1–3], we arrived at the conclusion that if there is a sudden change of the refractive index from positive to negative when we have a broadcasting procedure, a better way to take this phenomenon into account is to formulate the problem through an integral equation. Indeed, we have described the way in which this change is triggered and how the so-called evanescent waves [4–6] are liberated when their confinement is broken. However, the question remains valid about the visualization from the point of view of the traditional formulation of the Maxwell equations. Indeed, there is a qualitative difference between these two points of views because the integral formulation brings their own boundary conditions immersed on their kernel; meanwhile, Maxwell equations need the imposing of the particular boundary conditions directly. Now we have a new goal, that is to create a set of equations with the following properties: first, that preserve the advantage of the integral formulation when studying the changes in the refractive index of the propagation media and second, that we can impose some kind of boundary and initial conditions as in the pure Maxwell equations formulation. Then, we introduce in this chapter Maxwell-Fredholm equations with the aim to increase the knowledge of the broadcasting properties under circumstances like time reversal, change on refractive index, presence of random obstacles, and so on; we do not only obtain new kind of hybrid equations with these properties but we can apply to new kind of electromagnetic problems involving special propagating and broadcasting characteristics which occurs when an electromagnetic beam is strongly deflected in a media with a very high anisotropy and a negative refraction index. The first step to get the hybrid equations is to leave the time domain and transform our generalized Fredholm **integral equations** [7–10] into a system of algebraic equations through a Fourier transform. We must emphasize that the resonant behaviour associated with the transformation of the evanescent waves will be considered when we build the new equations, specifically when we establish the fact that we take a homogeneous equation and we employ the free Green function. In the other hand, we underline the role played of the resonance properties like orthogonally [11–14] and recall how we can overcome the fact that their frequencies cannot be used directly as a mathematical base, so we build the named information packs. After we obtain the Maxwell-Fredholm equations, we apply them to the problem of the beam bending inside a left-hand material. As a part of our procedure, we first show the equivalence of the two formulations that is Maxwell differential equations and generalized Fredholm integral equations. Then, we properly obtain the hybrid equations and apply them to the specific problem mentioned above, a very strong deflection of an electromagnetic beam.

## 2. Resonances and the Fredholm's eigenvalue

First of all, we recall the generalized homogeneous Fredholm's equations (GHFE) [7–10] taken from a theorem we have proved [7]:

$$f^{m}(\mathbf{r},\omega)=\eta(\omega)\int_{0}^{\infty}K_{n}^{m(\circ)}\left(\omega;\mathbf{r},\mathbf{r}^{'}\right)f^{n}\left(\mathbf{r}^{'},\omega\right)dr^{'} \tag{1}$$

Now, we have also proved that when the physical system can be considered as a discrete one, Eq. (1) can be written as:

$$f^{m}(\mathbf{r}_{i},\omega)=\eta(\omega)\mathbf{A}_{i,j}^{m,n}\mathbf{G}_{n}^{m(\circ)}(\omega;\mathbf{r}_{i},\mathbf{r}_{j})f^{m}(\mathbf{r}_{i},\omega) \tag{2}$$

By supposing that we can take a diagonal kernel $K_{n}^{m(\circ)}\left(\omega;\mathbf{r},\mathbf{r}^{'}\right)$, and that the interaction matrix is:

$$\mathbf{A}_{i,j}^{m,n} = \delta(\mathbf{r}_i - \mathbf{r}_j) \tag{3}$$

Now, we take $f^m(\mathbf{r},\omega) = E^m(\mathbf{r},\omega)$ in Eq. (1) and by applying the differential operator $rot = \nabla\times$ over the non-apostrophe variable $\mathbf{r}$, and we obtain the following equation:

$$\nabla \times \mathbf{E}(\mathbf{r},\omega) = \eta(\omega) \int_0^\infty \nabla \times \mathbf{K}^{(\circ)}\left(\omega; \mathbf{r}, \mathbf{r}'\right) \mathbf{E}\left(\mathbf{r}', \omega\right) dr' \tag{4}$$

Now, we use Maxwell equation:

$$rot\mathbf{E}(\mathbf{r},\omega) = -i\omega\boldsymbol{\mu}\mathbf{H}(\mathbf{r},\omega) \tag{5}$$

and, in order to transform Eq. (4), we use the relation:

$$\nabla \times \mathbf{E}(\mathbf{r},\omega)\delta(\mathbf{r}) = \mathbf{D}(\mathbf{r}) \times \mathbf{E}(\mathbf{r},\omega) - i\omega\boldsymbol{\mu}\mathbf{H}(\mathbf{r},\omega) \tag{6}$$

In Eq. (6), the vector $\mathbf{D}(\mathbf{r})$ is defined by:

$$\mathbf{D}(\mathbf{r}) = \delta(y)\delta(z)\widehat{\mathbf{i}} + \delta(x)\delta(z)\widehat{\mathbf{j}} + \delta(x)\delta(y)\widehat{\mathbf{k}} \tag{7}$$

By substituting Eqs. (5)–(7) in Eq. (4), we have after using the derivative properties of the delta function:

$$-i\frac{\eta(\omega)}{\omega} \int_0^\infty \mathbf{D}(\mathbf{r}) \times \mathbf{E}\left(\mathbf{r}', \omega\right) dr' + \eta(\omega) \int_0^\infty \mathbf{K}^{(\circ)}\left(\omega, \mathbf{r}'\right) \boldsymbol{\mu}\mathbf{H}\left(\mathbf{r}', \omega\right) dr' \tag{8}$$

In Eq. (8), the first term seems to be the current of magnetic monopoles, that is, a source term, so must be zero, and the final equation is:

$$\mathbf{H}(\mathbf{r},\omega) = \eta(\omega) \int_0^\infty \mathbf{K}^{(\circ)}\left(\omega, \mathbf{r}'\right) \mathbf{H}\left(\mathbf{r}', \omega\right) dr' \tag{9}$$

So, we can see that it is equivalent to use Maxwell equations or the generalized Fredholm equations. We have shown that the generalized homogeneous fredholm equation (GHFE) can be written in the following compact algebraic form:

$$\mathbf{E}_e^m(\omega) = \left[\eta_e(\omega)\mathbf{G}^{(\circ)}(\omega)\mathbf{A}\right]_n^m \mathbf{E}_e^n(\omega) \tag{10}$$

$$\mathbf{H}_e^m(\omega) = \left[\eta_e(\omega)G^{(\circ)}(\omega)A\right]_n^m \mathbf{H}_e^n(\omega) \tag{11}$$

We can apply operator $rot = \nabla\times$ to Eqs. (10) and (11) and by using Maxwell equations obtaining in terms of the kernels the equations we name Maxwell-Fredholm:

$$rot\mathbf{E}_e(\omega) = -i\omega\boldsymbol{\mu}e^{-ih(\omega_e)}\mathbf{K}^{(\circ)}(\omega)\mathbf{H}_e(\omega) \tag{12}$$

$$rot\mathbf{H}_e(\omega) = i\omega\varepsilon e^{ih(\omega_e)}\mathbf{K}^{(\circ)}(\omega)\mathbf{E}_e(\omega) \tag{13}$$

$$\eta_e(\omega) = e^{ih(\omega_e)} \tag{14}$$

In Eqs. (12) and (13), we must remember that the left-hand side is computed at the final sites, meanwhile, the right-hand term is computed at the initial sites.

Also, we remember that Eq. (1) can be written for the electric field $\mathbf{E}(\mathbf{r},\omega)$ as:

$$\mathbf{E}(\mathbf{r},\omega) = \eta(\omega)\int_0^\infty \mathbf{K}^{(\circ)}\left(\omega,\mathbf{r}'\right)\mathbf{E}\left(\mathbf{r}',\omega\right)dr' \tag{15}$$

which is a form identical to Eq. (9).

At this point, it is important to emphasize that Eqs. (9) and (15) are homogeneous generalized integral equations that properly allow us to follow the behaviour of a left-hand material media, that is a media with a negative refractive index; and the Maxwell-Fredholm Eqs. (12) and (13) also have a structure guided for the same purpose.

## 3. The role of orthogonality properties

There is a very important property of the resonant solutions for the generalized Fredholm equations, that is, the orthogonality between different resonances [7]. Indeed, we are giving an alternative point of view as the established in the work of Li et al. [5] or by Kong et al. [6], concerning the physical interpretation of a resonance. If the resonances would constitute a band of resonant states, we could use these properties directly as a mathematical base to represent any kind of desired broadcasting signal, but the set of resonant solutions is made of punctual frequencies that only permit a pedestrian kind of information transmission, perhaps like a telegraph mode in which even a single frequency can be used as a succession of signal, non-signal intervals. This last kind of information is very far from the goals of an efficient broadcasting. Nevertheless, if we use some results we have obtained previously like the definition of information packs, we can reach our desired results. Suppose that we want to send a signal represented by the function $S(t)$ and that we know that the propagating media bring us a set of resonances one of which we can name $\rho$ so that the associated resonant frequency will be known as $\omega_\rho$. Now we can project the original signal over a sub-space generated with the aid of Communication theory [15–20] by the rule:

$$S_\rho(t) = \sum_{-\infty}^{\infty} P_{m,\rho}\frac{\sin\left[\pi\left(2\omega_\rho t - m\right)\right]}{\pi\left(2\omega_\rho t - m\right)} \tag{16}$$

In expression (16) the span coefficients are:

$$P_{m,\rho} = S\left(\frac{m}{2\omega_\rho}\right) \tag{17}$$

http://dx.doi.org/10.5772/intechopen.76115

So, we have a collection of signals that are projections of the original $S(t)$. Then, we can emit simultaneously the different projections $S_\rho(t)$ and when they arrive to their destination, we can decode and rebuild the original $S(t)$. There is a limitation that comes also from communication theory about the frequencies appeared in every pack, that is, these frequencies cannot be major than the respective resonant frequency $\omega_\rho$. During the broadcasting, the orthogonality properties of the resonances and the structure of the information packs guarantee that there is no interference between the different projections.

In this chapter, we do not show how we can apply Eq. (16) explicitly, but we suppose that the signal we enter through the initial electric and magnetic fields comes from the building of information packs. In this manner, we are using resonances in two different ways, first by using the Maxwell-Fredholm equations created for an explicitly homogeneous situation and second, by the projection of the original signal over the sub-spaces generated with the rules (16) and (17).

## 4. Academic example

We have obtained a new type of algebraic equations named the Maxwell-Fredholm equations as our principal goal, in which we incorporate the resonant behaviour and we can apply them in the following academic special case, in which we suppose the media has very large left-handed material properties also with a very large inhomogeneity that force a light beam to follows a circular trajectory and we describe the phenomena with the aid of parabolic coordinates. We also suppose that we know the tensor $\boldsymbol{\varepsilon}$ [21, 22] in an appropriate form that directly operates over a space of parabolic coordinates $(\xi, \eta, \varphi)$. The specific relation between the new and old systems is depicted in the following equations:

$$\hat{\mathbf{i}} = \frac{\xi}{\sqrt{2}|\xi|}\left(\hat{\boldsymbol{\xi}}_0 + \hat{\boldsymbol{\eta}}_0\right) \tag{18}$$

$$\hat{\mathbf{j}} = \frac{\xi}{|\xi|\sqrt{4+\pi^2}}\left(\sqrt{2}\hat{\boldsymbol{\xi}}_0 + \sqrt{2}\hat{\boldsymbol{\eta}}_0 + \pi\hat{\boldsymbol{\varphi}}_0\right) \tag{19}$$

$$\hat{\mathbf{k}} = 2\left(\xi\hat{\boldsymbol{\xi}}_0 - \sqrt{\frac{1}{4} - \xi^2}\hat{\boldsymbol{\eta}}_0\right) \tag{20}$$

We underline that the vectors $\hat{\boldsymbol{\xi}}_0$, $\hat{\boldsymbol{\eta}}_0$ and $\hat{\boldsymbol{\varphi}}_0$ are functions of the coordinates $(\xi, \eta, \varphi)$.

Also we have:

$$x = \sqrt{\xi\eta}\cos(\varphi) \tag{21}$$

$$y = \sqrt{\xi\eta}\sin(\varphi) \tag{22}$$

$$z = \frac{1}{2}(\xi - \eta) \tag{23}$$

And, the rotational of a vector $\mathbf{V}$ in parabolic coordinates is:

$$\nabla \times \mathbf{V}(\xi, \eta, \varphi) =$$

$$\frac{1}{(\xi^2+\eta^2)\xi\eta}\left\{(\xi^2+\eta^2)^{1/2}\widehat{\boldsymbol{\xi}}_0\left[\xi V_\varphi + \xi\eta\frac{\partial V_\varphi}{\partial \eta} - (\xi^2+\eta^2)^{1/2}\frac{\partial V_\eta}{\partial \varphi}\right]\right.$$

$$-(\xi^2+\eta^2)^{1/2}\widehat{\boldsymbol{\eta}}_0\left[\eta V_\varphi + \xi\eta\frac{\partial V_\varphi}{\partial \xi} - (\xi^2+\eta^2)^{1/2}\frac{\partial V_\xi}{\partial \varphi}\right)]$$

$$\left.+\xi\eta\widehat{\boldsymbol{\varphi}}_0\left[\xi(\xi^2+\eta^2)^{-1/2}V_\eta + (\xi^2+\eta^2)^{1/2}\frac{\partial V_\eta}{\partial \xi} - \eta(\xi^2+\eta^2)^{-1/2}V_\xi - (\xi^2+\eta^2)^{1/2}\frac{\partial V_\xi}{\partial \eta}\right]\right\} \tag{24}$$

Let us take Eq. (13) and make in the left-hand term:

$$rot\mathbf{H}_e^{'}(\omega) = i\omega\boldsymbol{\varepsilon}^{'}\mathbf{E}_e^{'}(\omega) \tag{25}$$

So we arrive to the equation:

$$i\omega\mathbf{E}_e^{'}(\omega) = i\omega\boldsymbol{\varepsilon}e^{ih(\omega_e)}\mathbf{K}^{(\circ)}(\omega)\mathbf{E}_e(\omega) \tag{26}$$

Now, we suppose that the electric field points towards the unitary vector $\widehat{\boldsymbol{\xi}}_0$ that implies that Eq. (26) becomes (see **Figure 1**):

$$\mathbf{E}_e^{'}(\omega) = \boldsymbol{\varepsilon}e^{ih(\omega_e)}\mathbf{K}^{(\circ)}(\omega)\widehat{\boldsymbol{\xi}}_0 E_\xi(\omega) \tag{27}$$

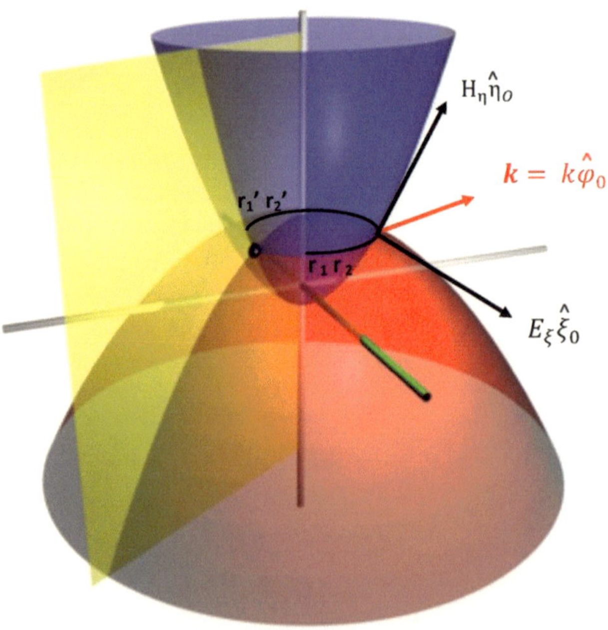


**Figure 1.** Electric and magnetic fields in parabolic coordinates and the beam trajectory.

Defining the permittivity tensor:

$$\boldsymbol{\varepsilon} \tag{28}$$

In principle, there is a dependence on the frequency $\omega$ but, for convenience, we bequeath this to the kernel, in order to easy look the contribution of the tensor $\boldsymbol{\varepsilon}$, which operates on the column vectors in the $(\xi, \eta, \varphi)$ space bending the beam trajectory (see **Figure 1**):

$$\boldsymbol{\varepsilon} = \varepsilon \begin{pmatrix} 0 & -1 & 0 \\ 1 & 0 & 0 \\ 0 & 0 & -1 \end{pmatrix} \tag{29}$$

In terms of this last tensor, Eq. (26) can be written as:

$$\mathbf{E}_e^{'}(\omega) = e^{ih(\omega_e)}\boldsymbol{\varepsilon}\mathbf{K}^{(\circ)}(\omega)\widehat{\boldsymbol{\xi}}_0 E_\xi(\omega) \tag{30}$$

For simplicity, we propose that we have only two punctual emitters with the kernel given by:

$$\mathbf{K}^{(\circ)} = \begin{bmatrix} {}^1\mathbf{K}(\circ) & \mathbf{0} \\ \mathbf{0} & {}^2\mathbf{K}(\circ) \end{bmatrix} \tag{31}$$

On matrix (23), the elements are:

$${}^{1,2}\mathbf{K}^{(\circ)} = \begin{bmatrix} \dfrac{\sin\left[(\omega - \omega_p)\delta\right]}{(\omega - \omega_p)\delta} & 0 & 0 \\ 0 & 1 & 0 \\ 0 & 0 & 1 \end{bmatrix} \tag{32}$$

As we have said, we suppose that the electric fields at the two initial points only have a $\xi$ component, for example:

$$E_\xi(\mathbf{r}_1) = E_\xi(\mathbf{r}_2) = E_0 \cos(\omega_0) \tag{33}$$

In Eq. (25), we impose the condition that the $rot\mathbf{H}_e^{'}(\omega)$ does not have $\varphi$ or $\eta$ components. Also:

$$H_\xi^{'} = 0 \tag{34}$$

and

$$H_\varphi^{'} = 0 \tag{35}$$

which means that $H_\eta^{'}$ satisfy the partial differential equation:

$$\frac{\partial H_\eta^{'}}{\partial \xi} = -\frac{\xi}{\xi^2 + \eta^2} H_\eta^{'} \tag{36}$$

and then

$$H_{\eta}^{'} = C_0 \frac{\varphi}{\xi\eta} e^{-\frac{1}{2}\ln\left(\xi^2+\eta^2\right)} \tag{37}$$

In (29), $C_0$ is a constant determined by the field at the starting point in Eq. (25).

The field $\mathbf{H}_e^{'}$ has the components:

$$H_{\eta}^{'} = C_0 \frac{\varphi}{\xi\eta} e^{-\frac{1}{2}\ln\left(\xi^2+\eta^2\right)} \tag{38}$$

$$H_{\xi}^{'} = 0 \tag{39}$$

$$H_{\varphi}^{'} = 0 \tag{40}$$

Now, because of Eqs. (31)–(33):

$$\mathbf{E}_e^{\prime}(\omega) = \boldsymbol{\varepsilon} e^{ih(\omega_e)} \begin{bmatrix} {}^1\mathbf{K}(\circ) & 0 \\ 0 & {}^2\mathbf{K}(\circ) \end{bmatrix} \widehat{\xi}_0 E_{\xi}(\omega) \tag{41}$$

So, the electric field at the final points is:

$$E_{\xi}^{'}(\mathbf{r}_{1,2}) = \frac{1}{\varepsilon^{'}} C_0 \frac{\varphi^{'}}{\xi_{1,2}^{\prime}\eta_{1,2}^{\prime}} e^{-\frac{1}{2}\ln\left(\xi_{1,2}^{\prime\,2}+\eta_{1,2}^{\prime\,2}\right)} = \varepsilon e^{ih(\omega_e)} \frac{\sin\left(\omega-\omega_p\right)\delta}{\left(\omega-\omega_p\right)\delta} \cos\left(\omega_0\right) E_0 \tag{42}$$

The value of $E_0$ is really a function of $(\xi, \eta, \varphi)$ that is, we know that:

$$E_0\left(\xi_{1,2}, \eta_{1,2}, \varphi_{1,2}\right) = D_0 \frac{\varphi e^{-\frac{1}{2}\ln\left(\xi_{1,2}^{\,2}+\eta_{1,2}^{\,2}\right)}}{\xi_{1,2}\eta_{1,2}} \tag{43}$$

and we obtain from (34):

$$\frac{1}{\varepsilon^{'}} C_0 \frac{\varphi^{'}}{\xi_{1,2}^{\prime}\eta_{1,2}^{\prime}} e^{-\frac{1}{2}\ln\left(\xi_{1,2}^{\prime\,2}+\eta_{1,2}^{\prime\,2}\right)} = \varepsilon e^{ih(\omega_e)} \frac{\sin\left(\omega-\omega_p\right)\delta}{\left(\omega-\omega_p\right)\delta} \cos\left(\omega_0\right) D_0 \frac{e^{-\frac{1}{2}\ln\left(\xi_{1,2}^{\,2}+\eta_{1,2}^{\,2}\right)}}{\xi\eta} \tag{44}$$

We can see that in parabolic coordinates there is a strong dependence on the specific values of the vectors $(\xi, \eta, \varphi)$ and $\left(\xi^{'}, \eta^{'}, \varphi^{'}\right)$. But we can say that all the initial conditions depend on the non-primed variables and put in a named constant factor $E_0$:

That is

$$\frac{1}{\varepsilon^{'}} C_0 \frac{\varphi^{'}}{\xi_{1,2}^{\prime}\eta_{1,2}^{\prime}} = \varepsilon e^{ih(\omega_e)} \frac{\sin\left(\omega-\omega_p\right)\delta}{\left(\omega-\omega_p\right)\delta} \cos\left(\omega_0\right) D_0 E_0 \tag{45}$$

From (37), we can see that we must calculate $C_0$ for every selected $\left(\xi^{'}, \eta^{'}, \varphi^{'}\right)$.

Then the fields can be calculated through the Eqs. (38)–(43) and (45).

http://dx.doi.org/10.5772/intechopen.76115

## 5. Conclusions

We have seen how it is possible to use the hybrid Maxwell-Fredholm equations to understand some kind of problems like the bending of a light beam inside a left-hand material described with an extremely deflective tensor $\boldsymbol{\varepsilon}$. So a light beam begin his trajectory measured at the two points $\mathbf{r}_1$, $\mathbf{r}_2$ and we determine with the aid of the Maxwell-Fredholm equations the electromagnetic fields at the final points $\mathbf{r}_1{}'$ and $\mathbf{r}_2{}'$ given by the Eqs. (33)–(45). Explicitly, we have shown how we can point towards the permittivity tensor as responsible for the beam deflection because in the kernel of the Maxwell-Fredholm equations only appear the free Green function $\mathbf{G}^{(\circ)}$. In this chapter, we have showed how we can add several tools to observe the behaviour of general devices and we have obtained that the Maxwell-Fredholm equations results in an appropriate tool in some interesting physical situations as the academic example illustrates. It is shown how we can select an appropriate system of coordinates as in the specific case of parabolic coordinates that we used in the example. Even if we do not establish an explicit link between the building of the named information packs and proper use of the Maxwell-Fredholm equations, we suppose that the initial signal is indeed a mix of different information packs and then it is possible to conclude that implementation of both cause a better achievement of signal transmission. In addition, we can say in advance that in a near future, the Maxwell-Fredholm equations could be used in extremely different problems like the nanofluid flux [23] providing we can establish a left-hand materials propagation condition for the electromagnetic field inside the nanofluid and others like a system of split ring resonators [24].

## Author details

Juan Manuel Velázquez-Arcos*, Alejandro Pérez-Ricardez, Ricardo Teodoro Páez-Hernández, Samuel Alcántara Montes and Jaime Granados-Samaniego

*Address all correspondence to: jmva@correo.azc.uam.mx

Universidad Autónoma Metropolitana, San Pablo, CDMX, México

## References

[1] Velázquez-Arcos JM, Pérez-Ricardez A, Páez-Hernández RT, Granados-Samaniego J. In: Awrejcewicz J, editor. Resonances and Exceptional Broadcasting Conditions, RESONANCE, Chapter XII. The Lodz University of Technology, Poland: IntechOpen; 2017. pp. 233-243

[2] Velázquez-Arcos JM. Fredholm's alternative breaks the confinement of electromagnetic waves. AIP Advances. 2013;**3**:092114. DOI: 10.1063/1.4821336

[3] Velázquez-Arcos JM, Granados-Samaniego J, Vargas CA. The confinement of electromagnetic waves and Fredholm's alternative. In: 2013 International Conference on Electromagnetics

in Advanced Applications (ICEAA); 9–13 September 2013; 2013. pp. 411-414. DOI: 10.1109/ICEAA.2013.6632268

[4] Smith DR, Pendry JB, Wiltshire MCK. Metamaterials and negative refractive index. Science. 2004;**305**(5685):788-792. DOI: 10.1126/science.1096796

[5] Li Z, Zhao R, Koshchny T, Kafesaki M, BoratayAlici K, Colak E, Caglayan H, Ozbay E, Soukolis CM. Chiral metamaterials with negative refractive index based on four "U" split ring resonators. Applied Physics Letters. 2010;**97**:0810901. DOI: 10.1063/1.3457448

[6] Kong X-K, Liu S-B, Zhang H-F, Bian B-R, Li H-M, et al. Evanescent wave decomposition in a novel resonator comprising unmagnetized and magnetized plasma layers. Physics of Plasmas. 2013;**20**:043515. DOI: 10.1063/1.4802807

[7] Velázquez-Arcos JM. Fredholm's equations for subwavelength focusing. Journal of Mathematical Physics. 2012;**53**(10):103520. DOI: 10.1063/1.4759502

[8] Velázquez-Arcos JM, Vargas CA, Fernández-Chapou JL, Salas-Brito AL. On computing the trace of the kernel of the homogeneous Fredholm equation. Journal of Mathematical Physics. 2008;**49**:103508

[9] de la Madrid R. Numerical calculation of the decay widths, the decay constants, and the decay energy spectra of the resonances of the delta-shell potential. Nuclear Physics A. 2017;**962**:24-45

[10] Velázquez-Arcos JM, Pérez-Martínez F, Rivera-Salamanca CA, Granados-Samaniego J. On the application of a recently discovered electromagnetic resonances to communication systems. IJETAE. 2013;**3**(1):466-471. Website: www.ijetae.com; ISSN: 2250-2459

[11] Velázquez-Arcos JM, Granados-Samaniego J. Wave propagation under confinement break. IOSR Journal of Electronics and Communication Engineering. 2016;**11**(2):42-48. DOI: 10.9790/2834-1102014248

[12] Velázquez-Arcos JM, Granados-Samaniego J, Navarrete-González TD. Top efficiency on communication theory. IOSR Journal of Electronics and Communication Engineering. 2016;**11**(6):01-11. DOI: 10.9790/2834-1106020111

[13] Velázquez-Arcos JM, Granados-Samaniego J, Vargas CA. Communication theory and resonances on electromagnetic systems. In: 2012 International Conference on Electromagnetics in Advanced Applications (ICEAA); 2–7 September 2012; Cape Town: IEEE; 2012. pp. 392-395. DOI: 10.1109/ICEAA.2012.6328657

[14] Velázquez-Arcos JM, Granados-Samaniego J. Recent technologies and recording time detection on time reversal signals. International Journal on Recent and Innovation Trends in Computing and Communication. November 2014;**2**(11):3447-3450

[15] Shannon CE. A mathematical theory of communication. The Bell System Technical Journal. 1948;**27**:379-423; 623-656

[16] Nyquist H. Certain factors affecting telegraph speed. Bell System Technical Journal. April 1924;**3**(2):181-352

[17] Nyquist H. Certain topics in telegraph transmission theory. AIEE Transactions. 1928;**47**(2): 617-644

[18] Hartley RVL. The interpolation, extrapolation and smoothing of stationary time series. In: NDRC Report. Wiley; 1949

[19] Wiener N. The Ergodic theorem. Duke Mathematical Journal. 1939;**5**(1):1-18

[20] Hopf E. Remarks on causality statistics and probability. Journal of Mathematical Physics. 1934;**13**:51-102

[21] Jackson JD. Classical Electrodynamics. Third ed. Wiley; 2017

[22] Clemmow PC. An Introduction to Electromagnetic Theory. Cambridge University Press; 1973

[23] Sheikholeslami M, Abelman S. Two-phase simulation of nanofluid flow and heat transfer in an annulus in the presence of an axial magnetic field. IEEE Transactions on Nanotechnology. 2015;**14**(3):561-569. DOI: 10.1109/TNANO.2015.2416318

[24] Hernández-Bautista F, Vargas CA, Velázquez-Arcos JM. Negative refractive index in split ring resonators. Revista Mexicana de Fisica. 2013;**59**(1):139-144. ISSN: 0035-00IX